Physical Principles of Remote Sensing

SECOND EDITION

Physical Principles of Remote Sensing explains remote sensing of the Earth's surface and atmosphere from space using electromagnetic radiation. The main emphasis of the book is on the physical and mathematical principles underlying the techniques, but examples of applications are drawn from a range of the environmental sciences.

This second edition of a popular book has been very substantially revised and expanded. The topics covered include overviews of electromagnetic propagation in free space and in matter, surface and volume scattering, the interaction of radiation with the atmosphere, the main classes of sensor (photographic, electro-optical, passive microwave, laser profiling and lidar, radar altimetric, microwave scatterometer and imaging radar), satellite orbits for remote sensing, and an introduction to image processing. Additions for this new edition include a discussion of the radiative transfer equation, atmospheric sounding techniques and interferometric radar, an expanded list of problems (now including solutions), and a discussion of GPS (the Global Positioning System). The discussions of all the main types of sensor are illustrated with up-to-date examples of real instruments. SI units are used throughout the book.

This book forms the basis of an introductory course in remote sensing highlighting physical and mathematical principles. The main readership will be undergraduate students, graduate students and researchers in remote sensing, geography, cartography, surveying, meteorology, earth sciences and environmental sciences generally, as well as physicists, mathematicians and engineers who wish to know about, or retrain into, the area of Earth remote sensing. It will also form a valuable reference source for researchers using remote sensing techniques.

W. G. Rees is Assistant Director of Research at the Scott Polar Research Institute, University of Cambridge. His research is principally concerned with developing and applying spaceborne remote sensing methods to the study of polar environments, especially snow, ice, and tundra vegetation. He has published extensively in the *International Journal of Remote Sensing*, and other polar and remote sensing journals. Dr Rees has published several books: the first edition of *Physical Principles of Remote Sensing* (1990, Cambridge University Press); *The Remote Sensing Data Book* (1999, Cambridge University Press); *Physics by Example* (1994, Cambridge University Press) and *Essential Quantum Physics* (Landshoff, Metherell and Rees, 1997, Cambridge University Press). He also teaches physics and remote sensing at undergraduate and graduate levels at Cambridge University. He is a Fellow of the Institue of Physics, a Fellow of the Royal Astronomical Society, and a Member of the Remote Sensing Society.

Physical Principles of Remote Sensing

SECOND EDITION

W. G. REES
Scott Polar Research Institute, University of Cambridge

CAMBRIDGE
UNIVERSITY PRESS

CAMBRIDGE UNIVERSITY PRESS
Cambridge, New York, Melbourne, Madrid, Cape Town, Singapore, São Paulo, Delhi

Cambridge University Press
The Edinburgh Building, Cambridge CB2 8RU, UK

www.cambridge.org
Information on this title: www.cambridge.org/9780521660341

First published 2001
Eighth printing 2009

Printed in the United Kingdom at the University Press, Cambridge

A catalog record for this publication is available from the British Library.

Library of Congress Cataloging in Publication Data

Rees, Gareth (William Gareth), 1959–
Physical principles of remote sensing / W. G. Rees. – 2nd ed.
p.   cm.
Includes bibliographical references (p.).
ISBN 0-521-66034-3 (hardback) – ISBN 0-521-66948-0 (pb)
1. Remote sensing.   I. Title.
G70.4.R44   2001
621.36′78 – dc21
00-045535

ISBN 978-0-521-66034-1 hardback
ISBN 978-0-521-66948-1 paperback

For Christine

Contents

Colour plates 1–10 and colour versions of figures 5.3 and 5.5 can be found between pages xvi and 1.

Preface

There are many books that explain the subject of remote sensing to those whose backgrounds are primarily in the environmental sciences. This is an entirely reasonable fact, since those people continue to be the main users of remotely sensed data. However, as the subject grows in importance, the need for a significant number of people to understand not only what remote sensing systems do, but how they work, will grow with it. This was already happening in 1990, when the first edition of *Physical Principles of Remote Sensing* appeared, and since then increasing numbers of physical scientists, engineers and mathematicians have moved into the field of environmental remote sensing. It is for such readers that this book, like its first edition, has been written. That is to say, the reader for whom I have imagined myself to be writing is educated to a reasonable standard (although not necessarily to first degree level) in physics, with a commensurate mathematical background. I have, however, found it impossible to be strictly consistent about this, because of the wide range of disciplines within and beyond physics from which the material has been drawn, and I trust that readers will be understanding when they find the treatment either too simple or over their heads.

This book attempts to follow a logical progression, more or less following the flow of information from the remotely sensed object to the user of the data. The first four chapters lay the general foundations. Chapter 1 sets the subject in context. Chapter 2 is a non-rigorous treatment of electromagnetic wave propagation in free space, which can be regarded as a compendium of necessary results. It will represent, I hope, mostly revision to most readers, although it assumes little or no previous knowledge of Fourier transforms or of Fraunhofer diffraction theory. Chapter 3 discusses the interaction of electromagnetic radiation with smooth and rough surfaces and with inhomogeneous materials like soil and snow, and chapter 4 discusses the interaction of radiation with the atmosphere and ionosphere. By this stage of the book, our information is, as it were, travelling upwards towards the sensor.

Chapters 5 to 9 discuss the sensors themselves, beginning with the more familiar passive sensors and going on to consider active systems. These chapters explain, so far as is consistent with the level of the book, the functioning of the sensors, important operational constraints, and some of the more important applications derived from them. These chapters also include brief descriptions of real instruments on existing or forthcoming satellite missions.

The platforms on which the sensors are supported are discussed in chapter 10. After a short discussion of remote sensing from aircraft, the chapter devotes itself to satellite orbits. Finally, chapter 11 presents an introduction to the data processing aspects of remote sensing, particularly digital image processing and analysis. There are two appendices. The first provides a brief introduction to the global positioning system (GPS) and its capabilities, since this is increasingly an integral part of remote sensing field work and indeed of some remote sensing systems themselves. The second contains tables of data frequently needed in remote sensing. A short list of problems is included at the end of most chapters, with hints and solutions given at the end of the book. Most of these problems are straightforward (I have tried to indicate which are for 'enthusiasts'), designed to extend and consolidate the reader's understanding of the material. Some problems will require material from more than one chapter for their solution.

It will perhaps be useful to indicate those features of the book that have been preserved from the first edition and those that are new. The underlying rationale has not changed. It has still been my intention to keep the book as short as possible, consistent with clarity, although this edition is significantly longer than the first because it includes new material. In particular, the book now includes a discussion of atmospheric sounding, which has in turn required a somewhat lengthier treatment of atmospheric propagation and the inclusion of an explicit discussion of the radiative transfer equation. Chapter 9 includes a treatment of SAR interferometry, and chapter 11 contains a significantly expanded discussion of image processing techniques. As before, the aim has been to teach principles of remote sensing rather than to present a lot of technical or engineering detail. However, I have now included brief discussions of real sensor systems or surveys of types of sensor in the relevant chapters, and have tried to keep these reasonably up-to-date. The list of problems has been expanded, and solutions, or partial solutions, have been provided to them. The book's bibliography has also been expanded somewhat and brought up to date, although I have still attempted to keep it short enough so as not to overwhelm the reader with an enormous list of references. Some selection and omission has therefore been necessary, and I hope my colleagues will forgive me if my selection does not tally with theirs.

As in the first edition, I have deliberately avoided the rigorous consistency in the use of symbols that demands that a given symbol be used to represent only a single physical quantity. Because of the wide scope of remote sensing, this would lead to an unforgivably confusing proliferation of symbols with many sub- and superscripts. Consistency of symbols is therefore confined to sections of the text that deal with a single topic, except for a few 'universal' symbols such as h for the Planck constant and ω for angular frequency, which are used throughout the book. SI units are used consistently, although a table in appendix 2 gives equivalents for some common non-SI units.

This book arose from a course of undergraduate lectures delivered first at the Scott Polar Research Institute, and later at the Cavendish Laboratory, both at the University of Cambridge. I am grateful to both departments for letting me try out my ideas. Many people are owed thanks for their contributions to the writing of the first edition. It is difficult to single out individuals, but I particularly wish to thank Professor Andrew Cliff, Dr Bernard Devereux, Michael Gorman and Christine Rees. If the second edition is an improvement on the first, much of the credit should go to the constructive criticisms of the users and reviewers of the first edition and of the many students who have come my way since 1990. As always, Cambridge University Press has provided advice and encouragement whenever it was needed.

W. G. R.
Cambridge

Acknowledgements

Permission to reproduce copyright and other material from the following sources is gratefully acknowledged: Australian Board of Meteorology, Australian Centre for Remote Sensing, Cambridge University Committee for Aerial Photography, Dundee University, EUMETSAT, European Space Agency, Japan Meteorological Agency, National Aeronautics and Space Administration (USA), National Remote Sensing Centre (United Kingdom), R&D Center ScanEx (Russia), Sovzond (Russia), SSC Satellitbild (Sweden), and Dr T. Wahl.

Plate 1. A satellite aerial photograph. The image was recorded by the Metric Camera, carried by the Space Shuttle at an altitude of approximately 250 km, and covers an area of about 190 km × 130 km. It shows the Himalaya mountains, including Everest (at top left, slightly clouded) and the Ganges plain (bottom right). (© European Space Agency. Reproduced with permission.)

Plate 2. False-colour infrared aerial photograph. The image was recorded from an altitude of approximately 760 m, and covers an area of about 1.1 km². It shows the Tay reed beds on the Firth of Tay, Scotland. (Cambridge University Collection: Copyright reserved.)

Plate 3. HCMM (thermal infrared) image of Death Valley, USA. The image was acquired on 31 May 1978, during the daytime, and has been colour coded with blue representing the lowest emission of radiation and red the highest. The image has been superimposed on a Landsat image of the same area to provide a geographical base. It was obtained from a height of approximately 620 km and has a coverage of about 70 km × 100 km. It shows the Panamint (left) and Amargosa (centre) ranges, with Death Valley between them. (Reproduced by courtesy of the National Remote Sensing Centre, UK, and the National Aeronautics and Space Administration, USA.)

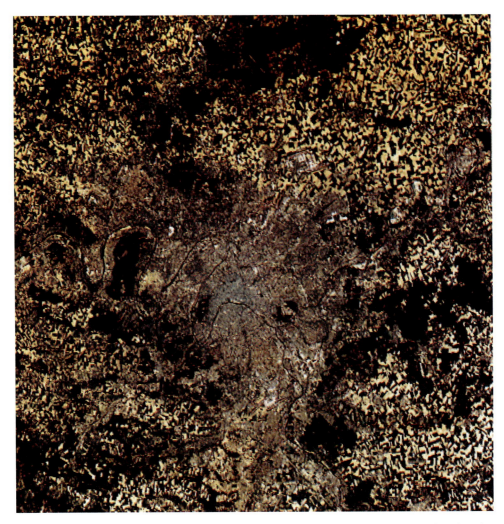

Plate 4. Extract from an enhanced true-colour Landsat Thematic Mapper image of Paris. The image was obtained from a height of approximately 700 km, and has a coverage of about 90 km × 80 km. (© European Space Agency. Reproduced with permission.)

Plate 5. Sea-surface temperature deduced from passive microwave (NIMBUS SMMR) data. The upper figure shows the average temperature for November 1978 to February 1979, the lower figure for July to October 1979. (Reproduced by courtesy of the National Aeronautics and Space Administration, USA.)

Plate 6. Three-band colour composite of NIMBUS-7 passive microwave data from the northern hemisphere on 1 January 1980, showing its potential for monitoring sea-ice concentrations. The red, green and blue bands correspond to microwave radiances at 6, 18 and 37 GHz, respectively, all in vertical polarisation. Blue areas are open water and black corresponds to areas for which no data are available. (Data courtesy of NASA Goddard Space Flight Center.)

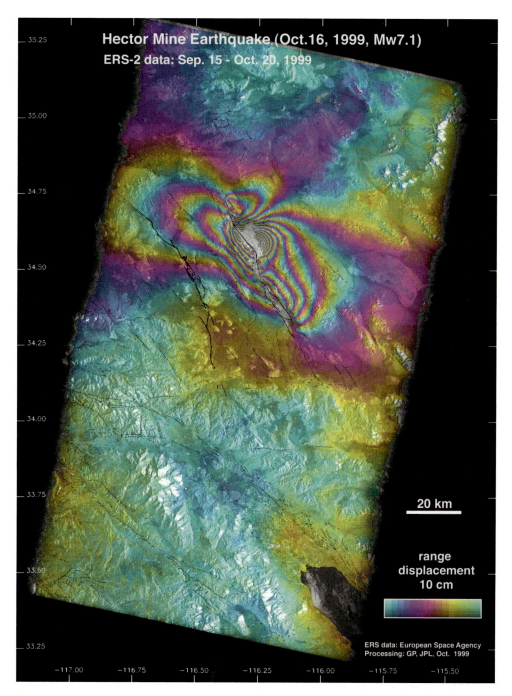

Plate 7. Interferometric image of the Hector Mine earthquake area in California. The image has been constructed from two ERS-2 spaceborne SAR images, on 15 September and 20 October, 1999. One full cycle of the colour scale represents a movement of 10 cm in the direction towards the radar. Grey areas are zones of low phase coherence. (Reproduced by courtesy of the Jet Propulsion Laboratory, from the web site at

http://www-radar.jpl.nasa.gov/sect323/InSar4crust/

and with acknowledgement of the authors: Frédéric Crampé, Gilles Peltzer, Paul Rosen and Mark Simons.)

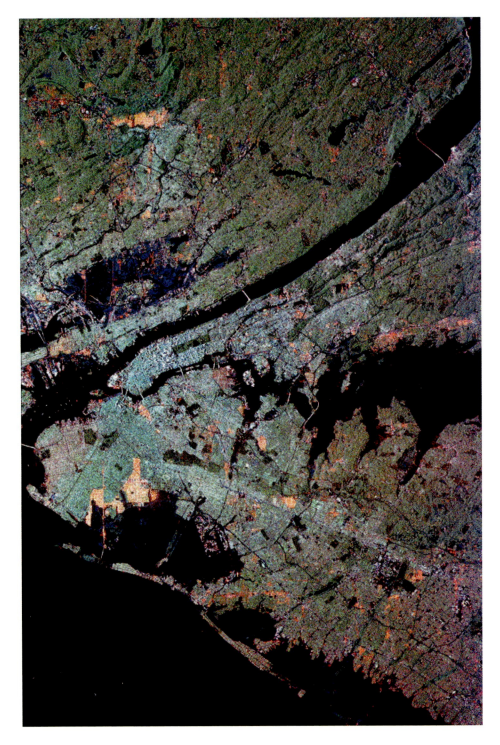

Plate 8. Multifrequency and polarisation SAR image of the New York City metropolitan area. The image covers an area of 75 km × 49 km and was acquired from the SIR-C/X-SAR instrument carried on the Space Shuttle *Endeavour*, on 10 October 1994. The red band of this image corresponds to the LHH backscattering coefficient; green is LHV and blue is CHV. (Reproduced by courtesy of the Jet Propulsion Laboratory from the web site at

http://southport.jpl.nasa.gov/imagemaps/html/srl-nyc.html)

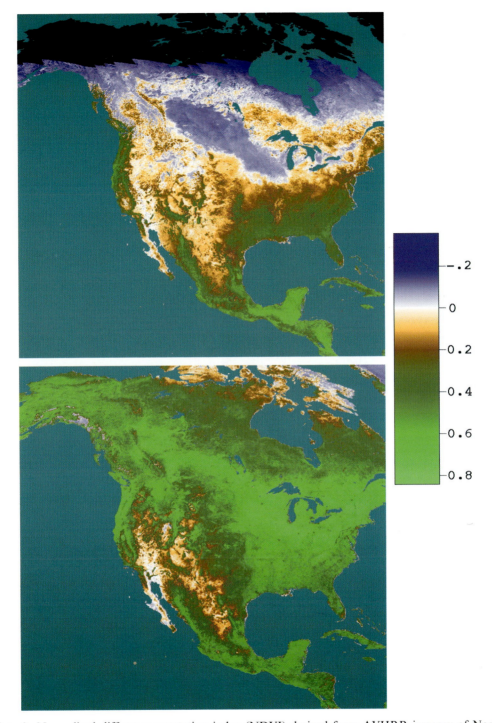

Plate 9. Normalised difference vegetation index (NDVI) derived from AVHRR imagery of North America. The top image shows the average NDVI for January 1999 and the lower image shows the average for July 1999. Sea areas have been masked in cyan, and areas of no data are shown in black. (The author thanks the Distributed Active Archive Center (Code 902.2) at the Goddard Space Flight Center, Greenbelt, MD 20771, for producing the data in their present form and distributing them. The original data products were produced under the NOAA/NASA Pathfinder program, by a processing team headed by Ms. Mary James of the Goddard Global Change Data Center; and the science algorithms were established by the AVHRR Land Science Working Group, chaired by Dr. John Townshend of the University of Maryland. Goddard's contributions to these activities were sponsored by NASA's Mission to Planet Earth program.)

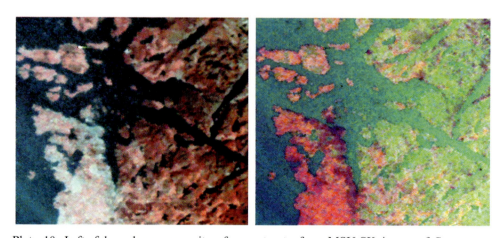

Plate 10. Left: false-colour composite of an extract of an MSU-SK image of Stavanger, Norway. The extract covers an area of approximately 38 km × 38 km, and has been assigned the following colours. Red: 0.7–0.8 μm; green: 0.6–0.7 μm; blue: 0.5–0.6 μm. (Reproduced by courtesy of Eurimage.) Right: false-colour composite of the principal components of this image. Red, green and blue correspond to the first, second and third principal components, respectively.

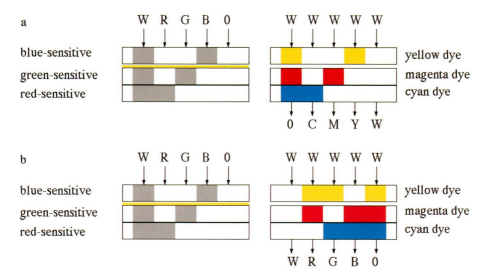

Figure 5.3 (a) Normal (negative) colour film; (b) reversal (positive) colour film. The diagrams on the left show the film after exposure to light, the shaded areas indicating the activated regions of the emulsion layers. The diagrams on the right show the regions in which yellow, magenta and cyan dyes are formed after development. The symbols for colour are W = white, R = red, G = green, B = blue, C = cyan, M = magenta, Y = yellow, 0 = no light = black.

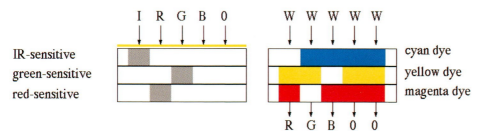

Figure 5.5 Reversal FCIR film, showing (left) activated regions after exposure, (right) the regions in which dye is formed after development. The symbols for the 'colours' are as for figure 5.3, with the addition of I for infrared radiation.

1

Introduction

1.1 Definition and origins of remote sensing

'Remote sensing' is, broadly but logically speaking, the collection of information about an object without making physical contact with it. This is a simple definition, but too vague to be really useful,[1] so for the purposes of this book we restrict it by confining our attention to the Earth's surface and atmosphere, viewed from above using electromagnetic radiation. This narrower definition excludes such techniques as seismic, geomagnetic and sonar investigations, as well as (for example) medical and planetary imaging, all of which could otherwise reasonably be described as remote sensing, but it does include a broad and reasonably coherent set of techniques, nowadays often described by the alternative name of *Earth Observation*. These techniques, which now have a huge range of applications in the 'civilian' sphere as well as their obvious military uses, make use of information impressed in some way on electromagnetic radiation ranging from ultraviolet to radio frequencies.

The origins of remote sensing can plausibly be traced back to the fourth century BC and Aristotle's *camera obscura* (or, at least, the instrument described by Aristotle in his *Problems*, but perhaps known even earlier). Although significant developments in the theory of optics began to be made in the seventeenth century, and glass lenses were known much earlier than this, the first real advance towards our modern conception of remote sensing came in the first half of the nineteenth century with the invention of photography. For the first time, it became possible to record an image permanently and objectively. Also during the nineteenth century, forms of electromagnetic radiation were discovered beyond the visible part of the spectrum – infrared radiation by Herschel, ultraviolet by Ritter, and radio waves by Hertz – and in 1863

---

[1] See Campbell (1996, p. 4) for a summary of the main definitions of remote sensing that have been adopted over the last few decades. The term 'remote sensing' itself was first used by the U.S. Office of Naval Research in the 1960s (see Cracknell and Hayes, 1991, p. 1).

Maxwell developed the electromagnetic theory on which so much of our under-standing of these phenomena depends.

Aerial photography followed almost immediately on the discovery of the photographic method. The first aerial photograph, unfortunately no longer in existence, was probably made in 1858 by Gaspard Félix Tournachon, taken from a balloon at an altitude of about 80 m. Kites were also soon used, and by 1890 the usefulness of aerial photography was so far recognised that Batut had published a textbook on the subject.

The next step towards what we now recognise as remote sensing was taken with the development of practicable aeroplanes in the early twentieth century. Again, the potential applications were quickly recognised and aerial photographs were recorded from aeroplanes from 1909. Aerial photography was used during the First World War for military reconnaissance, and, during the period between the two World Wars, civilian uses of this technique began to be developed, notably in cartography, geology, agriculture and forestry. Cameras, film and aircraft underwent significant improvements, and stereographic mapping attained an advanced state of development. Also during this period, John Logie Baird, the inventor of television, performed early work on the development of airborne scanning systems capable of transmitting images to the ground. This work was highly confidential, having been carried out on behalf of the French Air Ministry. It was ended by the war and forgotten about until 1985 (Newton, 1989).

The Second World War brought substantial developments to remote sensing. Photographic reconnaissance reached a high state of development – the German invasion of Britain, planned for September 1940, was forestalled by the observation of concentrations of ships along the English Channel. Infrared-sensitive instruments and radar systems were developed. In particular, the Plan Position Indicator used by night bombers was an imaging radar that presented the operator with a 'map' of the terrain, and thus it represented the ancestor of the imaging radar systems discussed in chapter 9.

By the 1950s, false-colour infrared film, originally developed for military use, was finding applications in vegetation mapping, and high-resolution imaging radars were being developed. As these developments continued through the 1960s, sensors began to be placed in space. This was originally part of the programme to observe the Moon, but the advantages of applying the same techniques to observation of the Earth were soon recognised and the first multispectral spaceborne imagery of the Earth was acquired from *Apollo 6*. Although there were earlier unmanned remote sensing satellites,[2] the opening of the modern era of spaceborne remote sensing ought probably to be dated to July 1972 with the successful operation of ERTS, the Earth Resources Technology Satellite, by the U.S. National Aeronautics and Space Administration (NASA). The ERTS was renamed Landsat-1, and the Landsat programme is still continuing – at the time of writing (November 1999), Landsat-7 is producing large quantities of data.

[2] The first was TIROS-1, launched in April 1960.

Since the launch of ERTS in 1972, the number and diversity of spaceborne and airborne remote sensing systems has grown dramatically. A larger range of variables can be measured, and consistent and systematic datasets can be constructed for progressively longer periods of time. The explosive growth in the quantity of data being generated has been matched by growth in the availability of computing resources and the facilities for data storage.

1.2   Applications

The enormous growth in the availability of remotely sensed data over the last four decades has been matched by a fall in the real cost of the data. Nevertheless, it is still clear that use of the data must offer some tangible advantages to justify the cost of acquiring and analysing them. These advantages derive from a number of characteristics of remote sensing. Probably the most important of these is that data can be gathered from a large area of the Earth's surface, or a large volume of the atmosphere, in a short space of time, so that a virtually instantaneous 'snapshot' can be obtained. For example, scanners carried on geostationary meteorological satellites such as METEOSAT can acquire an image of approximately one quarter of the Earth's surface in less than half an hour. When this aspect is combined with the fact that airborne or spaceborne systems can acquire data from locations that would be difficult (slow, expensive, dangerous, politically inconvenient . . .) to measure *in situ*, the potential power of remote sensing becomes apparent. Of course, further advantages derive from the fact that most remote sensing systems generate calibrated digital data that can be fed straight into a computer for analysis.

Remote sensing finds a very wide range of applications, obviously including the area of military reconnaissance in which many of the techniques had their origins. In the non-military sphere, most applications can loosely be categorised as 'environmental', and we can distinguish a range of environmental variables that can be measured. In the atmosphere, these include temperature, precipitation, the distribution and type of clouds, wind velocities, and the concentrations of gases such as water vapour, carbon dioxide, ozone, and so on. Over land surfaces, we can measure tectonic motion, topography, temperature, albedo (reflectance) and soil moisture content, and determine the nature of the land cover in considerable detail, for example by characterising the type of vegetation and its state of health or by mapping man-made features such as roads and towns. Over ocean surfaces, we can measure the temperature, topography (from which the Earth's gravitational field, as well as ocean tides and currents, can be inferred), wind velocity, wave energy spectra, and colour (which is often related to biological productivity by plankton). The 'cryosphere', that part of the Earth's surface covered by snow and ice, can also be studied, giving data on the distribution, condition and dynamical behaviour of snow, sea ice, icebergs, glaciers and ice sheets.

This list of measurable variables, while not complete, is large enough to indicate that there is a correspondingly large number of disciplines to which remote sensing data can be applied. While by no means exhaustive, a list of applications could include the following disciplines: agriculture and crop monitoring, archaeology, bathymetry, cartography, climatology, civil engineering, coastal erosion, disaster monitoring and prediction, forestry, geology, glaciology, oceanography, meteorology, pollution monitoring, snow resources, soil characterisation, urban mapping, water resource mapping and monitoring. It is not really possible to present a detailed cost–benefit analysis in this introduction, partly because, at least until recently, most spaceborne remote sensing operations have been part of national or international space programmes and so their costs have been to some extent hidden.[3] Perhaps it is sufficient to point out that the data available from remote sensing, particularly from spaceborne observations, often cannot be obtained in any other way; that our current understanding of the global climate system is very largely based on spaceborne observations; and that the use of remotely sensed data for disaster warning has already saved many thousands of human lives.

1.3   A systems view of remote sensing

We stated rather briefly in section 1.1 that remote sensing involves the collection of information, carried by electromagnetic radiation, about the Earth's surface or atmosphere. Let us try to expand this statement a little.

First, where does the radiation come from? One major classification of remote sensing systems is into passive systems, which detect naturally occurring radiation, and active systems, which emit radiation and analyse what is sent back to them. Passive systems can be further subdivided into those that detect radiation emitted by the Sun (this radiation consists mostly of ultraviolet, visible light and near-infrared radiation), and those that detect the thermal radiation emitted by all objects that are not at absolute zero (i.e. all objects). For objects at typical terrestrial temperatures, this thermal emission occurs mostly in the infrared part of the spectrum, at wavelengths of the order of $10\,\mu m$ (the so-called thermal infrared region), although measurable quantities of radiation also occur at longer wavelengths, as far as the microwave part of the spectrum. Active systems can, in principle, use any type of electromagnetic radiation. In practice, however, they are restricted by the transparency of the Earth's atmosphere. This is shown schematically in figure 1.1. The figure shows that there are two main 'windows' in the atmosphere. The first of these includes the visible and infrared parts of the spectrum, between wavelengths of about $0.3\,\mu m$ and $10\,\mu m$, although it does also contain a number of opaque regions; and the second more or less corresponds to the microwave region, between wavelengths of a few millimetres and a few metres. Thus, we can expect that any active system designed to penetrate the

---

[3] However, a reasonable cost estimate for a typical remote sensing satellite mission might be $100 million.

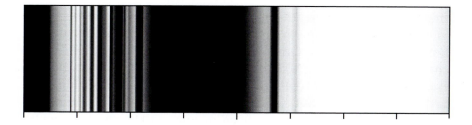

Figure 1.1. Transparency of the Earth's atmosphere as a function of wavelength (schematic). Black regions are opaque, white regions transparent.

Earth's atmosphere will operate in one of these two 'window' regions. Figure 1.2 summarises the main types of remote sensing system on the basis of the classifications we have just outlined.

The sensor, whether it is part of a passive or an active instrument, detects electromagnetic radiation after it has interacted with or been emitted by the 'target' material. In what way can this radiation contain useful information about the target? There are essentially only two variables to describe the radiation that is received: (1) *How much* radiation is detected?[4] (2) *When* does it arrive? The time-structure of the detected radiation is obviously relevant only in the case of active systems where the time-structure of the emitted radiation can be controlled. In this case, it is possible to determine the distance from the sensor to the target, and this is the principle behind various ranging systems such as the laser profiler, lidar, radar altimeter and other types of radar system. In all other cases, the only information we have is the quantity of radiation received at the sensor. If the radiation arises from

	Passive systems			Active systems	
	Reflected sunlight	Thermal emission		Visible/IR	Microwave (radio)
		Infrared	Microwave (radio)		
Non-imaging		Thermal infrared radiometry (6)	Passive microwave radiometry (7)	Laser profiling (8)	Radar altimetry (8) Microwave scatterometry (9)
Imaging	Aerial photography (5) Visible/near-infrared imaging (6)	Thermal infrared imaging (6)	Passive microwave radiometry (7)		Real aperture radar (9) Synthetic aperture radar (9)
Sounding	Ultraviolet backscatter sounding (6)	Thermal infrared sounding (6)	Passive microwave sounding (7)	Lidar (9)	

Figure 1.2. A simple taxonomy of remote sensing systems. The numbers in parentheses refer to the chapters of this book.

---

[4] The 'quantity' of radiation, however, may need to be qualified with a statement of its polarisation state.

thermal emission, the quantity is characteristic of the temperature of the
target material and its emissivity, a property that describes its efficiency at
emitting thermal radiation. Otherwise (in the cases of both passive and active
systems that measure reflected radiation), the amount of radiation that is
received is determined by the amount illuminating the target material, and
the target's reflectivity. Thus, we can see that the information about a target
material that is directly observable from remote sensing observations is actually rather limited: we can measure its range, its reflectivity, and a combination of its temperature and emissivity. However, these can be measured at
different times, over a range of wavelengths and, sometimes, at different
polarisation states, and this increase in the diversity of the variables at our
disposal is responsible for the large range of indirect observables that was
sketched out in section 1.2.

The foregoing discussion has not included the effects of the Earth's atmosphere, except to point out that atmospheric opacity limits the scope for observation of the Earth's surface to the two main atmospheric 'windows'. In fact, as
almost any electromagnetic wave propagates through the atmosphere, its characteristics will be somewhat modified. This modification may be troublesome,
requiring correction, or advantageous, depending on whether we are more
interested in studying the Earth's surface or the atmosphere itself. In general,
we can say that if the observation is made at a wavelength at which the atmosphere is opaque, the measured signal will be characteristic of the atmosphere;
whereas if the atmosphere is transparent, the data will be characteristic of the
surface below.

Once the data have been collected by the sensor, they must be retrieved
and analysed. In most, though not all, cases, the data will form an image, by
which we mean a two-dimensional representation of the two-dimensional
distribution of radiation intensity. Although the familiar photograph can
serve as a prototype of an image, the images with which we have to deal
in remote sensing are normally digital, so they can conveniently be analysed
by computer, and need not be confined to the visible part of the electromagnetic spectrum. For example, an image might represent the radar reflectivity
in one or more frequencies or polarisation states, or the thermal emission, as
well as the visible or near-infrared reflectivity. Image processing forms an
integral part of remote sensing. Typically, this involves several steps. The
first is to correct the image so that it has a known geometric correspondence
to the Earth's surface and a known calibration, with atmospheric propagation effects removed. At this stage, the image may also be enhanced in
various ways, for example by suppressing noise, to increase its intelligibility.
The major goal of image processing, however, is the extraction of useful
information from the sensor data, based on the brightness values of the
image (probably in a number of spectral bands, at a number of different
dates, in different polarisation states, etc.) and also on the spatial context.
Using the analogy of a colour photograph, we can say that information can
be extracted on the basis of colour, texture, shape and spatial context. In the

majority of cases, it is necessary or at least desirable to 'train' the process of extracting information from the image using data from known locations. The process can therefore be seen as one of extrapolation from areas that are already known, for example on the basis of field work, to much wider areas. The extrapolation need not be confined to the spatial domain, however, and the analysis of time-series of images for change detection is also an important application of remote sensing.

1.4   Further reading, and how to obtain data

The field of remote sensing is now well served with textbooks, and the interested (or puzzled) reader should be able to find alternative treatments of most of the topics discussed in this book. The 1990s have seen the publication of, amongst other general or introductory texts in remote sensing, *Fundamentals of Remote Sensing and Airphoto Interpretation* by Avery and Berlin (1992), *Introduction to environmental Remote Sensing* by Barrett and Curtis (1992), *Introduction to Remote Sensing* by Campbell (1996), *Introduction to Remote Sensing* by Cracknell and Hayes (1991), *Images of the Earth: a guide to Remote Sensing* by Drury (1998), *Remote Sensing and Image Interpretation* by Lillesand and Kiefer (1994), and *Remote Sensing: principles and interpretations* by Sabins (1996). In addition, useful collections of remote sensing data have been published by, for example, Kramer (1996), Rees (1999) and Ryerson (1998). Amongst older but still very valuable works on remote sensing it is useful to mention *Principles of Remote Sensing* by Curran (1985), *Introduction to the physics and techniques of Remote Sensing* by Elachi (1987), and *Satellite Remote Sensing: an introduction* by Harris (1987).

Scientific journals also represent an important source of information. Articles in the scientific literature are usually aimed at specialists, but the more general reader can often also extract a useful understanding from them, and the journals sometimes also publish review articles. In this book I have provided references to both books and journal articles. The principal English language journals in remote sensing are the *IEEE Transactions on Geoscience and Remote Sensing*, the *International Journal of Remote Sensing*, *Photogrammetric Engineering and Remote Sensing*, and *Remote Sensing of Environment*.

Finally, a few remarks about the Internet may be useful. This can represent a very powerful means of obtaining up-to-date information of all sorts, for example on the operational status of a particular remote sensing satellite, or the latest results from a research group, or access to remote sensing data (indeed, some of the illustrations used in this book have been obtained in this way) or the software needed to process it. As anyone who has grappled with the Internet will know only too well, the problem is usually to locate the information one needs. The well-known 'search engines' can be extremely helpful, as can the collections of 'links' assembled by public-spirited individuals and

organisations. I will mention just one source of such links, and that is the web
site of the Remote Sensing Society. The interested reader should easily be able
to build up a much larger list using this as a starting point. At the time of going
to press, its URL (uniform resource locator, i.e. its 'address' on the Internet)
was

    http://www.the-rss.org

2

Electromagnetic waves in free space

2.1 Electromagnetic waves

The propagation of electromagnetic radiation as waves is a consequence of the form of Maxwell's equations, as Maxwell himself realised.[1] One form in which these equations can be written, for free space, is

$$\nabla \cdot \mathbf{E} = 0 \tag{2.1.1}$$

$$\nabla \cdot \mathbf{B} = 0 \tag{2.1.2}$$

$$\nabla \times \mathbf{E} = -\dot{\mathbf{B}} \tag{2.1.3}$$

$$\nabla \times \mathbf{B} = \varepsilon_0 \mu_0 \dot{\mathbf{E}} \tag{2.1.4}$$

In these expressions, $\mathbf{E}$ and $\mathbf{B}$ are the electric and magnetic field vectors, respectively, of the wave, and ε_0 and μ_0 are the *electric permittivity* and the *magnetic permeability* of free space.

It can easily be confirmed that the plane wave

$$E_x = E_0 \cos(\omega t - kz) \tag{2.2}$$
$$E_y = 0$$
$$E_z = 0$$
$$B_x = 0$$
$$B_y = \frac{E_0}{c} \cos(\omega t - kz) \tag{2.3}$$
$$B_z = 0$$

satisfies the equations (2.1.1) to (2.1.4), provided that the wave speed

$$c = \frac{\omega}{k} = \frac{1}{\sqrt{\varepsilon_0 \mu_0}} \tag{2.4}$$

The constant c is the speed of light, and of all electromagnetic waves, in free space. It has a value of $2.99792458 \times 10^8 \, \mathrm{m\,s^{-1}}$. (This value is very well deter-

---

[1] It is assumed that the reader is more or less familiar with the theory of electromagnetism. If not, the range of suitable textbooks is very wide. I still find that volume 2 of the *Feynman Lectures on Physics* (Feynman et al., 1964) offers one of the most illuminating approaches.

mined, and in fact now defines the metre in terms of the second. Values of important constants such as c are given in the appendix.)

Note that we have used the *angular frequency* ω and the *wavenumber* k, rather than the more familiar cyclic frequency f and wavelength λ. The former are usually more useful, and we shall use them often. They are related to frequency and wavelength, respectively, by

$$\omega = 2\pi f \tag{2.5}$$

and

$$k = \frac{2\pi}{\lambda} \tag{2.6}$$

In principle, the frequency of an electromagnetic wave can take any value, and the whole range of possible frequencies is called the *electromagnetic spectrum*. Different regions of the spectrum are conventionally given names such as light, radio waves, ultraviolet radiation, and so on, usually referring to the manner in which the radiation is generated or detected. The electromagnetic spectrum is shown schematically in figure 2.1.

Returning to the electromagnetic wave specified by equations (2.2) and (2.3), E_0 is the *amplitude* of the electric field, and E_0/c is the amplitude of the magnetic field, although since these two amplitudes are related by the factor c it is common to speak of E_0 as the amplitude of the wave. The wave carries energy in its direction of propagation, which is the positive z-direction, and the *flux density* (power crossing unit area normal to the propagation direction) is given by

$$F = \frac{E_0^2}{2Z_0} \tag{2.7}$$

where Z_0 is the *impedance of free space*, defined by

$$Z_0 = \sqrt{\frac{\mu_0}{\varepsilon_0}} \tag{2.8}$$

It has a value of approximately 377 Ω.

2.2   Polarisation

The wave specified by equations (2.2) and (2.3) is not the most general electromagnetic wave propagating in the z-direction. We can find another such wave by simply rotating our coordinate system by $90°$ about the z-axis, to give

$$E_y = E_0 \cos(\omega t - kz) \tag{2.9}$$

$$B_x = -\frac{E_0}{c} \cos(\omega t - kz) \tag{2.10}$$

all other components being zero. If we now add the waves represented by equations (2.2) and (2.9), giving them different amplitudes and phases, we obtain an expression for a general wave propagating in the z-direction:

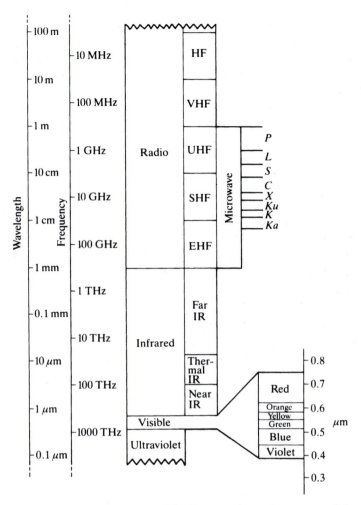

Figure 2.1. The electromagnetic spectrum. The diagram shows those parts of the electromagnetic spectrum that are important in remote sensing, together with the conventional names of the various regions of the spectrum. The letters (*P*, *L*, *S*, etc.) used to denote parts of the microwave spectrum are in common use in remote sensing, being standard nomenclature amongst radar engineers in the USA. Note that this nomenclature varies somewhat in other countries, particularly in military usage. Note also that various terminologies are in use for the subdivisions of the infrared (**IR**) part of the spectrum. That adopted here defines the thermal infrared band as lying between 3 and $15\,\mu$m, since this region contains most of the power emitted by black bodies at terrestrial temperatures.

$$E_x = E_{0x}\cos(\omega t - kz - \phi_x) \qquad\qquad (2.11)$$
$$E_y = E_{0y}\cos(\omega t - kz - \phi_y) \qquad\qquad (2.12)$$
$$E_z = 0$$

Note that we do not need to specify the components of the magnetic field **B**, since they are defined uniquely by the components of the electric field **E**. The two fields are always perpendicular to one another, and to the propagation direction, and the ratio of the amplitude of the electric field to that of the magnetic field is always equal to c.

The values of E_{0x}, E_{0y}, ϕ_x and ϕ_y determine the way in which the direction of the electric field (and hence also of the magnetic field) varies with time. This is termed the *polarisation* of the radiation and, as we shall see later, it is important to consider it in discussing the operation of a remote sensing system.

If the effect of the variables in equations (2.11) and (2.12) is to cause the electric field vector **E** to remain pointing in the same direction, the radiation is said to be *plane polarised*. This is illustrated in figure 2.2. Clearly, this requires that the phase difference $\phi_y - \phi_x = 0$, π or $-\pi$. (We need only consider values of this phase difference in the range $-\pi$ to $+\pi$, since a value outside this range can be expressed as a value within it by adding or subtracting some integral multiple of 2π.) Although in principle the direction of the polarisation could be specified using either the electric or the magnetic field, it is conventional to use the electric field, so the example in figure 2.2 would be described as x-polarised.

If, instead of being confined to a fixed direction, the electric field vector rotates in the xy-plane with a constant amplitude, the radiation is said to be *circularly polarised* (figure 2.3). If the sense of the rotation is clockwise when viewed along the propagation direction the polarisation is called *right-hand circular* (RHC), and if anticlockwise it is *left-hand circular* (LHC). Clearly, circular polarisation requires that

$$E_{0x} = E_{0y}$$

and right-hand polarisation requires that

$$\phi_y - \phi_x = \frac{\pi}{2}$$

For left-hand polarisation,

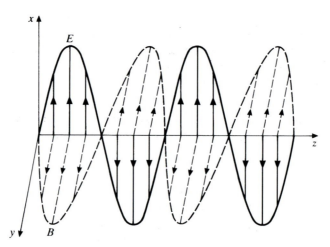

Figure 2.2. Plane-polarised radiation. The wave is propagating in the z-direction and is polarised with the electric field parallel to the x-axis and the magnetic field parallel to the y-axis. The arrows represent the instantaneous magnitudes and directions of the fields.

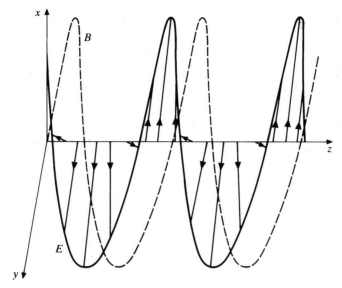

Figure 2.3. Right-hand circularly polarised radiation. The notation is the same as in figure 2.2, although the magnetic field vectors have been omitted for clarity. They are, as always, oriented perpendicularly to the electric field vectors.

$$\phi_y - \phi_x = -\frac{\pi}{2}$$

The only other kind of 'pure' polarisation (completely polarised radiation) is *elliptically polarised* radiation, in which the path traced by the electric field vector in the *xy*-plane is an ellipse. This corresponds to a phase difference of $\pm\pi/2$, but different amplitudes for the *x* and *y* components of the field. In general, the polarisation of an electromagnetic wave will be a mixture of these various types (elliptical polarisation is itself a combination of linear and circular polarisation), and may also include a *randomly polarised* component in which the direction of the electric field vector changes randomly on a time-scale too short to measure. This kind of radiation is often called *unpolarised* radiation, although this is a somewhat misleading name since it suggests that the electric field vector does not point in any direction.

There are a number of notations for specifying the polarisation state of electromagnetic radiation. One of the most common is the *Stokes vector*, the four components of which can be defined in terms of equations (2.11) and (2.12) as follows:

$$S_0 = \langle E_{0x}^2 \rangle + \langle E_{0y}^2 \rangle \tag{2.13.1}$$

$$S_1 = \langle E_{0x}^2 \rangle - \langle E_{0y}^2 \rangle \tag{2.13.2}$$

$$S_2 = \langle 2E_{0x}E_{0y}\cos(\phi_y - \phi_x)\rangle \tag{2.13.3}$$

$$S_3 = \langle 2E_{0x}E_{0y}\sin(\phi_y - \phi_x)\rangle \tag{2.13.4}$$

The angle brackets $\langle\,\rangle$ in these expressions denote time-averages.

Examples of some Stokes vectors are given below. In each case, the Stokes vector has been normalised so that $S_0 = 1$.

$$
\begin{array}{llll}
[1 & 0 & 0 & 0] & \text{Random polarisation} \\
[1 & 1 & 0 & 0] & x\text{-Polarised linear} \\
[1 & -1 & 0 & 0] & y\text{-Polarised linear} \\
[1 & 0 & 1 & 0] & +45° \text{ Linear} \\
[1 & 0 & -1 & 0] & -45° \text{ Linear} \\
[1 & 0 & 0 & 1] & \text{Right-hand circular} \\
[1 & 0 & 0 & -1] & \text{Left-hand circular} \\
[1 & 0.6 & 0 & 0.8] & \text{Right-hand elliptical, } E_{0x}/E_{0y} = 2
\end{array}
$$

The *degree of polarisation* of an electromagnetic wave is defined as the fraction of the total power that is contained in polarised components. It is given in terms of the components of the Stokes vector by

$$
\sqrt{\frac{S_1^2 + S_2^2 + S_3^2}{S_0}}
$$

It can be verified that in all of the examples above, with the exception of the first, the degree of polarisation is 1. The total flux density of the radiation is proportional to S_0, and in fact is given by

$$
F = \frac{S_0}{2Z_0} \tag{2.14}
$$

The Stokes components of two electromagnetic waves of the same frequency, travelling in the same direction, can be added provided that the two waves are *incoherent* (i.e. that there is a randomly changing phase difference between them). This allows us to 'decompose' a Stokes vector into its polarised components, together with a randomly polarised component if necessary. If a remote sensing system responds only to one polarisation state (this is a common situation for microwave systems), we will need to consider the component of the incident radiation that has that polarisation state. For example, randomly polarised radiation can be decomposed into incoherent *x*- and *y*-polarised components:

$$
[1 \quad 0 \quad 0 \quad 0] = \frac{1}{2}[1 \quad 1 \quad 0 \quad 0] + \frac{1}{2}[1 \quad -1 \quad 0 \quad 0]
$$

so that an instrument capable of detecting only *y*-polarised radiation will collect half of the power available from a randomly polarised wave. Most natural sources of radiation are randomly polarised, although, as we shall see, scattering and reflection may change the state of polarisation.

2.3   Spectra and the Fourier transform

Up to this point we have said nothing about the frequency (or wavelength) of the radiation, other than that electromagnetic radiation may, in principle, have any frequency we wish. It will often happen, however, that we wish to describe a particular radiation field in which a number (possibly a continuous distribution) of frequencies is present. This can be done by specifying the complete waveform, which obviously contains all the necessary information, or the *spectrum* of the radiation – the amplitudes of the various frequency components that are present in the waveform. These two methods are equivalent, and it is important to know how to convert from one description to another. The conversion is achieved using the *Fourier transform*, and since this is of great importance in many aspects of remote sensing it is worth deriving the theory.

It will be convenient to use the *complex exponential* notation to describe sinusoidal or cosinusoidal components, since it greatly simplifies the following analysis. Using this notation, we express a variation having angular frequency ω and amplitude A as

$$A \exp(i\omega t) \tag{2.15}$$

where i is the square root of -1 and 'exp' is the exponential function; that is,

$$\exp(x) \equiv e^x$$

By allowing A to take complex values, and adopting the convention that it is the *real part* of equation (2.15) that corresponds to the variation of the physical quantity, we can represent both sinusoidal and cosinusoidal components. We can see this by writing A in terms of its real and imaginary parts, expanding $\exp(i\omega t)$ as $\cos(\omega t) + i \sin(\omega t)$, and taking the real part of equation (2.15):

$$\mathrm{Re}((\mathrm{Re}(A) + i\,\mathrm{Im}(A))(\cos(\omega t) + i\,\sin(\omega t))) = \mathrm{Re}(A)\,\cos(\omega t) - \mathrm{Im}(A)\,\sin(\omega t)$$

Let us suppose that some time-varying quantity (e.g. the electric field amplitude at a given location as an electromagnetic wave passes through it) is written as a function of time $f(t)$, and that it is also possible to express it as the sum of components of various angular frequencies ω. If the distribution of frequencies is continuous, the amount of each frequency present can be expressed by a density function $a(\omega)$, such that the total amplitude of the components having frequencies in the range ω to $\omega + d\omega$ ($d\omega$ being very small) is $a(\omega)\,d\omega$. Thus, the contribution from this range of frequencies is written as

$$a(\omega)\,d\omega\,\exp(i\omega t)$$

and the sum of the contributions from all frequencies can be obtained by integrating this expression:

$$f(t) = \int_{-\infty}^{\infty} a(\omega)\,\exp(i\omega t)\,d\omega \tag{2.16}$$

So far, this is merely an assertion. We have neither proved that the distribution $a(\omega)$ uniquely represents $f(t)$, nor shown how to find $a(\omega)$ given $f(t)$. It is beyond our scope to find a rigorous answer to the former problem, so we shall content ourselves with answering the latter.

If we multiply equation (2.16) by $\exp(i\omega't)$, where ω' is an arbitrary angular frequency, we obtain

$$f(t)\exp(i\omega't) = \int_{-\infty}^{\infty} a(\omega)\,\exp(i[\omega+\omega']t)\,d\omega$$

Next, we integrate this with respect to t, giving

$$\int_{-\infty}^{\infty} f(t)\,\exp(i\omega't)\,dt = \int_{-\infty}^{\infty}\int_{-\infty}^{\infty} a(\omega)\,\exp(i[\omega+\omega']t)\,d\omega\,dt$$

$$= \int_{-\infty}^{\infty} a(\omega)\int_{-\infty}^{\infty} \exp(i[\omega+\omega']t)\,d\omega\,dt$$

Now

$$\int_{-\infty}^{\infty} \exp(i\alpha t)\,dt$$

is a function of α which is zero everywhere but at $\alpha = 0$, where it is infinite. The area underneath a graph of this function is, however, finite, and has a value of 2π. This can be written as

$$\int_{-\infty}^{\infty} \exp(i\alpha t)\,dt = 2\pi\,\delta(\alpha)$$

where $\delta(\alpha)$ is the *Dirac delta-function*. Thus, we have

$$\int_{-\infty}^{\infty} f(t)\,\exp(i\omega't) = 2\pi\,a(-\omega')$$

which can be rewritten, by changing the symbols and rearranging the expression, as

$$a(\omega) = \frac{1}{2\pi}\int_{-\infty}^{\infty} f(t)\,\exp(-i\omega t)\,dt \qquad (2.17)$$

This is very similar to equation (2.16) and shows that, apart from a change of sign and scale, $a(\omega)$ is obtained from $f(t)$ in exactly the same way as $f(t)$ is obtained from $a(\omega)$. The integral transforms defined by equations (2.16) and (2.17) are called Fourier transforms, although we should note that some authors increase the symmetry between (2.16) and (2.17) still further by writing

$$f(t) = \frac{1}{\sqrt{2\pi}} \int\limits_{-\infty}^{\infty} a(\omega)\,\exp(i\,\omega\,t)\,d\omega$$

$$a(\omega) = \frac{1}{\sqrt{2\pi}} \int\limits_{-\infty}^{\infty} f(t)\,\exp(-i\,\omega\,t)\,dt$$

Let us apply the Fourier transform to a practical example. Suppose we have a waveform $f(t)$ that consists of a single angular frequency ω_0, which is turned on for a finite time T (figure 2.4). What is its spectrum $a(\omega)$? We might think that since we only used one frequency to construct $f(t)$, the spectrum would consist of a single spike or delta-function at that frequency. However, this cannot be correct since the spectrum $a(\omega)$ has to contain *all* the information contained by $f(t)$, including the fact that the waveform drops abruptly to zero for $|t| > T/2$. Using equation (2.17), then, we find that

$$a(\omega) = \frac{1}{2\pi} \int\limits_{-T/2}^{T/2} \cos(\omega_0 t)\,\exp(-i\,\omega k t)\,dt$$

$$= \frac{1}{2\pi}\left[\frac{\sin(\omega_0 - \omega)T/2}{\omega_0 - \omega} + \frac{\sin(\omega_0 + \omega)T/2}{\omega_0 + \omega}\right]$$

This is evidently the sum of two functions, each of the form $(\sin x)/x$, centred at frequencies ω_0 and $-\omega_0$. The function $(\sin x)/x$, often called sinc(x), is shown in figure 2.5. (Note that some authors define sinc(x) to be $(\sin \pi x)/(\pi x)$).

Thus, the complete spectrum of the waveform whose time dependence was shown in figure 2.4 is shown by figure 2.6. It can be seen that the delta-functions that we initially expected at $\omega = \pm\omega_0$ have been spread out over a range $2\delta\omega$ in frequency, where

$$\delta\omega \approx \frac{2\pi}{T} \tag{2.18}$$

or $\delta f \approx 1/T$. This is in fact a general result of fundamental importance: in order to represent a waveform of duration δt, we need a range of frequencies of at least $\pm 1/\delta t$. It is a form of 'uncertainty principle'. Defining exactly what is

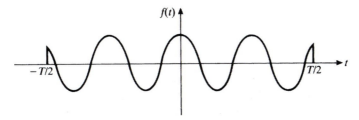

Figure 2.4. A truncated cosine wave. The Fourier transform of this function is shown in figure 2.6.

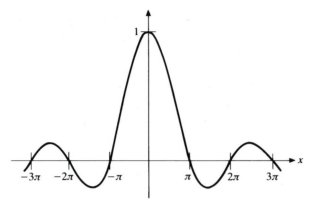

Figure 2.5. The function sinc(x), defined as (sin x)/x.

usefully meant by 'length' and 'range' is not always obvious, so we will leave equation (2.18) in its approximate form, although a more exact formulation of the result is possible.

A related result – which is, however, quite exact – is the *Nyquist sampling theorem*. This states that if a signal is to be sampled at regular intervals, the sampling frequency must exceed some minimum value if it is to be possible to reconstruct the original signal unambiguously from the samples. This frequency is the *Nyquist frequency*, and it is twice the bandwidth of the signal. The bandwidth is defined as the range of frequencies f over which the signal spectrum is non-zero. If the signal is undersampled – that is, sampled at a rate below the Nyquist frequency – aliases are introduced which, amongst other undesirable effects, degrade the signal-to-noise ratio. The practical implications of the Nyquist theorem are many, but it clearly finds an important application in the design of electronic systems in which a signal is first filtered to define a bandwidth, and then sampled at regular intervals.

2.4   The Doppler effect

If a source of electromagnetic radiation of frequency f is in motion with respect to an observer (e.g. a sensor), the observer will in general detect the radiation at

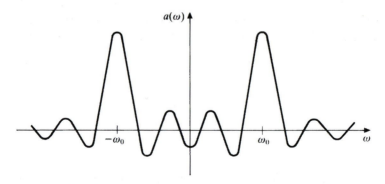

Figure 2.6. The Fourier transform of the function shown in figure 2.4.

a different frequency f'. If the source is approaching the observer, or equivalently if the observer is approaching the source, f' will be greater than f, and conversely. This is known as the Doppler effect, and is analogous to the similar (and familiar) effect observed with sound waves. However, whereas the Doppler effect for sound is not the same for the source approaching the observer and for the observer approaching the source, the effect is symmetrical in this manner for electromagnetic radiation in free space. The result has to be derived using Einstein's Special Theory of Relativity, so it will merely be stated here.

If the source S approaches the observer O with a velocity v directed at an angle θ to the line of sight, as shown in figure 2.7, the Doppler shift is given by

$$\frac{f'}{f} = \frac{\sqrt{1 - \dfrac{v^2}{c^2}}}{1 - \dfrac{v \cos \theta}{c}} \tag{2.19}$$

where c is the speed of light. However, in all cases that will concern us, the relative speed v will be very much smaller than the speed of light, in which case a very good approximation to equation (2.19) is given by

$$\frac{f'}{f} = 1 + \frac{v \cos \theta}{c} \tag{2.20}$$

For example, if a satellite is travelling away from an observer on the Earth with a speed of $7\,\mathrm{km\,s^{-1}}$ at an angle of $10°$ to the line of sight (thus, $\theta = 170°$), and it emits a signal with a frequency of exactly $5\,\mathrm{GHz}$, the received frequency will be $4.999885\,\mathrm{GHz}$. In other words, the frequency has been shifted downwards by $115\,\mathrm{kHz}$. The error in calculating this shift using the approximate equation (2.20) is only about $1\,\mathrm{Hz}$ and may be ignored.

Although it is small, a consideration of the Doppler effect is important for some radar systems, particularly the synthetic aperture radar systems discussed in chapter 9.

2.5   Describing angular distributions of radiation

We have already seen how to describe electromagnetic radiation that contains a range of frequencies or a range of polarisations. Up to this point, however, we have considered only collimated radiation: that is, radiation travelling in a

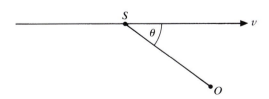

Figure 2.7. The Doppler effect. The source of electromagnetic radiation is located at S, travelling with velocity v. The observer is located at O.

single direction. It is clear that we will also need to be able to describe radiation distributed over a range of directions in space. The radiometric quantities introduced in this section are also discussed by Curran (1985).

Let us begin by considering a plane surface that is illuminated by radiation from a variety of directions. To specify a particular direction of incident radiation we will need two angles: θ, the angle between the propagation direction and the normal to the surface element; and ϕ, the azimuthal angle, measured around the normal in the plane of the surface (see figure 2.8). Now we consider an element dA of this surface, and radiation incident from the range of directions between θ and $\theta + d\theta$ and between ϕ and $\phi + d\phi$ (figure 2.9). The solid angle (unit: steradian; symbol: sr) defined by this range of directions is

$$d\Omega = \sin\theta \, d\theta \, d\phi \qquad (2.21)$$

and it is clear that the power incident on the element dA from this range of directions must be proportional to dA and $d\Omega$ as well as to a term that defines the strength of the radiation. This relationship can be expressed as

$$dP = L \cos\theta \, dA \, d\Omega \qquad (2.22)$$

where dP is the contribution to the power incident on the area dA from solid angle $d\Omega$ in the direction (θ, ϕ), and L is the *radiance* of the incident radiation in that direction. From this definition, it follows that the SI unit of radiance is $\mathrm{W\,m^{-2}\,sr^{-1}}$.

The inclusion of the factor $\cos\theta$ in equation (2.22) seems perverse at first sight. However, it gives the radiance the valuable property that, if the medium through which the radiation propagates does not scatter or absorb and has a constant refractive index (and these conditions are obviously all met if we are still considering radiation in free space), then the radiance is constant along any ray. The concept of radiance is of prime importance in considering measurements made by optical and near-infrared remote sensing systems, discussed in chapter 6.

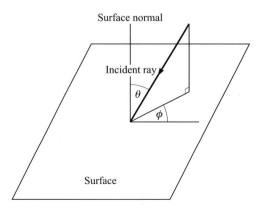

Figure 2.8. Definition of the angles θ and ϕ to describe the angular distribution of radiation.

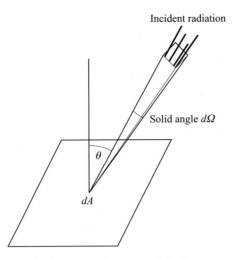

Figure 2.9. Geometrical construction to explain the concept of radiance.

The *irradiance E* at the surface is defined as the total incident power per unit area, and its SI unit is $\mathrm{W\,m^{-2}}$. It is found by integrating equation (2.22) over all the directions for which $\theta \leq \pi/2$; namely, the hemisphere of directions from which the surface can be illuminated:

$$E = \int_{\theta=0}^{\pi/2} \int_{\phi=0}^{2\pi} L_{\mathrm{incident}} \cos\theta \, d\Omega \qquad (2.23)$$

Although the radiance may be a function of direction, the irradiance clearly cannot be.

We can use the same ideas to describe radiation emitted or reflected from a surface. Since the concept of radiance describes radiation in space, the same terminology will suffice for both incoming and outgoing radiation. All we need to do is to 'label' the radiation so that we know in which direction it is propagating, and terms such as 'upwelling' and 'downwelling' radiation are frequently used for this purpose. The outgoing analogue of irradiance is termed *radiant exitance*, and given the symbol M:

$$M = \int_{\theta=0}^{\pi/2} \int_{\phi=0}^{2\pi} L_{\mathrm{outgoing}} \cos\theta \, d\Omega \qquad (2.24)$$

For *isotropic* radiation, the radiance is independent of direction. In this case, the relationship between the radiance and the exitance is given by

$$M = L \int_{\theta=0}^{\pi/2} \int_{\phi=0}^{2\pi} \cos\theta \, d\Omega = \pi L \qquad (2.25)$$

2.6   Thermal radiation

Thermal radiation is emitted by all objects above absolute zero (−273.15°C – see box) and is, at first or second hand, the radiation that is detected by the great majority of passive remote sensing systems.

THE ABSOLUTE TEMPERATURE SCALE

In describing thermal radiation, it is convenient to use the absolute scale in which temperatures are measured in *Kelvin* (K). The relationship between a temperature T in Kelvin and a temperature t in degrees Celsius is

$$T = t + 273.15-$$

In general, a hot object (by which, for the present, we mean one that is not at absolute zero) will distribute its emission over a range of wavelengths in a continuous spectrum. To describe this radiation we can use the same radiometric quantities that were defined in section 2.5, but we need to modify the definitions to include the variation with wavelength or frequency. This is done by defining the *spectral radiance* L_λ such that the radiance ΔL contained in a small range of wavelengths $\Delta \lambda$ is given by

$$\Delta L = L_\lambda \Delta \lambda \tag{2.26}$$

In other words, L_λ is just the differential of L with respect to λ, or more strictly the absolute value (modulus) of this differential:

$$L_\lambda = \left| \frac{\partial L}{\partial \lambda} \right| \tag{2.27}$$

It is clear that the SI unit of spectral radiance is $\mathrm{W\,m^{-2}\,sr^{-1}\,m^{-1}}$, although the unit $\mathrm{W\,m^{-2}\,sr^{-1}\,\mu m^{-1}}$ is also commonly used.

The spectral radiance can also be defined in terms of the frequency f:

$$L_f = \left| \frac{\partial L}{\partial f} \right| \tag{2.28}$$

so that its unit is $\mathrm{W\,m^{-2}\,sr^{-1}\,Hz^{-1}}$, and the relationship between the definitions (2.27) and (2.28) is therefore given by

$$\frac{L_\lambda}{L_f} = \left| \frac{df}{d\lambda} \right| = \frac{c}{\lambda^2} = \frac{f^2}{\lambda} \tag{2.29}$$

where c is the speed of light.

All of the radiometric quantities defined in section 2.5, not just the radiance, can similarly be defined spectrally.

If we make a closed cavity with opaque walls, and hold the cavity at an absolute temperature T, the electromagnetic radiation inside it is known as *black-body radiation*. The spectral radiance of this radiation was calculated by Planck, using quantum mechanics (see e.g. Longair, 1984), during the early years of the twentieth century. It is

$$L_f = \frac{2hf^3}{c^2(e^{hf/kT} - 1)} \qquad (2.30)$$

which may also be expressed, using equation (2.29), as

$$L_\lambda = \frac{2hc^2}{\lambda^5(e^{hc/\lambda kT} - 1)} \qquad (2.31)$$

In these equations, h is the Planck constant and k is the Boltzmann constant. Equation (2.31) is plotted in figure 2.10 for two different temperatures. Note the steep rise at short wavelengths, and the long tail at long wavelengths.

The radiation inside a closed cavity may not seem particularly interesting or relevant, but we may observe it by making a small hole in the cavity and letting some of it escape. In this case, equations (2.30) or (2.31) describe the radiation emerging from the hole, and from any *black body* (perfect emitter of thermal radiation) at temperature T.

At sufficiently long wavelengths, equation (2.30) can be approximated as

$$L_f \approx \frac{2kTf^2}{c^2} = \frac{2kT}{\lambda^2} \qquad (2.32)$$

This is called the *Rayleigh–Jeans approximation*, and corresponds to the right-hand half of figure 2.10 where the graphs can be approximated as straight lines with a slope of -2. The condition for this approximation to be valid is

$$\frac{hc}{\lambda kT} \ll 1$$

or equivalently

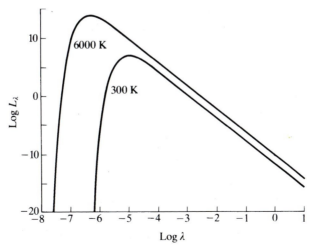

Figure 2.10. Black-body radiation according to the Planck law. The graphs show $\log_{10} L_\lambda$ in units of $\mathrm{W\,m^{-3}\,sr^{-1}}$ plotted against $\log_{10} \lambda$ in metres.

$$\frac{hf}{kT} \ll 1$$

For $T = 280\,\text{K}$, this gives $f \ll 6000\,\text{GHz}$ or $\lambda \gg 50\,\mu\text{m}$, so the approximation is valid for microwave and radio frequencies for objects at typical terrestrial temperatures.

We can integrate the Planck formula (either equation (2.30) or (2.31), it does not matter which) to calculate the total radiance of black-body radiation over all wavelengths:

$$L = \int_0^\infty L_\lambda \, d\lambda = \frac{2\pi^4 k^4}{15 c^2 h^3} T^4 \tag{2.33}$$

Since the radiation is isotropic, the total radiant exitance M is found, using equation (2.25), to be

$$M = \pi L = \frac{2\pi^5 k^4}{15 c^2 h^3} T^4$$

This is normally written more compactly as

$$M = \sigma T^4 \tag{2.34}$$

where $\sigma = 2\pi^5 k^4 / 15 c^2 h^3 \approx 5.67 \times 10^{-8}\,\text{W}\,\text{m}^{-2}\,\text{K}^{-4}$ is called the Stefan–Boltzmann constant, and equation (2.34) is called Stefan's law. It shows how much power is emitted by a black body at temperature T, integrated over all wavelengths. If we want to know how this power is distributed in wavelength, we can of course use equation (2.30) directly, but it may be sufficient merely to know the wavelength λ_{max} at which L_λ reaches a maximum. This is found by differentiating equation (2.30), which shows that

$$\lambda_{\text{max}} = \frac{A}{T} \tag{2.35}$$

where A is a constant whose value is about $2.898 \times 10^{-3}\,\text{K}\,\text{m}$. Equation (2.35) is called Wien's law, or Wien's displacement law. For example, the Sun is a good approximation to a black body at a temperature of 5800 K, so the peak spectral radiance occurs at $\lambda_{\text{max}} \approx 0.50\,\mu\text{m}$, in the middle of the visible spectrum – where we expect it to be. If, on the other hand, we consider a black body at a temperature of 280 K, which is fairly typical of temperatures on the Earth's surface, we find $\lambda_{\text{max}} \approx 10.3\,\mu\text{m}$, in the thermal infrared region of the electromagnetic spectrum.

We may also occasionally need to calculate the radiance or radiant exitance of a black body over a finite range of wavelengths. This can be simplified a little by integrating with respect to a dimensionless variable. Specifically, we can put

$$\int_{\lambda_1}^{\lambda_2} M_\lambda \, d\lambda = \sigma T^4 (f(x_1) - f(x_2)) \tag{2.36}$$

where the dimensionless variables x_1 and x_2 are defined by

$$x_1 = \frac{hc}{\lambda_1 kT}, \qquad x_2 = \frac{hc}{\lambda_2 kT}$$

and the function $f(x)$ is defined by

$$f(x) = \frac{15}{\pi^4} \int_0^x \frac{z^3 \, dz}{e^z - 1} \tag{2.37}$$

This integral cannot be evaluated analytically, although numerical integration using computer programs such as *Mathematica, Maple,* or *Matlab* is straight-forward. In cases where no such program is available, the function $f(x)$ is tabulated in table 2.1, and the box (see e.g. Houghton, 1986) shows how the function may be approximated.

We remarked earlier that a small hole in the wall of a cavity behaves as a black body. This is not a particularly plausible model for real materials, so we introduce the idea of the *emissivity* ε to relate the actual radiance of a body at temperature T to the black-body value. (Note that emissivity and dielectric constant are both conventionally denoted by the symbol epsilon, which has potential for confusion. The usage is too well established, however, for us to introduce a different notation, and we will rely on the context, or an explicit statement, to differentiate between them.) The emissivity is often dependent on

Table 2.1. Integral of the Planck distribution function
The table gives values of the function $f(x)$ defined in the text.

x	$f(x)$	x	$f(x)$
0	0	1.4	0.08040
0.10	0.00005	1.6	0.11023
0.12	0.00009	1.8	0.14402
0.14	0.00013	2.0	0.18115
0.16	0.00020	2.5	0.28403
0.18	0.00028	3.0	0.39302
0.20	0.00038	3.5	0.49938
0.25	0.00073	4.0	0.59703
0.30	0.00124	4.5	0.68251
0.35	0.00193	5.0	0.75453
0.40	0.00282	6.0	0.86016
0.45	0.00394	7.0	0.92443
0.50	0.00529	8.0	0.96084
0.60	0.00879	9.0	0.98039
0.70	0.01341	10.0	0.99045
0.80	0.01923	12.0	0.99788
0.90	0.02629	14.0	0.99956
1.0	0.03462	16.0	0.99991
1.2	0.05506	18.0	0.99998
		20.0	1.00000

> ### APPROXIMATIONS TO THE PLANCK INTEGRAL
>
> The function defined by equation (2.37) can be expanded as
>
> $$f(x) = 1 - \frac{15}{\pi^4} \sum_{m=1}^{\infty} e^{-mx} \left(\frac{x^3}{m} + \frac{3x^2}{m^2} + \frac{6x}{m^3} + \frac{6}{m^4} \right)$$
>
> although for small values of x the convergence is rather slow. For $x \geq 3$, the first three terms are sufficient to give five significant figures in the value of $1 - f(x)$. For $x \leq 0.5$, the power series
>
> $$f(x) \approx \frac{15}{\pi^4} \left(\frac{x^3}{3} - \frac{x^4}{8} + \frac{x^5}{60} - \frac{x^7}{5040} \right)$$
>
> gives an accuracy of five significant figures.

wavelength, so in general we should write it as $\varepsilon(\lambda)$, and we can define it through

$$L_\lambda = \varepsilon(\lambda) L_{\lambda,p} \tag{2.38}$$

where we have now written $L_{\lambda,p}$ for the black-body radiance defined by equation (2.31) (the 'p' stands for 'Planck'). A simple thermodynamic argument shows that a body which is a good emitter (high ε) must also be a good absorber of radiation – in fact the two factors must be equal (this is Kirchhoff's law of radiation). We can see this quite easily by realising that any body at temperature T must be in equilibrium with black-body radiation whose spectrum corresponds to the same temperature. If, say, the body absorbs better than it emits, it will heat up, and thus cannot in fact be at equilibrium. Thus, the reflectivity is given by $1 - \varepsilon$. It also follows from this argument that the emissivity must lie between 0 and 1. The factors that determine emissivity are discussed in more detail in sections 3.5.2 and 3.5.3.

It is often convenient, especially when discussing passive microwave systems (chapter 7), to define the *brightness temperature* of a body that is emitting thermal radiation. This is the temperature of the equivalent black body that would give the same radiance at the wavelength under consideration. By combining equations (2.31) and (2.38), we can see that at wavelength λ, a body with temperature T and emissivity ε has a brightness temperature T_b that is given by

$$\varepsilon \frac{2hc^2}{\lambda^5(e^{hc/\lambda kT} - 1)} = \frac{2hc^2}{\lambda^5(e^{hc/\lambda kT_b} - 1)}$$

The solution of this equation for T_b is

$$T_b = \frac{hc}{\lambda k \ln\left(1 + \dfrac{1}{\varepsilon}(e^{hc/\lambda kT} - 1)\right)} \tag{2.39}$$

but at sufficiently long wavelengths (high frequencies) this can be approximated very simply, using the Rayleigh–Jeans approximation, as

$$T_b = \varepsilon T \tag{2.40}$$

We saw earlier that a black body at a typical terrestrial temperature of 280 K will radiate maximally at a wavelength of 10.3 μm. How well does it radiate at other wavelengths? Specifically, let us calculate the fraction of the total radiant exitance that is emitted in four wavelength ranges: 0.5–0.6 μm, 1.55–1.75 μm, 10.5–12.5 μm and 1.52–1.56 cm. These have been chosen to be typical of remote sensing measurements in the optical, near-infrared, thermal infrared and passive microwave regions, respectively. Using the methods described from equation (2.36) onwards, we find that these fractions are approximately 6×10^{-33}, 7×10^{-10}, 0.12 and 1×10^{-10}, respectively. This illustrates the very rapid fall in the Planck function at shorter wavelengths and the much slower decline at longer wavelengths. It also shows that, while objects at normal terrestrial temperatures do not emit thermal radiation in the form of visible light (which is a fact of everyday experience), small but potentially measurable quantities of radiation are emitted in the microwave region. In fact, it is possible to build receivers sensitive enough to detect this microwave radiation, and this forms the basis of the passive microwave radiometry techniques that will be discussed in chapter 7.

2.6.1   Characteristics of solar radiation

By way of illustration, we will apply some of the results of sections 2.5 and 2.6 to characterise radiation from the Sun. To a fairly good approximation the Sun can be taken to be a *grey body* (i.e. it has a constant emissivity over the range of wavelengths at which emission is significant) with an effective temperature T of about 5800 K and an emissivity of 0.99. It can be assumed to be a sphere of radius $r = 6.96 \times 10^8$ m located a distance $D = 1.50 \times 10^{11}$ m from the Earth.

From equations (2.34) and (2.38), we can write the Sun's radiant exitance, integrated over all wavelengths, as

$$M = \varepsilon \sigma T^4 = 6.35 \times 10^7 \, \mathrm{W\,m^{-2}}$$

The total power radiated by the Sun is obtained by multiplying this by the Sun's surface area:

$$P = 4\pi r^2 \varepsilon \sigma T^4 = 3.87 \times 10^{26} \, \mathrm{W}$$

By considering a sphere of radius D centred on the Sun, we can see that the irradiance at the Earth (but above the Earth's atmosphere, so we that we do not need to consider atmospheric absorption) is given by

$$E = \frac{P}{4\pi D^2} = 1.37 \times 10^3 \, \mathrm{W\,m^{-2}}$$

This value is often called the *mean exoatmospheric irradiance*. We can calculate the corresponding exoatmospheric radiance L by considering the range of directions over which this radiation is distributed. Seen from a distance D, the Sun subtends a solid angle

$$\Delta\Omega = \frac{\pi r^2}{D^2}$$

which is much less than 1, so a sufficiently accurate estimate is given by

$$L = \frac{E}{\Delta\Omega} = \frac{\varepsilon\sigma T^4}{\pi} = 2.02 \times 10^7 \, \text{W}\,\text{m}^{-2}\,\text{sr}^{-1}$$

This radiance is confined to the range $\Delta\Omega$ of solid angle. Outside this range, the radiance is of course zero.

We can also calculate the exoatmospheric radiance spectrally, taking equation (2.31) as our starting point and following the same procedure. We find that

$$L_\lambda = \frac{2\varepsilon hc^2}{\lambda^5(e^{hc/\lambda kT} - 1)}$$

which of course is just the Planck formula for the radiance, modified by equation (2.37) to take account of the emissivity. For example, at a wavelength of $0.5\mu\text{m}$ this gives $L_\lambda = 2.65 \times 10^{13}\,\text{W}\,\text{m}^{-2}\,\text{sr}^{-1}\,\text{m}^{-1}$. The corresponding spectral irradiance is obtained by multiplying this by $\Delta\Omega$ to give $E_\lambda = 1.79 \times 10^9\,\text{W}\,\text{m}^{-2}\,\text{m}^{-1}$, which can also be expressed in less standard but more common units as $1.79 \times 10^3\,\text{W}\,\text{m}^{-2}\,\mu\text{m}^{-1}$ or even as $179\,\text{mW}\,\text{cm}^{-2}\,\mu\text{m}^{-1}$.

2.7 Diffraction

We conclude this review of the propagation of electromagnetic radiation in free space by discussing diffraction. Diffraction can be roughly defined as the changes that occur to the direction of electromagnetic radiation when it encounters an obstructing obstacle of some kind. It could therefore be argued that, since the radiation is interacting with matter (the obstacle), the phenomenon should be discussed in chapter 3. However, it is more convenient to treat it here since (a) we will assume that, until the radiation encounters the obstacle and after it has left it, it is propagating in free space, and (b) our approach will build upon the discussion of Fourier transforms developed in section 2.3. The treatment presented here, which will lead to results that are of fundamental importance in understanding the *spatial resolution* of remote sensing systems, will be very brief. Much fuller treatments can be found in any textbook on optics: for example, Hecht (1987) or Lipson et al. (1995).

We shall begin by considering plane parallel radiation (i.e. radiation travelling in a single direction) incident on a very long slit, of width w, in an infinite opaque screen. The slit has its long axis parallel to the x-axis of a Cartesian coordinate system, and the centre of the slit is located at the origin of this coordinate system. We wish to determine the amplitude of the electric field at the point P shown in figure 2.11.

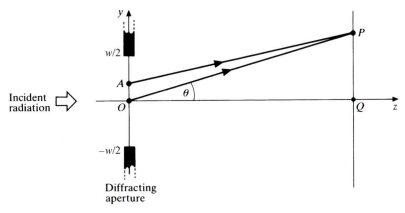

Figure 2.11. Geometry of Fraunhofer diffraction at a slit.

If the distance z is sufficiently large (we will discuss later how large it needs to be), the rays OP and AP may be regarded as parallel, and AP is shorter than OP by $y \sin \theta$. The phase difference between the two rays is thus $ky \sin \theta$, where k is the wavenumber of the radiation. If this condition, that the phase difference in a given direction varies linearly with the position in the slit, is met, what we are describing is termed *Fraunhofer diffraction*. The complex amplitude at P contributed by an element of the slit of width dy, located at A, is thus proportional to

$$\exp(iky \sin \theta) \, dy$$

(We are ignoring the reduction of amplitude with distance due to geometrical spreading, as well as one or two other effects.) The total amplitude at P is found by integrating this expression over the entire slit:

$$a(\theta) = \int_{-w/2}^{w/2} \exp(iky \sin \theta) \, dy$$

This expression looks very similar to the Fourier transform defined in equation (2.16). We can make the correspondence exact by introducing the idea of an *amplitude transmittance function* $f(y)$ for the plane of the screen, which defines the fraction of the incident amplitude that is transmitted. For the slit we have been discussing, $f(y) = 1$ for $-w/2 < y < w/2$, and 0 everywhere else. Using $f(y)$ to characterise any general one-dimensional aperture distribution, the expression for the complex amplitude in the direction θ becomes

$$a(\theta) = \int_{-\infty}^{\infty} f(y) \exp(iky \sin \theta) \, dy \qquad (2.41)$$

which is clearly a Fourier transform, though it is often called the *Fraunhofer diffraction integral*.

In section 2.3 we identified time t and angular frequency ω as a pair of conjugate variables related by the Fourier transform; here, the corresponding variables are y and $(k \sin \theta)$. Again, a form of uncertainty principle applies. Evaluating the integral (2.41) for our slit of width a, we find that

$$a(\theta) \propto \operatorname{sinc}\left(\frac{kw \sin \theta}{2}\right)$$

This is a function that has the same shape as figure 2.5, and it first falls to zero when $\sin \theta = \pm 2\pi/kw = \pm \lambda/w$. If $w \gg \lambda$, $\sin \theta$ will be much less than 1, so that we can put $\sin \theta \approx \theta$, and hence

$$\delta\theta \approx \frac{\lambda}{w} \tag{2.42}$$

This is the result that corresponds to equation (2.18), and it shows that if a beam of plane parallel radiation of wavelength λ passes through an aperture of width w, it will spread into a diverging beam whose angular width will be of the order of λ/w radians.

Equation (2.41) applies to one-dimensional diffraction; namely, to the case in which the amplitude transmission function depends only on y. For the two-dimensional case in which the amplitude transmission function must be written as $f(x, y)$, the diffraction integral becomes

$$a(\theta_x, \theta_y) = \int\limits_{-\infty}^{\infty} \int\limits_{-\infty}^{\infty} f(x, y) \exp(ikx \sin \theta_x) \exp(iky \sin \theta_y) \, dx \, dy \tag{2.43}$$

This double integral is rather hard to solve in general, although there are two special cases that should be mentioned. The first is when $f(x, y)$ can be factorised into two independent parts: $f(x, y) = g(x)h(y)$. The double integral can then be factorised into the product of two single integrals of the form of equation (2.41). This approach allows us to calculate, for example, the diffraction pattern of rectangular apertures. The second special case is when the amplitude transmission function has circular symmetry. In this case, it is simpler to use polar coordinates. We shall need only one result for general reference, and that is the diffraction pattern of a uniform circular aperture of diameter D. The amplitude of the diffracted wave in this case is given by

$$a(\theta_r) \propto \frac{J_1\left(\dfrac{kD \sin \theta_r}{2}\right)}{\left(\dfrac{kD \sin \theta_r}{2}\right)} \tag{2.44}$$

where $J_1(x)$ is the first-order Bessel function and θ_r is the radial angle. The function in equation (2.43) is sketched in figure 2.12. $J_1(x)$ first falls to zero when $x = 3.832$, so the first zero occurs when $\sin(\theta_r) = 7.66/kD = 1.22\lambda/D$.

Now we return to the comment we made regarding figure 2.11, that the distance z must be large enough for the two rays OP and AP to be regarded

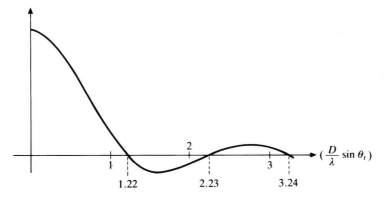

Figure 2.12. The Fraunhofer diffraction pattern of a circular aperture of diameter D. θ_r is the radial angle (i.e. the angle from the normal to the plane of the aperture).

as parallel. How large is this? We assume conventionally that the Fraunhofer description is valid if the phase differences computed by its use are accurate to within $\pi/2$ radians. Inspection of figure 2.11 shows that this is equivalent to putting

$$AQ - OQ < \frac{\lambda}{4}$$

and since OA takes a maximum value of $w/2$ we may use Pythagoras' theorem to derive

$$\sqrt{\left(\frac{w}{2}\right)^2 + z^2} - z < \frac{\lambda}{4}$$

Now if $w/2 \ll z$, we can use the binomial approximation to simplify this condition to

$$\frac{w^2}{8z} < \frac{\lambda}{4}$$

or

$$z > \frac{w^2}{2\lambda} \equiv z_F \tag{2.45}$$

The distance z_F is often called the *Fresnel distance*, after A. Fresnel who made many important discoveries in physical optics in the early nineteenth century, and if the condition (2.44) is not satisfied a more rigorous form of diffraction theory, known as *Fresnel diffraction*, must be used. The region in which $z < z_F$ is often called the *near field*, and $z > z_F$ is the *far field*.

As was mentioned earlier, one important practical implication of diffraction is that it limits the spatial resolution of a remote sensing system. Without developing a rigorous theory for this phenomenon, we can see the principles involved by considering a very simple system (figure 2.13) consist-

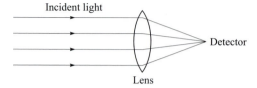

Figure 2.13. Ray diagram for a simple model of a remote sensing system.

ing of a lens arranged so that it focusses plane parallel light onto an extremely small detector. Figure 2.13 is a ray diagram, so it includes the phenomena of geometric optics but not the effects of diffraction. As it stands, the diagram implies that all of the radiation that reaches the detector was originally travelling in the same direction; namely, that it subtended an angular width of zero. In fact, a range of incident directions will contribute to the signal that reaches the detector. By imagining that the light is propagating in the opposite direction to that shown in the figure, from the detector to the lens, we can see that the effect of the finite aperture represented by the lens will be to spread the outgoing light into a cone with an angular width of the order of λ/D, where D is the diameter of the lens. Thus, in general, we expect that diffraction will limit the angular resolution of any remote sensing system to $\approx \lambda/D$, where D is the width of the lens, antenna, mirror, or whatever is at the 'front end' of the system to define the spatial extent of the wavefront captured by the system. Other parts of the system may further degrade the resolution, of course.

We can illustrate this calculation with two examples. The first is a spaceborne optical sensor operating at a wavelength of 0.5μm and with a lens diameter of 5 cm. Using the formula λ/D we find that the angular resolution is limited to about 10^{-5} radians (about 2 seconds of arc), which corresponds to a spatial resolution of about 10 m at a distance of 1000 km. The Fresnel distance (equation (2.45)) is 2.5 km, so the simple λ/D calculation is valid. This is in fact typical of the spatial resolution of many spaceborne optical remote sensing systems. The second example is a passive microwave radiometer operating at a wavelength of 3 cm and with an antenna diameter of 1 m. In this case, the angular resolution is about 0.03 radians (1.7 degrees), corresponding to a spatial resolution of 30 km at a distance of 1000 km (the Fresnel distance is 17 m). Thus, we can see why the passive microwave systems to be discussed in chapter 7 have very much poorer angular resolution than the optical and infrared systems described in chapter 6.

Finally, it should be noted that some microwave remote sensing systems have been designed to circumvent the diffraction limit. The methods by which this is possible are discussed in chapters 8 and 9.

PROBLEMS

1. The electric field of an electromagnetic wave in free space is given by

$$E_x = 0$$
$$E_y = E \cos(\omega t - kx)$$
$$E_z = 2E \cos(\omega t - kx)$$

where $E = 1 \, \text{kV m}^{-1}$. Find the corresponding magnetic field and the flux density of the radiation.

2. When radiation having Stokes vector $\mathbf{S} = [S_0, S_1, S_2, S_3]$ is incident on an antenna that receives only linearly x-polarised radiation, the detected power is proportional to $\mathbf{S} \cdot \mathbf{P}$ where $\mathbf{P} = [1, 1, 0, 0]$. Show how the detected power varies with the polarisation state for radiation of a given flux density.

3. Prove equation (2.32).

4. (For mathematical enthusiasts.) Show that the Fourier transform of the Gaussian function

$$f(t) = \exp\left(-\frac{(t - t_0)^2}{2\sigma^2}\right)$$

is proportional to

$$\exp\left(-\frac{i\omega t_0}{2} - \frac{\omega^2 \sigma^2}{2}\right)$$

and interpret this result.

5. Calculate the ratio of the spectral radiances of black bodies at $300 \, \text{K}$ and $6000 \, \text{K}$ at (a) $1 \, \text{GHz}$, (b) $1000 \, \text{GHz}$, (c) $1 \, \mu\text{m}$ and (d) $0.1 \, \mu\text{m}$.

3

Interaction of electromagnetic radiation with matter

The interaction of electromagnetic radiation with matter is evidently fundamental to remote sensing. The subject is a vast one, embracing many areas of physics, and a fully systematic treatment would require at least a book in itself. In this chapter, therefore, we will attempt to provide an overview that will be sufficient to gain an understanding of the operation of remote sensing systems. In order to keep the chapter to a manageable length, we reserve a discussion of the interaction of electromagnetic radiation with the Earth's atmosphere to chapter 4. Nevertheless, this is still a long chapter, and it is also the most technical in the book. It is not necessary to understand all of the material in this chapter in order to follow the subsequent material.

3.1   Propagation through homogeneous materials

To describe propagation in a uniform homogeneous material, we need to introduce two properties of the medium: its *relative electric permittivity* ε_r and its *relative magnetic permeability* μ_r. The relative electric permittivity, which is also known as the *dielectric constant*, is the ratio of ε, the electric permittivity of the material, to ε_0, the permittivity of free space; and the relative magnetic permeability is the ratio of μ, the magnetic permeability μ_0, to the permeability of free space. Thus, we can put

$$\mu = \mu_r \mu_0 \tag{3.1}$$

and

$$\varepsilon = \varepsilon_r \varepsilon_0 \tag{3.2}$$

Both ε_r and μ_r are pure numbers; that is, they are dimensionless. Note that some authors use the symbols ε and μ to represent the relative, rather than the absolute, values of the permittivity and the permeability.

An electromagnetic wave can propagate in such a medium. For consistency with chapter 2, we will write the equation for the electric field in exactly the same form as for the free-space wave defined in equation (2.1):

$$E_x = E_0 \cos(\omega t - kz) \tag{3.3}$$

However, the equation for the magnetic field now differs from equation (2.2), becoming

$$B_y = \frac{E_0 \sqrt{\varepsilon_r \mu_r}}{c} \cos(\omega t - kz) \tag{3.4}$$

The ratio of the amplitudes of the electric and magnetic fields has become

$$\frac{c}{\sqrt{\varepsilon_r \mu_r}}$$

instead of c as it is for an electromagnetic wave propagating in free space.

In order for equations (3.3) and (3.4) to represent a valid solution of Maxwell's equations, the angular frequency ω and the wave number k must be related by

$$\frac{\omega}{k} = \frac{c}{\sqrt{\varepsilon_r \mu_r}} \tag{3.5}$$

This is the wave speed, or more precisely the *phase velocity* of the wave (see section 3.1.3), which we denote by the symbol v. The *refractive index n* of the medium is defined as c/v; thus,

$$n = \sqrt{\varepsilon_r \mu_r} \tag{3.6}$$

Clearly, free-space propagation is the special case $n = 1$.

Most of the discussion of polarisation presented in chapter 2 can be applied equally well to the case of a medium in which the refractive index n is not equal to 1. The electric and magnetic fields are still perpendicular to one another and to the direction of propagation, though of course the ratio of the amplitudes is now v rather than c. The Stokes parameters are still defined by equations (2.13). However, the flux density of the radiation (the power transmitted per unit area normal to the propagation direction) is

$$F = \frac{S_0}{2Z} \tag{3.7}$$

where Z, the impedance of the medium, is given by

$$Z = Z_0 \sqrt{\frac{\mu_r}{\varepsilon_r}} \tag{3.8}$$

3.1.1 Complex dielectric constants: absorption

We have seen how the behaviour of an electromagnetic wave in a uniform homogeneous medium is controlled by the refractive index n, defined by equation (3.6). For most of the media we shall need to consider, the relative magnetic permeability μ_r can be taken as 1 (these are so-called 'non-magnetic materials'), and we can therefore focus our attention on the dielectric constant ε_r.

For the case of a uniform homogeneous medium that does not absorb any energy from an electromagnetic wave propagating through it, the dielectric constant must be a real number. However, if the medium does absorb energy from the wave, we must use a complex number to represent the dielectric constant. The conventional way of doing this is to put

$$\varepsilon_r = \varepsilon' - i\varepsilon'' \tag{3.9}$$

where $i^2 = -1$ and ε' and ε'', respectively, are referred to as the real and imaginary parts of the dielectric constant. (Strictly speaking, ε'' is the negative of the imaginary part.) Another commonly encountered way of writing the complex dielectric constant is to put

$$\varepsilon_r = \varepsilon'(1 - i \tan \delta) \tag{3.10}$$

where $\tan \delta$ is called the *loss tangent*. This notation is clearly equivalent to equation (3.9), and is also equivalent to making the refractive index complex (we are still assuming that $\mu_r = 1$):

$$n = m - i\kappa \tag{3.11}$$

which we can see as follows. From equation (3.6), and taking $\mu_r = 1$, we know that

$$n^2 = \varepsilon_r$$

Squaring equation (3.11), and equating it to equation (3.9), gives

$$n^2 = m^2 - \kappa^2 - 2im\kappa = \varepsilon' - i\varepsilon''$$

and so the two descriptions are equivalent provided that

$$\varepsilon' = m^2 - \kappa^2 \tag{3.12.1}$$

$$\varepsilon'' = 2m\kappa \tag{3.12.2}$$

We can also see how a complex dielectric constant represents propagation with absorption by considering an x-polarised wave propagating in the z-direction. The electric field can be written using the complex exponential notation

$$E_x = E_0 \exp(i[\omega t - kz]) \tag{3.13}$$

where E_0 is a constant. This is equivalent to equation (3.3). From equations (3.5) and (3.6) we know that

$$k = \frac{\omega n}{c}$$

which can be rewritten using equation (3.11) as

$$k = \frac{\omega}{c}(m - i\kappa)$$

Substituting this expression for k into equation (3.13), and rearranging, we obtain

$$E_x = E_0 \exp\left(-\frac{\omega \kappa z}{c}\right) \exp\left(i\left[\omega t - \frac{\omega m z}{c}\right]\right)$$

This clearly represents a simple harmonic wave whose amplitude decreases exponentially with the distance z. Since we are usually more interested in the flux density F of the wave, and this is proportional to the square of the amplitude, we can put

$$F = F_0 \exp\left(-2\frac{\omega\kappa z}{c}\right) \qquad (3.14)$$

where F_0 is a constant.

Equation (3.14) shows that the flux density of the electromagnetic wave is reduced by a factor of $e\,(\approx 2.718)$ as the wave advances a distance l_a through the medium, where l_a, the *absorption length*, is given by

$$l_a = \frac{c}{2\omega\kappa} \qquad (3.15)$$

If there are no other factors influencing the intensity of the radiation (i.e. we can ignore scattering and emission for the present – although we will discuss these phenomena in section 3.4), the absorption length provides an order-of-magnitude estimate for the distance that radiation will propagate through the material before its intensity is significantly reduced. For example, after travelling two absorption lengths, the intensity is reduced by a factor of e^2, which means that it has been reduced to about 14% of its original value. After five absorption lengths, the intensity is only 0.7% of its original value, and so on. Figure 3.1 shows the variation with frequency of the absorption length in various materials.

3.1.2   Dielectric constants and refractive indices of real materials

The dielectric constant of a real material is normally dependent on the frequency, and the variation can be quite complicated (e.g. see Feynman et al., 1964, volume 2, for a very clear discussion of the physical principles that determine the dielectric constants of real materials). However, over a limited frequency range a simple physical model can often give a satisfactory description. Although the foregoing theory has been developed in terms of the angular frequency ω, it is more usual to specify the frequency f, or, especially for optical and infrared radiation, the *free-space wavelength* λ_0. This is the wavelength that electromagnetic radiation of the same frequency would have if it were propagating in free space, and is given by equations (2.4) and (2.6) as

$$\lambda_0 = \frac{2\pi c}{\omega} = \frac{c}{f} \qquad (3.16)$$

The wavelength of the radiation in the medium is given by

$$\lambda = \frac{\lambda_0}{m} \qquad (3.17)$$

3.1.2.1   Gases
The dielectric constant of a gas is given to reasonable accuracy, provided that the radiation is not strongly absorbed by the gas, by

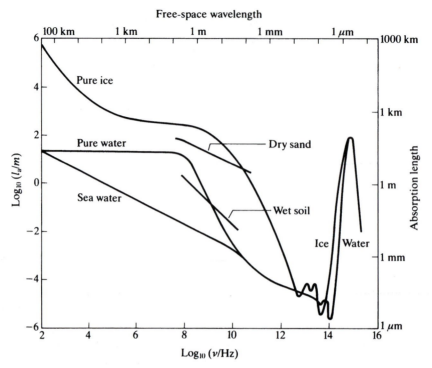

Figure 3.1. Absorption lengths (schematic) of various materials. Note that the absorption lengths are strongly influenced by such factors as temperature and the content of trace impurities, especially at low frequencies.

$$\varepsilon_r = 1 + \frac{N\alpha}{\varepsilon_0} \qquad (3.18)$$

where N is the number density of the gas molecules (i.e. the number of molecules per unit volume) and α is the *polarisability* of the gas molecule. Since the term $N\alpha/\varepsilon_0$ is much smaller than 1 for a gas, the refractive index is given, to a good approximation, by

$$n = 1 + \frac{N\alpha}{2\varepsilon_0} \qquad (3.19)$$

The quantity α/ε_0 has the dimensions of a volume, and is normally similar to the actual physical volume of the molecule. Table 3.1 shows typical values of this quantity for various gases at optical ($\lambda_0 = 589\,\text{nm}$) and radio (typically 1 MHz) frequencies.

3.1.2.2   *Solids and liquids that are electrical insulators*
Simple *non-polar* materials are characterised by a constant (possibly complex) value of ε_r. Simple *polar* materials can be described by the *Debye equations* (3.20), which represent a resonant phenomenon with a time-constant (relaxation time) τ:

Table 3.1. Values of α/ε_0 (where α is the molecular polarisability) of various gases at optical and radio frequencies

The values are given in units of $10^{-30}\,m^3$

Gas	Optical	Radio
Air	21.7	21.4
Carbon dioxide	33.6	36.8
Hydrogen	9.8	10.1
Oxygen	20.2	19.8
Water vapour	18.9	368

$$\varepsilon' = \varepsilon_\infty + \frac{\varepsilon_p}{1 + \omega^2\tau^2} \qquad (3.20.1)$$

$$\varepsilon'' = \frac{\omega\tau\varepsilon_p}{1 + \omega^2\tau^2} \qquad (3.20.2)$$

In these equations, ε_∞ is the dielectric constant at 'infinite' frequency (in practice, at frequencies much greater than $1/\tau$), and ε_p is the polar contribution to the dielectric constant. For example, pure water follows the Debye equations fairly closely between 1 MHz and 1000 GHz, with values (at 20°C) of $\varepsilon_\infty = 5.0$, $\varepsilon_p = 75.4$, $\tau = 9.2 \times 10^{-12}$ s. The corresponding variation of ε' and ε'' is shown in figure 3.2.

3.1.2.3 Metals

The electrical properties of metals are dominated by the very high densities of delocalised electrons. In general, the dielectric constant of a metal can be written as

$$\varepsilon' = 1 - \frac{\sigma\tau}{\varepsilon_0(1 + \omega^2\tau^2)} \qquad (3.21.1)$$

$$\varepsilon'' = \frac{\sigma}{\varepsilon_0\omega(1 + \omega^2\tau^2)} \qquad (3.21.2)$$

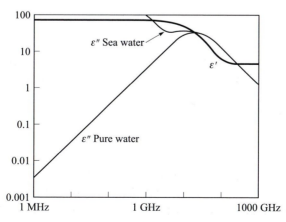

Figure 3.2. Real and imaginary parts of the dielectric constants of pure water and sea water.

In these expressions, σ is the electrical *conductivity* of the metal, and

$$\tau = \frac{m_e \sigma}{N e^2} \qquad (3.22)$$

where m_e is the mass of the electron, e is the charge on the electron, and N is the number density of delocalised electrons in the metal. Equations (3.21) can be simplified for the cases where $\omega \gg 1/\tau$ and where $\omega \ll 1/\tau$. For metals, τ has a value of typically 10^{-15} to 10^{-14} s, so these cases correspond to radio frequencies and optical or ultraviolet frequencies, respectively. At low (radio) frequencies, we obtain

$$\varepsilon_r \approx -\frac{i\sigma}{\varepsilon_0 \omega} \qquad (3.23)$$

From equations (3.12) we see that this corresponds to real and imaginary components of the refractive index of

$$m = \kappa = \sqrt{\frac{\sigma}{2\varepsilon_0 \omega}}$$

and hence from equation (3.15) the absorption length is given by

$$l_a = c \sqrt{\frac{\varepsilon_0}{2\sigma \omega}}$$

For example, let us consider electromagnetic radiation at a frequency of 5 GHz ($\omega = 3.14 \times 10^{10}\,\text{s}^{-1}$) propagating in stainless steel ($\sigma = 1.0 \times 10^6\,\Omega^{-1}\,\text{m}^{-1}$). We find that the absorption length is $3.6\,\mu\text{m}$, which shows that the material is opaque to radio frequency radiation unless it is extremely thin.

At high (optical or ultraviolet) frequencies, the dielectric constant of a metal can be approximated as

$$\varepsilon_r \approx 1 - \frac{N e^2}{\varepsilon_0 m_e \omega^2} \qquad (3.24)$$

which is real and very slightly less than 1. Thus, at sufficiently high frequencies, metals become transparent.

Equation (3.24) also describes the dielectric constant of a *plasma*, a state of matter in which all the atoms have been ionized. Because the mass of the electron is so much smaller than that of any other charged particle, the latter may effectively be ignored in considering the response of the material to an electromagnetic wave. It is clear from equation (3.24) that a plasma is transparent (ε_r is real) for angular frequencies higher than

$$\omega_p = \sqrt{\frac{N e^2}{\varepsilon_0 m_e}} \qquad (3.25)$$

and absorbs radiation at angular frequencies below this value. ω_p is called the *plasma frequency*, and considerations of the properties of plasmas will be important when we discuss the ionosphere.

3.1.3   Dispersion

We noted earlier that, in a number of cases of practical importance, the dielectric properties (and hence refractive index) of a medium vary with frequency. Such media are said to be *dispersive*, and a wave propagating in such a medium is called a *dispersive wave*. It is usual to characterise this behaviour by expressing the angular frequency ω as a function of the wavenumber k, and this relationship is called the *dispersion relation*.

We saw in equation (3.5) that the wave velocity v is given by

$$v = \frac{\omega}{k} \tag{3.26}$$

This is true even if ω varies with k, and v (the wave velocity or *phase velocity*) is the speed at which the crests and troughs of the wave move in the propagation direction. However, if we modulate the wave in some way, for example by breaking it up into pulses, it is this modulation that carries information, and we therefore need to know the speed at which the modulating function travels. This is called the *group velocity*, and it is given by

$$v_g = \frac{d\omega}{dk} \tag{3.27}$$

Only in the case of a non-dispersive wave, for which ω is proportional to k, will equations (3.26) and (3.27) be, in general, equal to one another.

Figure 3.3 illustrates the idea of a dispersive wave. It shows a sinusoidal wave that has been modulated by a Gaussian envelope to give a pulse. The pulse is travelling to the right, and is shown at four equally spaced intervals of time. The circle drawn on each position of the pulse shows the progress of a particular crest of the wave, and it can be seen that this crest is travelling more slowly than the envelope itself. Thus, in this particular case, the phase velocity is less than the group velocity. Figure 3.3 also illustrates another consequence of a wave travelling in a dispersive medium, namely the spreading (elongation) of the envelope as time progresses. Whether this phenomenon occurs depends on the precise form of the dispersion relation.

It will sometimes happen in practice that data about the dispersion relation for a particular medium will be given not as $\omega(k)$, but as $n(\lambda_0)$, the dependence of the refractive index on the free-space wavelength. In this case, equation (3.27) may conveniently be expressed as

$$\frac{c}{v_g} = n - \lambda_0 \frac{dn}{d\lambda_0} \tag{3.28}$$

As an example, we can consider the dispersion of visible light in air. In the optical region, the refractive index of dry air at atmospheric pressure and 15°C can be approximated as

$$n = 1 + \frac{1}{a + b/\lambda_0^2}$$

where $a = 3669$ and $b = 2.1173 \times 10^{-11}$ m^2. Applying equation (3.28) gives

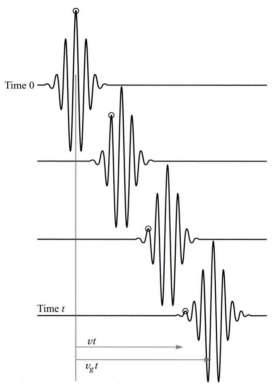

Figure 3.3. Dispersion of a modulated wave. The figure shows a sinusoidal wave that has been modulated by a Gaussian envelope, at four successive instants of time. After time t, a particular wave crest (shown by the circle) has moved a distance vt, whereas the peak of the modulating function has moved a distance $v_g t$. In this case, v is the phase velocity and v_g is the group velocity.

$$\frac{v_g}{c} = 1 - \frac{\lambda_0^2(a\lambda_0^2 + b)}{a(a+1)\lambda_0^4 + (1-2a)\lambda_0^2 + b^2}$$

Figure 3.4 shows the phase velocity v and the group velocity v_g, calculated from these formulae, for free-space wavelengths between 0.4 and 0.7 μm.

Returning to equation (3.24), which describes the dielectric constant of a metal (at sufficiently high frequencies) or a plasma, we note that above the plasma frequency the dielectric constant is real and less than 1. This means that the phase velocity v is greater than c, the speed of light, and at first sight this seems to contradict Einstein's principle that nothing can travel faster than c. However, as we remarked earlier, *information* is carried at the group velocity, not the phase velocity. It is easy to show, using equation (3.27), that if the dielectric constant is given by equation (3.24), then the relationship between v and v_g is

$$vv_g = c^2 \tag{3.29}$$

and so the group velocity is indeed less than c.

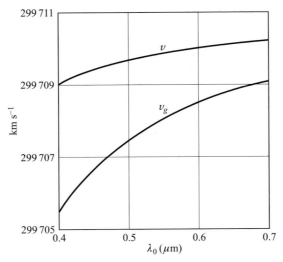

Figure 3.4. Phase and group velocities for light propagating in dry air at standard atmospheric pressure and 15°C.

3.2 Plane boundaries

In this section, we will review the phenomena of reflection and transmission when electromagnetic radiation encounters a plane boundary between two uniform homogeneous media (figure 3.5). We will call these media 1 and 2. The radiation is travelling in medium 1 towards the boundary with medium 2, and makes an angle θ_1 with the normal to the boundary. In general, some of the radiation will be reflected back into medium 1, again at an angle θ_1 but on the opposite side of the normal, and some will be refracted across the boundary so that it makes an angle θ_2 in medium 2. *Snell's law* relates the angles θ_1 and θ_2 through

$$n_1 \sin \theta_1 = n_2 \sin \theta_2 \tag{3.30}$$

where n_1 and n_2 are the refractive indices of the two media, and it also states that the incident, reflected and refracted rays, and the normal to the boundary, all lie in the same plane.

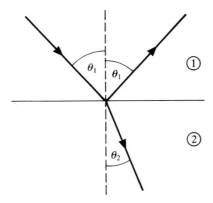

Figure 3.5. Reflection and refraction at a plane boundary between two media.

We will also need to know the reflection and transmission coefficients r and t. The reflection coefficient is defined as the electric field amplitude of the reflected radiation, expressed as a fraction of the electric field amplitude of the incident radiation, and similarly for the transmission coefficient. Since the values of these coefficients depend on the polarisation of the incident radiation, we will need to specify each coefficient for two orthogonal polarisations, giving a total of four coefficients. (The coefficients for any other polarisation state can be calculated by resolving the state into the components for the two states we have chosen, as shown in section 2.2.) The two polarisations that are usually chosen are called parallel and perpendicular, denoted by the symbols ∥ and ⊥ (figure 3.6). The term 'parallel polarisation' means that the electric field vector of the radiation is parallel to the plane containing the incident, reflected and refracted rays (and the normal to the boundary), and 'perpendicular polarisation' means that the electric field vector is perpendicular to this plane. Sometimes, especially in describing microwave systems, the terms 'horizontal polarisation' and 'vertical polarisation' are used instead. To understand this notation, it is necessary to think of the boundary as being horizontal, and to realise that 'vertically' polarised radiation merely has a vertical component. Provided that the two media are homogeneous, parallel-polarised incident radiation will give rise to parallel-polarised reflected and refracted radiation, and no perpendicularly polarised radiation. The converse of this statement is also true, so that perpendicularly polarised incident radiation does not produce any parallel-polarised components.

Now that we have defined our terms, we can proceed to state the formulae for the reflection and transmission coefficients. These are calculated, in terms of the impedances Z_1 and Z_2 of the media, by solving Maxwell's equations at the boundary:

$$r_\perp = \frac{Z_2 \cos \theta_1 - Z_1 \cos \theta_2}{Z_2 \cos \theta_1 + Z_1 \cos \theta_2} \qquad (3.31.1)$$

$$t_\perp = \frac{2Z_2 \cos \theta_1}{Z_2 \cos \theta_1 + Z_1 \cos \theta_2} \qquad (3.31.2)$$

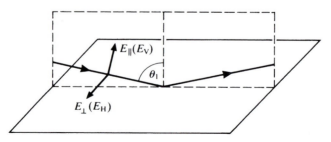

Figure 3.6. Parallel and perpendicular (vertical and horizontal) polarisations of radiation incident at and reflected from a plane boundary between two media.

$$r_\parallel = \frac{Z_2 \cos \theta_2 - Z_1 \cos \theta_1}{Z_2 \cos \theta_2 + Z_1 \cos \theta_1} \tag{3.31.3}$$

$$t_\parallel = \frac{2Z_2 \cos \theta_1}{Z_2 \cos \theta_2 + Z_1 \cos \theta_1} \tag{3.31.4}$$

The full expressions of equations (3.31) become rather complicated in the case where both media have significant absorption coefficients (i.e. the complex form of their refractive indices has to be taken into account). However, in many cases of practical importance we may assume that medium 1 has a refractive index of 1 (a vacuum, or air, to a good approximation). If medium 2 is absorbing, the Fresnel reflection coefficients for non-magnetic media are given by the following formulae:

$$r_\perp = \frac{\cos \theta_1 - \sqrt{\varepsilon_{r2} - \sin^2 \theta_1}}{\cos \theta_1 + \sqrt{\varepsilon_{r2} - \sin^2 \theta_1}} \tag{3.32.1}$$

$$t_\perp = \frac{2 \cos \theta_1}{\cos \theta_1 + \sqrt{\varepsilon_{r2} - \sin^2 \theta_1}} \tag{3.32.2}$$

$$r_\parallel = \frac{\sqrt{\varepsilon_{r2} - \sin^2 \theta_1} - \varepsilon_{r2} \cos \theta_1}{\sqrt{\varepsilon_{r2} - \sin^2 \theta_1} + \varepsilon_{r2} \cos \theta_1} \tag{3.32.3}$$

$$t_\parallel = \frac{2 \sqrt{\varepsilon_{r2}} \cos \theta_1}{\sqrt{\varepsilon_{r2} - \sin^2 \theta_1} + \varepsilon_{r2} \cos \theta_1} \tag{3.32.4}$$

Note that these expressions must in general be evaluated using complex arithmetic. If medium 1 has $n = 1$ and medium 2 is non-absorbing, the reflection coefficients become

$$r_\perp = \frac{\cos \theta_1 - \sqrt{n_2^2 - \sin^2 \theta_1}}{\cos \theta_1 + \sqrt{n_2^2 - \sin^2 \theta_1}} \tag{3.33.1}$$

$$r_\parallel = \frac{\sqrt{n_2^2 - \sin^2 \theta_1} - n_2^2 \cos \theta_1}{\sqrt{n_2^2 - \sin^2 \theta_1} + n_2^2 \cos \theta_1} \tag{3.33.2}$$

We can see from equation (3.33.2) that $r_\parallel = 0$ when θ_1 takes the value θ_B, given by

$$\tan \theta_B = n_2 \tag{3.34}$$

This is called the *Brewster angle*. Parallel (vertically) polarised radiation incident on a surface at the Brewster angle cannot be reflected, and so must all be transmitted into the medium. Consequently, we can note that randomly polarised radiation incident from an arbitrary direction on a boundary between two media will in general, on reflection, be partially polarised, and if it is incident at the Brewster angle it will be completely plane-polarised. This is the simplest justification for the remark made in section 2.2 that the degree of polarisation is changed on reflection.

To illustrate equations (3.33) and the phenomenon of the Brewster angle, we will calculate the power reflection coefficients for light meeting an air–water interface. The refractive index of air can be taken as 1 and that of pure water as 1.333 with no imaginary part. Figure 3.7 shows the reflection coefficients as a function of the incidence angle θ_1. The figure shows that the value of $r_\parallel$ falls to zero near 50°, which is confirmed by calculating the Brewster angle from equation (3.34) as 53.1°. The figure also shows that when $\theta = 0$ the two reflection coefficients have the same value, which they must since for normally incident radiation there can be no distinction between parallel and perpendicular polarisations, and that when $\theta = 90°$ (grazing incidence) the power reflection coefficients are both 1 (i.e. all the radiation is reflected).

3.3   Scattering from rough surfaces

Scattering (reflection) of radiation from the Earth's surface is a fundamental process in most remote sensing situations. The exceptions to this principle are atmospheric sounding observations, and those passive observations (of thermal infrared or microwave emission) that do not respond to reflected sunlight. Thus, a consideration of the reflectance properties of real surfaces will be of considerable importance. In fact, as we saw in chapter 2, the thermal emissivity of a surface is directly related to its reflectance, so these properties will be important even in the case of passive microwave and thermal infrared remote sensing.

In section 3.2 we reviewed the behaviour of electromagnetic radiation when it is incident on a planar (i.e. perfectly smooth) boundary between two homogeneous media. In this section, we will consider radiation incident, from within

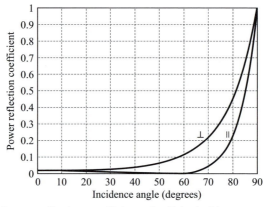

Figure 3.7. Power reflection coefficients for light incident on a water surface.

a vacuum (to which air is a reasonably good approximation), on a rough surface. The material below this surface will be assumed to be homogeneous, but we will consider what happens when it is *not* homogeneous in section 3.4.

3.3.1   *Description of surface scattering*

The first thing we need to do is to develop some of the terminology needed to describe rough-surface scattering. The treatment presented in this section is amplified in a number of works. Swain and Davis (1978) is a particularly useful source of information on quantitative descriptions of surface scattering in remote sensing, and Schanda (1986) also provides a helpful treatment. A more recent, and very full, discussion is provided by Hapke (1993).

Figure 3.8 shows a well-collimated beam of radiation of flux density F, measured in a plane perpendicular to the direction of propagation, incident on a surface at an angle θ_0. This angle is often called the *incidence angle*, and its complement $\pi/2 - \theta_0$ is the *depression angle*.[1] A proportion of the incident radiation will be scattered into the solid angle $d\Omega_1$, in a direction specified by the angle θ_1. For simplicity, the azimuthal angles have been omitted from figure 3.8. These will be denoted by ϕ_0 and ϕ_1, respectively.

The irradiance E at the surface is given by $F \cos \theta_0$. If we write $L_1(\theta_1, \phi_1)$, for the radiance of the scattered radiation in the direction (θ_1, ϕ_1), we can define the *bidirectional reflectance distribution function* (BRDF) R as

$$R = \frac{L_1}{E} \tag{3.35}$$

R has no dimensions, and its unit is sr^{-1}. It is also commonly represented by the symbols f or ρ. In considering radar systems (chapter 9), the BRDF is

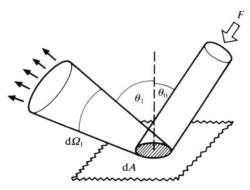

Figure 3.8. Radiation, initially of flux density F, is incident at angle θ_0 on an area dA and is then scattered into solid angle $d\Omega_1$ in the direction θ_1. The azimuthal angles ϕ_0 and ϕ_1 are omitted for clarity.

---

[1] This use of the term 'depression angle' assumes that the surface is horizontal. If not, it is probably safer to refer to the 'local incidence angle' and to avoid the term 'depression angle'.

usually replaced by the equivalent *bistatic scattering coefficient* γ, which has no units and is related to R by

$$\gamma = 4\pi R \cos\ \theta_1 \qquad (3.36)$$

In fact, most radar systems, and all those we shall consider in chapter 9, detect only the *backscattered* component of the radiation, which retraces the path of the incident radiation. In this case, $\theta_1 = \theta_0$ and $\phi_1 = \phi_0$, and the usual way of specifying the proportion of scattered radiation is through the (dimensionless) *backscattering coefficient* σ^0, defined by

$$\sigma^0 = \gamma \cos\ \theta_0 = 4\pi R \cos^2 \theta_0 \qquad (3.37)$$

The BRDF is a function of the incidence and scattered directions (σ^0 is a function of the incidence direction only, since the scattered direction is the same), so in principle it should be written as a function of its arguments: $R(\theta_0, \phi_0, \theta_1, \phi_1)$. This notation is useful since it allows us to state the reciprocity relation obeyed by the BRDF:

$$R(\theta_0, \phi_0, \theta_1, \phi_1) = R(\theta_1, \phi_1, \theta_0, \phi_0) \qquad (3.38)$$

but for compactness we will just write R, and take the arguments to be implied. In the majority of cases, the surface will lack azimuthally dependent features so that the dependence on ϕ_0 and ϕ_1 will simplify to a dependence on $(\phi_0 - \phi_1)$, and often the azimuthal dependence can be neglected altogether.

The *reflectivity r* of the surface is a function only of the incidence direction. It defines the ratio of the total power scattered to the total incident power. It is thus given by

$$r(\theta_0, \phi_0) = \frac{M}{E}$$

where M is the radiant exitance of the surface, and on substituting for M from equation (2.24) we find that

$$r(\theta_0, \phi_0) = \int\limits_{\theta_1=0}^{\pi/2} \int\limits_{\phi_1=0}^{2\pi} R \cos\ \theta_1 \sin\ \theta_1\ d\theta_1\ d\phi_1 \qquad (3.39)$$

The reflectivity is also commonly called the *albedo* (from the Latin for 'whiteness') of the surface, and it is related to the emissivity ε in the direction (θ_0, ϕ_0) through

$$r = 1 - \varepsilon \qquad (3.40)$$

We can define the *diffuse albedo* r_d, also called the *hemispherical albedo*, as the average value of r over the hemisphere of possible incidence directions. In this case, it represents the ratio of the total scattered power to the total incident power when the latter is distributed isotropically. Since the incident radiance is therefore constant, we may write it as L_0, so that the contribution dE to the irradiance from the direction (θ_0, ϕ_0) is $L_0 \cos\ \theta_0 \sin\ \theta_0\ d\theta_0\ d\phi_0$. The contribution dM that this makes to the radiant exitance in the direction (θ_1, ϕ_1) must therefore be given by $RL_0 \cos \theta_0 \sin \theta_0 \, d\theta_0 \, d\phi_0 \cos \theta_1 \sin \theta_1 \, d\theta_1 \, d\phi_1$. The radiant exitance is thus

$$M = L_0 \int_{\theta_0=0}^{\pi/2} \int_{\phi_0=0}^{2\pi} \int_{\theta_1=0}^{\pi/2} \int_{\phi_1=0}^{2\pi} R \cos \theta_0 \sin \theta_0 \cos \theta_1 \sin \theta_1 \, d\theta_0 \, d\phi_0 \, d\theta_1 \, d\phi_1$$

and the irradiance is

$$E = L_0 \int_{\theta_0=0}^{\pi/2} \int_{\phi_0=0}^{2\pi} \cos \theta_0 \sin \theta_0 \, d\theta_0 \, d\phi_0 = \pi L_0$$

Since the diffuse albedo is given by M/E in this case, we can write it (making use of equation (3.39) to simplify the formula a little) as

$$r_{\mathrm{d}} = \frac{1}{\pi} \int_{\theta_0=0}^{\pi/2} \int_{\phi_0=0}^{2\pi} r(\theta_0, \phi_0) \cos \theta_0 \sin \theta_0 \, d\theta_0 \, d\phi_0 \qquad (3.41)$$

3.3.2 Simple models of surface scattering

In this section, we will discuss a few of the important models of the BRDF of real surfaces. Further information can be found in, for example, Hapke (1993).

If the scattering surface is sufficiently smooth, it will behave like a mirror. This is called *specular scattering* or specular reflection (Latin *speculum* = a mirror). Radiation incident from the direction (θ_0, ϕ_0) will be scattered only into the direction $\theta_1 = \theta_0, \phi_1 = \phi_0 - \pi$, as illustrated schematically in figure 3.9a. The BRDF must therefore be a delta-function, and we can write it as

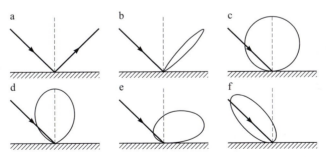

Figure 3.9. Schematic illustration of different types of surface scattering. The lobes are *polar diagrams* of the scattered radiation: the length of a line joining the point where the radiation is incident on the surface to the lobe is proportional to the radiance scattered in the direction of the line. (a) Specular reflection; (b) quasi-specular scattering; (c) Lambertian scattering; (d) Minnaert model ($\kappa = 2$); (e) Henyey–Greenstein model of forward scatter ($\Theta = 0.7$); (f) Henyey–Greenstein model of backscatter ($\Theta = -0.5$).

$$R = \frac{|r(\theta_0)|^2}{\cos \theta_0 \sin \theta_0} \delta(\theta_1 - \theta_0)\, \delta(\phi_1 - \phi_0 + \pi) \tag{3.42}$$

where $r(\theta_0)$ is the appropriate Fresnel amplitude reflection coefficient for radiation with incidence angle θ_0. Inserting this expression into equation (3.39), we find that the reflectivity r for radiation with incidence angle θ_0 is just

$$r = |r(\theta_0)|^2$$

as of course it must be, and from equation (3.41) the diffuse albedo is

$$r_d = 2 \int\limits_0^{\pi/2} |r(\theta_0)|^2 \cos \theta_0 \sin \theta_0\, d\theta_0$$

Specular scattering is one limiting case of surface scattering, and it arises when the surface is very smooth (later we will consider just how smooth it needs to be). The other important limiting case is that of an ideally rough surface, giving *Lambertian scattering*. This has the property that, for any illumination that is uniform across the surface, the scattered radiation is distributed isotropically, and so the BRDF has a constant value. This is illustrated schematically in figure 3.9b. From equations (3.39) and (3.41) it can easily be seen that, for such a surface,

$$R(\theta_0, \phi_0, \theta_1, \phi_1) = r(\theta_0, \phi_0) = \frac{r_d}{\pi} \tag{3.43}$$

Thus, for example, a Lambertian surface that scatters all of the radiation incident upon it has $R = 1/\pi$.

The scattering behaviour of real surfaces is often specified, not by using the BRDF, but instead by measuring the *bidirectional reflectance factor* (BRF). This is defined as the ratio of the flux scattered into a given direction by a surface under given conditions of illumination, to the flux that would be scattered in the same direction by a perfect Lambertian scatterer under the same conditions. The usefulness of this approach is that surfaces can be manufactured to have a BRF very close to unity for a fairly wide range of wavelengths and of incidence and scattering angles. The most common materials are barium sulphate, which, as a pressed powder and for $\theta < 45°$, has a BRF greater than 0.99 for wavelengths between 0.37 and 1.15 μm, and magnesium oxide, which has a BRF greater than 0.98 over roughly the same range of conditions.

Although the Lambertian model is simple and idealised, the scattering from many natural surfaces can often, to a first approximation at least, be described using it. A simple modification is provided by the *Minnaert model*, in which the BRDF is given by

$$R \propto (\cos \theta_0 \cos \theta_1)^{\kappa - 1} \tag{3.44}$$

where the parameter κ has the effect of increasing or decreasing the radiance scattered in the direction of the surface normal (figure 3.9d). Lambertian scattering is the special case of the Minnaert model with $\kappa = 1$.

The scattering from real rough surfaces can often, as we have just remarked, be described by the Lambert or Minnaert models. However, neither of these models accounts for the fact that real surfaces may also show additional back-scattering (where the radiation is scattered back into the incidence direction) or specular scattering. These can of course be incorporated by devising an empirical model that combines a Lambertian or Minnaert component with, for example, a 'quasi-specular' component. One common modification is to multiply the Lambert or Minnaert BRDF by the *Henyey–Greenstein* term

$$\frac{1 - \Theta^2}{(1 + 2\Theta \cos g + \Theta^2)^{3/2}} \tag{3.45}$$

where the parameter Θ represents the anisotropy of the scattering, with $0 < \Theta \le 1$ corresponding to forward scattering and $-1 \le \Theta < 0$ corresponding to backscattering. The scattering phase angle g is given by

$$\cos g = \cos \theta_0 \cos \theta_1 + \sin \theta_0 \sin \theta_1 \cos(\phi_1 - \phi_0) \tag{3.46}$$

Figure 3.9e and f illustrate typical BRDFs that incorporate the Henyey–Greenstein term.

3.3.3   The Rayleigh roughness criterion

We have distinguished between the behaviour of a perfectly smooth surface and a Lambertian surface that is in some sense perfectly rough. It is clear that in order to understand which of these simple models is likely to give a better model of scattering from a real surface, some measure of surface roughness must be developed. The usual approach is via the Rayleigh criterion, which we will develop in this section.

Figure 3.10 shows schematically the detailed behaviour when radiation is incident on an irregular surface at angle θ_0, and scattered specularly from it at the same angle. We consider two rays: one is scattered from a reference plane,

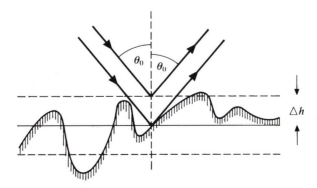

Figure 3.10. The Rayleigh criterion. Radiation is specularly reflected at an angle θ_0 from a surface whose r.m.s. height deviation is Δh. The difference in the lengths of the two rays is $2\Delta h \cos \theta_0$.

and the other from a plane at a height Δh above this reference plane. After scattering, the path difference between these two rays is $2\Delta h \cos \theta_0$, so the phase difference between them is

$$\Delta\phi = \frac{4\pi\,\Delta h \cos\,\theta_0}{\lambda}$$

where λ is the wavelength of the radiation. If we now let Δh stand for the root-mean-square (r.m.s.) variation in the surface height, $\Delta\phi$ becomes the root mean square variation in the phase of the scattered rays. A surface can be defined as smooth enough for scattering to be specular if $\Delta\phi$ is less than some arbitrarily defined value of the order of 1 radian. The conventional value is $\pi/2$, and this is called the Rayleigh criterion. Thus, for a surface to be smooth according to this criterion,

$$\Delta h < \frac{\lambda}{8 \cos\,\theta_0} \tag{3.47}$$

Note that other criteria have also been adopted for the value of $\Delta\phi$ at which the surface becomes effectively smooth. A common definition that provides for the possibility of some intermediate cases between rough and smooth is that if $\Delta\phi$ is greater than $\pi/2$ the surface is rough, and if $\Delta\phi$ is less than $4\pi/25$ (so that the numerical part of the denominator in equation (3.47) becomes 25 instead of 8) it is smooth.

Equation (3.47) evidently dictates that for a surface to be effectively smooth at normal incidence, irregularities must be less than about $\lambda/8$ (or perhaps $\lambda/25$) in height. Thus, for a surface to give specular reflection at optical wavelengths (say $\lambda = 0.5\,\mu\text{m}$), Δh must be less than about 60 nm. This is a condition of smoothness likely to be met only in certain man-made surfaces such as sheets of glass or metal. On the other hand, if the surface is to be examined using VHF radio waves (say $\lambda = 3\,\text{m}$), Δh need only be less than about 40 cm, a condition that could be met by a number of naturally occurring surfaces. A further aspect of equation (3.47) is the dependence on θ_0. The smoothness criterion is more easily satisfied at large values of θ_0 than at normal incidence, so that a moderately rough surface may be effectively smooth at glancing incidence. This fact is well known to anyone who has endured the glare of reflected sunlight from a low sun over an ordinary road surface. Although the scattering cannot really be described as specular in this case, the component of the BRDF in the specular direction is greatly enhanced.

3.3.4   *Models for microwave backscatter*

In this section, we shall discuss some of the commonest physical methods used to model the microwave backscatter from rough surfaces. This is a large and important area in which considerable research is still taking place, and the reader who wishes to pursue it in greater depth is recommended to study, for example, the books by Beckmann and Spizzichino (1963), Colwell (1983), Tsang et al. (1985) and Ulaby et al. (1981, 1986), and the more recent research literature. The mathematical development of these models is generally rather difficult, and in this section we can do little more than sketch their principles.

3.3.4.1 *The small perturbation model*

The most helpful way to begin to look at the problem of microwave scattering from a rough surface is through the small perturbation model. This is essentially a Fraunhofer diffraction approach to rough-surface scattering, in which the interaction of the incident radiation with the surface is used to calculate the outgoing radiation field in the vicinity of the surface. This field can then be regarded as having been produced from a uniform incident radiation field by a fictitious screen that changes both the amplitude and the phase, and the far-field radiation pattern is obtained by calculating the Fraunhofer diffraction pattern of this screen.

In order to see how this is applied, but without becoming too deeply immersed in mathematical detail, we will consider a surface $z(x, y)$ in which the height z depends only on x, and which scatters all radiation incident upon it (i.e. the diffuse albedo is 1). Figure 3.11 illustrates radiation incident on the surface at angle θ_0 and scattered from it at angle θ_1. The rays AO and OB, of length a and b respectively, provide a reference from which phase differences can be measured. P is a point on the surface, that has coordinates x, $z(x)$. Simple trigonometry shows that the length of the ray CP is

$$a + x \sin \theta_0 - z(x) \cos \theta_0$$

and the length of the ray PD is

$$b - x \sin \theta_1 - z(x) \cos \theta_1$$

so the phase of the wave at D relative to the reference ray can be written as

$$\phi(x) = k\alpha x - k\beta z(x)$$

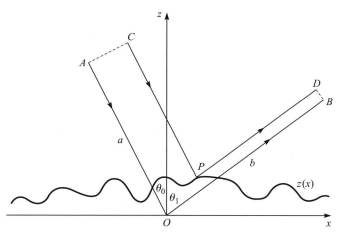

Figure 3.11. Model for developing the small perturbation model of rough-surface scattering.

where $\alpha = (\sin \theta_0 - \sin \theta_1)$, $\beta = (\cos \theta_0 + \cos \theta_1)$, and k is the wavenumber of the radiation. The total amplitude E scattered from the direction θ_0 into the direction θ_1 can therefore be written as

$$E = \int_{-\infty}^{\infty} e^{-i\phi(x)} \, dx = \int_{-\infty}^{\infty} e^{-ik\alpha x} \, e^{ik\beta z(x)} \, dx$$

This expression is clearly the Fourier transform of the function

$$e^{ik\beta z(x)}$$

which we can expand as a power series:

$$e^{ik\beta z(x)} = 1 + ik\beta z(x) - \frac{(k\beta z(x))^2}{2} \cdots \tag{3.48}$$

so that our expression for the scattered field amplitude becomes

$$E = \int_{-\infty}^{\infty} \left[1 + ik\beta z(x) - \frac{(k\beta z(x))^2}{2} \cdots \right] e^{-ik\alpha x} \, dx \tag{3.49}$$

The first term in this expression is a delta-function at $\alpha = 0$. Using the fact that $\alpha = (\sin \theta_0 - \sin \theta_1)$, we can see that this is just the specularly scattered component $\theta_1 = \theta_0$. We can write the second term as

$$ik\beta \int_{-\infty}^{\infty} z(x) \, e^{-ik\alpha x} \, dx$$

which is proportional to the Fourier transform of the surface height function $z(x)$. This suggests that it will be helpful to write the height function in terms of its Fourier transform $a(q)$, where q is the spatial frequency:

$$z(x) = \int_{-\infty}^{\infty} a(q) \, e^{iqx} \, dq$$

Thus, the second term in equation (3.49) becomes

$$ik\beta \int_{-\infty}^{\infty} \int_{-\infty}^{\infty} a(q) \, e^{i(q - k\alpha)x} \, dq \, dx$$

Using the definition of the Dirac delta-function that we met in section 2.3, we see that this can be written as

$$2\pi ik\beta \int_{-\infty}^{\infty} a(q)\delta(q - k\alpha) \, dq \tag{3.50}$$

This is the result that we need. It shows that the amplitude of the radiation scattered into the direction specified by α is proportional to the component of the surface height function with spatial frequency $q = k\alpha$.

There is another way of thinking about this result. The wave vector of the incident radiation has a horizontal component $k \sin \theta_0$ and the wave vector of the scattered radiation has a horizontal component $k \sin \theta_1$, so the term $k\alpha$ is just the change in this horizontal component. We can therefore say that the component of the scattered radiation amplitude is proportional to the spatial frequency component of the surface profile for which the spatial frequency corresponds to the change in the horizontal component of the radiation's wave vector.

Up to this point, we have assumed that the third and subsequent terms in the power-series expansion of equation (3.48) are negligible in comparison with the first two. If this is true, the phenomenon is known as *Bragg scattering*. Clearly, the condition that needs to be satisfied is

$$k\beta \, \Delta h \ll 1$$

which is equivalent to

$$\Delta h \ll \frac{\lambda}{2\pi(\cos \theta_0 + \cos \theta_1)}$$

Thus, the surface must be smooth according to the Rayleigh roughness criterion (3.47). The Bragg scattering mechanism is thought to be largely responsible for the reflection of microwave radiation from small-scale (of the order of 1 cm) roughness on water surfaces, especially where the structure of this roughness contains a dominant spatial frequency, in which case the Bragg scattering is said to be *resonant*.

We have also assumed that the surface height z depends only on the x-coordinate. A more general derivation for two-dimensional isotropic surfaces, which more or less follows the argument we have just presented, leads to the following expression for the backscattering coefficient σ^0:

$$\sigma^0_{pp} = 4k^4 L^2 (\Delta h)^2 \cos^4 \theta |f_{pp}(\theta)|^2 \exp(-k^2 L^2 \sin^2 \theta) \qquad (3.51)$$

In this expression, σ^0_{pp} is the backscattering coefficient for *pp*-polarisation (so that, for example, $p = \mathrm{H}$ means HH-polarisation, or radiation both incident and scattered in the horizontal polarisation state), θ is the incidence angle of the radiation, L is the correlation length[2] of the surface (i.e. the 'width' of the irregularities, which contains information about the shape of the spatial frequency spectrum of the surface), and $f_{pp}(\theta)$ is a measure of the surface reflectivity for radiation with incidence angle θ. For HH-polarised radiation we have

---

[2] In fact, equation (3.51) is based on the assumption that the surface has a Gaussian autocorrelation function. L is the distance over which the autocorrelation coefficient falls to a value of $1/e$. See box for more information.

$$f_{HH}(\theta) = \frac{\cos\theta - \sqrt{\varepsilon_r - \sin^2\theta}}{\cos\theta + \sqrt{\varepsilon_r - \sin^2\theta}} \qquad (3.52.1)$$

which is just the Fresnel reflection coefficient for radiation incident at angle θ from a vacuum onto the surface of a medium with a (complex) dielectric constant ε_r. For VV-polarised radiation, the corresponding formula is

$$f_{VV}(\theta) = (\varepsilon_r - 1)\frac{\sin^2\theta - \varepsilon_r(1 + \sin^2\theta)}{\left(\varepsilon_r\cos\theta + \sqrt{\varepsilon_r - \sin^2\theta}\right)^2} \qquad (3.52.2)$$

The conditions for the validity of equation (3.51) are usually given as

$$k\Delta h < 0.3 \qquad (3.53.1)$$

and

$$kL < 3 \qquad (3.53.2)$$

although these are somewhat approximate.

Figure 3.12 illustrates the backscatter predicted by the small perturbation model for a surface with $\varepsilon_r = 10$. In each case, the r.m.s. height variation Δh is the same. It can be seen that the effect of increasing kL, the spatial scale of the surface roughness features, is to increase the specular scattering ($\theta = 0$) at the

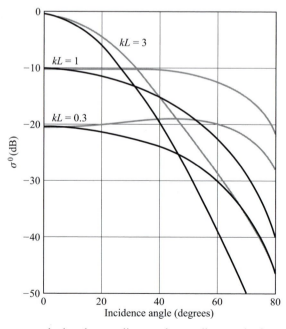

Figure 3.12. Backscatter calculated according to the small perturbation model for a surface having dielectric constant $\varepsilon_r = 10$. In each case, $k\,\Delta h = 0.3$ and the curves are labelled with the values of kL. The black curves are for HH-polarisation and the grey curves are for VV-polarisation.

THE AUTOCORRELATION FUNCTION

The r.m.s. height variation Δh is the simplest measure of the roughness of a surface, but it tells us nothing about the *scale* of the irregularities. The autocorrelation function provides this information.

Suppose (for simplicity) we consider a one-dimensional surface $z(x)$, where z is the height at position x. The mean height is $\langle z \rangle$, where the angle brackets denote an average over all values of x, and the r.m.s. height variation is defined by

$$\Delta h = \left\langle (z(x) - \langle z \rangle)^2 \right\rangle^{1/2}$$

The autocorrelation function is defined by

$$\rho(\xi) = \frac{\left\langle (z(x + \xi) - \langle z \rangle)(z(x) - \langle z \rangle) \right\rangle}{\Delta h^2}$$

and is measure of the similarity of the heights at two points separated by distance ξ. By the definition, $\rho(0) = 1$, and for most surfaces, $\rho(\infty) = 0$. Common models for the autocorrelation function are the Gaussian

$$\rho(\xi) = \exp\left(-\frac{\xi^2}{L^2}\right)$$

and the negative exponential

$$\rho(\xi) = \exp\left(-\frac{\xi}{L}\right)$$

In each case, L (the correlation length) is a measure of the width of the irregularities of the surface.

The extension of this idea to two dimensions is straightforward.

expense of the scattering at larger angles. Since the root-mean-square surface slope is of the order of $\Delta h/L$, increasing the value of kL while keeping the value of $k\,\Delta h$ constant corresponds to decreasing the r.m.s. surface slope, so it is not surprising that this increases the specular component of the scattering. The effect of varying $k\,\Delta h$, not shown in the figure, is very straightforward since equation (3.51) shows that the backscatter coefficient is just proportional to $(\Delta h)^2$. Thus, decreasing $k\,\Delta h$ by a factor of 10, say, would have the effect of shifting all the curves in the figure down by 20 dB without changing their shapes.

3.3.4.2 *The Kirchhoff model*

The approach that is adopted by the Kirchhoff model for scattering from randomly rough surfaces is to model the surface as a collection of variously oriented planes, each of which is locally tangent to the surface. This is called the *tangent plane approximation*. The scattered radiation field can then be calculated using the results for radiation incident on a plane interface. Two variants of the Kirchhoff model are in common use: the *stationary phase* (or *geometric optics*) *model*, which is valid for rougher surfaces, and the *scalar approximation*.

The backscatter coefficient is given by the stationary phase model as

$$\sigma_{HH}^0(\theta) = \sigma_{VV}^0(\theta) = \frac{|r(0)|^2 \exp\left(-\dfrac{\tan^2 \theta}{2m^2}\right)}{2m^2 \cos^4 \theta} \tag{3.54}$$

where $r(0)$ is the Fresnel reflection coefficient for normally incident radiation, and m is the root-mean-square surface slope. For a surface having a Gaussian autocorrelation function with correlation length L and root-mean-square height variation Δh,

$$m = \sqrt{2}\frac{\Delta h}{L} \tag{3.55}$$

The conditions for the validity of this model are

$$k \,\Delta h \cos \theta > 1.58 \tag{3.56.1}$$

$$kL > 6 \tag{3.56.2}$$

and

$$kL^2 > 17.3 \,\Delta h \tag{3.56.3}$$

The backscatter coefficient is given by the scalar approximation as

$$\sigma_{pp}^0(\theta) = k^2 L^2 \cos^2 \theta |r_p(\theta)|^2 \exp(-4k^2 \Delta h^2 \cos^2 \theta)$$

$$\times \sum_{n=1}^{\infty} \frac{(2k \,\Delta h \cos \theta)^{2n}}{n!n} \exp\left(-\frac{k^2 L^2 \sin^2 \theta}{n}\right) \tag{3.57}$$

where we have again assumed that the autocorrelation function of the surface is Gaussian. $r_p(\theta)$ is the Fresnel coefficient for p-polarised radiation incident at angle θ. The conditions for the validity of this model are

$$\Delta h < 0.18L \tag{3.58.1}$$

$$kL > 6 \tag{3.58.2}$$

and

$$kL^2 > 17.3 \,\Delta h \tag{3.58.3}$$

We can note that the second and third of these conditions are the same as for the stationary phase model.

Figure 3.13 illustrates the backscatter predicted by both variants of the Kirchhoff model for a surface with $\varepsilon_r = 10$. In each case, the value of kL has been kept constant. Two sets of curves show the predicted backscatter from a smooth ($k\,\Delta h = 3$) and rough ($k\,\Delta h = 10$) surface. Again, we see that smoother surfaces give enhanced scattering near the specular direction.

There are, in fact, further restrictions on the validity of the Kirchhoff model that we should note. As was mentioned above, the first step in constructing the model is to replace the surface by a set of planes or facets, each of which is locally tangent to the surface. It is clear that, for this programme to succeed, we must be able to define facets whose spatial extent is much greater than the wavelength λ (so that diffraction effects do not dominate), but whose deviation from the real surface is much less than λ (so that we do not incur large phase errors in modelling the surface). This is, in fact, a restriction on the local curvature of the surface, as we can see by the following simple one-dimensional argument.

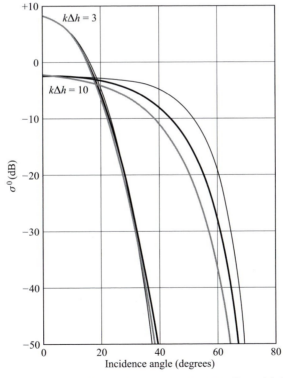

Figure 3.13. Backscatter calculated according to the Kirchhoff model for a surface having dielectric constant $\varepsilon_r = 10$. In each case, $kL = 30$, and the curves are labelled with the values of $k\,\Delta h$. The thin black curves are for the stationary phase model, the thick black curves are for the scalar approximation for HH-polarisation, and the thick grey curves are for the scalar approximation for VV-polarisation.

We will assume that, locally, the surface has a constant radius of curvature R, so that it forms part of a sphere or, in one dimension, a circle. Figure 3.14 shows a facet of length $2w$ tangent to this circle. The facet subtends an angle 2ψ at the centre of curvature, where $\psi = \tan^{-1}(w/R)$, and the maximum deviation of the facet, x, from the surface is $R(\sec \psi - 1)$. If $w \ll R$, this can be approximated as $x \approx w^2/2R$. Now we will assume that $w > \lambda$ (so that the facet is large enough for diffraction effects not to dominate) and that $x < \lambda/2$ (so that we do not incur large phase errors in approximating the surface by the facet). Thus, we obtain the condition that $R > \lambda$. In other words, in order for the surface to be adequately represented by facets, its radius of curvature must exceed a few wavelengths.

Another condition that must be satisfied is that the incidence or scattering angles should not be so large that one part of the surface can obscure another. If this occurs in practice, it can usually be dealt with by an appropriate modification of the model, or by specifying that the model is valid only up to some maximum angle.

3.3.4.3   *Other models*

It is clear that the models of backscatter from randomly rough surfaces that we have considered in the two preceding sections do not provide a complete description of the possible phenomena. Firstly, we can note that the ranges of validity of the three models that we have discussed, defined in equations (3.53), (3.56) and (3.58), do not cover all the possibilities. This fact is illustrated in figure 3.15, which shows the valid range of each model in terms of the dimensionless parameters $(\Delta h/\lambda)$ and (L/λ). Secondly, we can note from figure 3.13 that, even where two models are both apparently valid, they can give different predictions for the backscatter. This is to some extent a consequence of the approximations inherent in the models. Thirdly, we can observe that none of the three models that has been discussed has an explicit dependence on

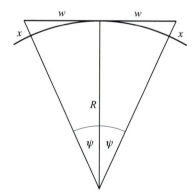

Figure 3.14. A one-dimensional facet of length $2w$ is tangent to a surface with radius of curvature R. The facet subtends an angle 2ψ at the centre of curvature, and its maximum deviation from the surface is x.

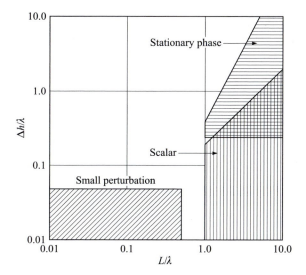

Figure 3.15. Range of validity of the small perturbation, stationary phase and scalar approximation models of rough-surface scattering. For the stationary phase model, the incidence angle has been taken as zero.

the imaginary part of the dielectric constant, which is in contradiction to the experimental evidence.

Many of these difficulties can be circumvented by the use of the *integral equation model*. This is still an approximation, but its grounding in the physics of the interactions between the radiation and the surface is more nearly fundamental. The integral equation model has a larger region of validity and is sensitive to both the real and the imaginary parts of the dielectric constant. It is thus useful for estimating these parameters from backscattering measurements. Unfortunately, its mathematical complexity is such that it is beyond our scope to discuss it any further here.

3.4 Volume scattering

In sections 3.2 and 3.3 we have considered the scattering of electromagnetic radiation from the boundary between two media, for example at the interface between a vacuum (to which air may be a good approximation) and some surface material. Unless the transmission coefficient of the interface is zero, some fraction of the radiation will also enter the material beyond the interface and hence have the possibility of interacting with the bulk of the material as well as with its surface. First, we discuss what happens if this material is homogeneous but absorbing. For simplicity, we will consider the situation shown in figure 3.16. Radiation is incident normally, from a vacuum, onto a parallel-sided slab of medium 1 of thickness d. Below this slab is an infinitely thick slab of medium 2.

The incident radiation has unit amplitude, so the ray A has amplitude r_{01}, defined as the amplitude reflection coefficient for radiation incident from med-

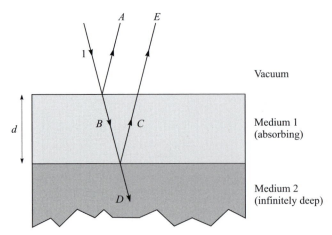

Figure 3.16. Radiation incident normally on a slab of medium 1 of thickness d, underlain by an infinitely thick slab of medium 2. The rays have been shown at an angle for clarity.

ium 0 (vacuum) onto medium 1. However, it is clear from figure 3.16 that there may be an additional contribution E to the radiation reflected from this system. The ray B has amplitude t_{01}, at the top of medium 1, so at the bottom of this slab it must have amplitude $t_{01}e^{-ikd}$, where k is the (complex) wavenumber of the radiation in medium 1. A fraction r_{12} of this is reflected at the interface between media 1 and 2, so the amplitude of the ray C at the bottom of medium 1 is $t_{01}r_{12}e^{-ikd}$, and the amplitude of this ray at the top of the slab is therefore $t_{01}r_{12}e^{-2ikd}$. Finally, we can write the amplitude of the ray E as $t_{01}r_{12}t_{10}e^{-2ikd}$.

Adding together the contributions from rays A and E, we see that the reflection coefficient has become

$$r_{01} + t_{01}r_{12}t_{10}e^{-2ikd}$$

We have ignored the ray D in this analysis. This is correct, because medium 2 is infinitely thick so there is no interface to reflect this radiation back into medium 1. We have also ignored the possibility that the ray C can be reflected back into medium 1, and in fact the radiation can bounce between the upper and lower surfaces of medium 1 indefinitely, with a little more radiation escaping upwards at each reflection from the upper surface. When this is taken into account, the formula for the reflection coefficient of the system becomes

$$r_{01} + \frac{t_{01}r_{12}t_{10}\exp(-2ikd)}{1 - r_{10}r_{12}\exp(-2ikd)} \tag{3.59}$$

Figure 3.17 illustrates the behaviour of this function (or rather, the power reflection coefficient which is the square of its magnitude) as a function of the slab thickness d, for the case where medium 1 has $\varepsilon_r = 10 - 2i$ and medium 2 has $\varepsilon_r = \infty$ (i.e. is perfectly reflecting). The frequency is 1 GHz. (These dielectric constants are not intended to correspond to any particular real materials, and are just for illustration.) The figure shows that, when medium 1 has a thickness of zero, the power reflection coefficient is equal to that of the inter-

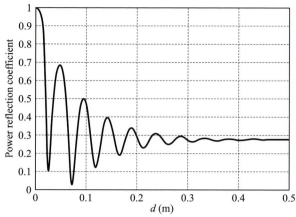

Figure 3.17. Power reflection coefficient at normal incidence and 1 GHz for a slab of material with $\varepsilon_r = 10 - 2i$ and thickness d overlying a perfectly reflecting surface.

face between media 1 and 2. When medium 2 is sufficiently thick, the power reflection coefficient is equal to that of the upper surface of medium 1. This is easy to understand from figure 3.16. If d is sufficiently large, the ray B is attenuated practically to zero as it travels down through medium 1, so there is nothing left to be reflected back as ray C. From equations (3.12) and (3.15) we can calculate the absorption length in medium 1 to be 0.076 m, so we can see from figure 3.17 that this situation is reached in practice once the depth of the layer is of the order of five absorption lengths. We also note from figure 3.17 that the reflection coefficient oscillates, with a diminishing amplitude, as d increases. These oscillations are due to interference between the emerging rays.

Although we have considered only a simple example, it is clear from the analysis we have just performed that, if there is *no* second medium underneath medium 1 (or, equivalently, if medium 1 is sufficiently thick), none of the radiation that enters medium 1 will be reflected back out of it. In this case, only surface scattering can occur. However, this assumes that the medium is homogeneous. If the medium is inhomogeneous – 'lumpy' – the inhomogeneities can *scatter* radiation. This phenomenon will be discussed in the next section.

3.4.1 The radiative transfer equation

To begin with, we will develop a model of the behaviour of radiation when both absorption and scattering are present. We have already encountered the phenomenon of absorption. In section 3.1.1 we defined the absorption length and noted (from equations (3.14) and (3.15)) that, for radiation propagating in the z-direction in a medium with absorption length l_a, the flux density [3]F varies according to

---

[3] Note that throughout this section we are considering only the *intensity* of radiation, not its *amplitude*, and we will not therefore have the added complcation of interference effects, such as those illustrated in figure 3.17. In effect, we are assuming that the scattering phenomena are *incoherent*.

$$F = F_0 \exp\left(-\frac{z}{l_a}\right)$$

where F_0 is a constant. This can also be written as a differential equation:

$$\frac{dF}{dz} = -\frac{F}{l_a}$$

in which case it is valid even if l_a varies with z, but it is more usual to write this expression as

$$\frac{dF}{dz} = -\gamma_a F \qquad (3.60)$$

where $\gamma_a = 1/l_a$ is the *absorption coefficient*. A helpful way to visualise the meaning of this equation is to consider radiation of flux density F incident normally on a thin slab of absorbing material of thickness Δz. Rearranging equation (3.60) very slightly, we see that $(\Delta F)/F = -\gamma_a \Delta z$, so that the *fraction* of the incident power that is absorbed by the slab is $\gamma_a \Delta z$. The fraction absorbed is proportional to the thickness, as we would expect (as long as the slab is thin), and γ_a is just the coefficient of proportionality.

Scattering can be defined as the deflection of electromagnetic radiation, without absorption, as a result of its interaction with particles (electrons, atoms, molecules or larger particles) or a solid or liquid surface. We will consider the nature of scattering by particles in the following sections, but for the present it will be enough to define the *scattering coefficient*. This is very similar to the definition of the absorption coefficient: a thin slab, of thickness Δz, scatters a fraction $\gamma_s \Delta z$ of the power incident upon it, where γ_s is the scattering coefficient. Of course, the scattered radiation is not 'lost', unlike the absorbed radiation, and we need to keep track of it.[4] In order to see how the radiative transfer equation works, we will first derive a simplified, one-dimensional version of it.

Figure 3.18 shows radiation propagating the $+z$ and $-z$ directions in three adjacent parallel slabs, each of thickness Δz. We have used the symbols F_+ and F_- to represent the flux densities propagating in the two directions. When radiation is incident on one of these slabs, a fraction $\gamma_a \Delta z$ is absorbed and a fraction $\gamma_s \Delta z$ is scattered. We assume that all of this scattered radiation is scattered backwards, so that the fraction of the radiation that is transmitted through the slab is $1 - (\gamma_a + \gamma_s) \Delta z$. It is clear from figure 3.18 that the the flux density F_+ in the middle slab is contributed to by the transmitted component of

---

[4] We are considering only *elastic* scattering, in which the wavelength of the radiation is unchanged by the scattering process. Thus, we are neglecting the phenomenon of *fluorescence*, in which radiation is absorbed at one wavelength and re-emitted at another, usually longer. Some minerals, especially sulphides, fluoresce in the visible part of the spectrum when excited by ultraviolet radiation. Plant material also shows a diagnostically useful fluorescence response. However, most fluorescence phenomena are too small to be measured accurately from airborne or (especially) spaceborne observations. Cracknell and Hayes (1991) give a useful discussion of 'fluorosensing'.

z-direction

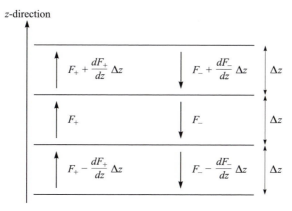

Figure 3.18. Radiation propagating in the $+z$ and $-z$ directions in three parallel slabs of thickness Δz.

the positive-direction flux in the lower slab, and the reflected (scattered) component of the negative-direction flux in the middle slab, so we must have

$$F_+ = \left(F_+ - \frac{dF_+}{dz} \Delta z \right)(1 - (\gamma_a + \gamma_s)\,\Delta z) + F_-\gamma_s\,\Delta z$$

Ignoring the terms in $(\Delta z)^2$ and rearranging, we find

$$\frac{dF_+}{dz} = -(\gamma_a + \gamma_s)F_+ + \gamma_s F_- \qquad (3.61.1)$$

This is an intuitively reasonable equation. It shows that radiation is being lost from the forward direction as a result of both absorption and scattering, but gained from the scattering of backward-travelling radiation. The corresponding equation for the backward-travelling radiation is

$$\frac{dF_-}{dz} = (\gamma_a + \gamma_s)F_- - \gamma_s F_+ \qquad (3.61.2)$$

We can note from these equations the significance of the *sum* of the absorption and scattering coefficients in describing the propagation of radiation where both phenomena are important, since it represents the loss of energy from the forward-propagating radiation. This combination of absorption and scattering is usually called *attenuation* (sometimes *extinction*), and the attenuation coefficient is defined as

$$\gamma_e = \gamma_a + \gamma_s \qquad (3.62)$$

Equation (3.61) can be used to identify some of the important consequences of volume scattering. We will consider an infinitely deep slab of material in which the absorption and scattering coefficients are constant, and radiation of unit flux density incident normally on the slab. To keep things as simple as possible, we will also assume that the reflection coefficient at the surface of this slab is zero, so that only volume scattering is important. This situation is described by equations (3.61.1) and (3.61.2) in the region $z \leq 0$, and subject to the boundary conditions that $F_-(0) = 1$, $F_+(-\infty) = F_-(-\infty) = 0$. It is not difficult to show that the solution is

$$F_- = \exp(\mu z)$$

$$F_+ = \frac{\gamma_a + \gamma_s - \mu}{\gamma_s} \exp(\mu z)$$

where

$$\mu = \sqrt{\gamma_a^2 + 2\gamma_a\gamma_s}$$

so the intensity reflection coefficient is

$$R = \frac{\gamma_a + \gamma_s - \sqrt{\gamma_a^2 + 2\gamma_a\gamma_s}}{\gamma_s} \tag{3.63}$$

This function, which depends only on the *ratio* of γ_s to γ_a, is shown in figure 3.19.

Figure 3.19 shows that, if the scattering coefficient is much larger than the absorption coefficient, the volume scattering will be large. This is the reason that many finely divided materials, such as snow, clouds and (for example) table salt, are white. The total optical absorption in a slab of pure ice 1 m thick is small, but if the ice is divided up into snow grains each of which is only 1 mm across, a ray of light will encounter 2000 ice–air interfaces as it traverses the snow layer. Scattering can occur at each of these interfaces, and although the amount of scattering at each interface is small, the cumulative effect is large. Provided the absorption coefficient is small across the whole of the visible spectrum, as it is for ice, water and sodium chloride, the material will therefore appear white.

Equations (3.61) can only strictly be applied to problems in which the scattering of radiation is all backwards, opposite to the direction of incidence. In general, scattered radiation will be distributed over all possible directions, and

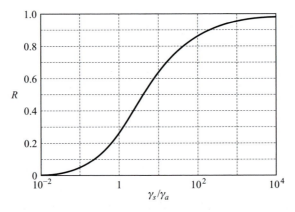

Figure 3.19. Dependence of the intensity reflection coefficient R for volume scattering on the ratio of the scattering coefficient to the absorption coefficient (one-dimensional model).

a simple one-dimensional approach to the problem is no longer possible. To describe this situation, we will need to use the *radiance* of the radiation, and to consider its distribution with direction as well as position. However, the essential principles remain the same. In three dimensions, if only absorption and scattering are involved, the radiative transfer equation becomes

$$\frac{dL_f(\theta, \phi)}{dz} = -(\gamma_a + \gamma_s)L_f(\theta, \phi) + \gamma_s J_f \qquad (3.64)$$

where $L_f(\theta, \phi)$ is the spectral radiance propagating in the direction (θ, ϕ), and dz measures distance in this same direction. J_f describes the radiation scattered into the direction (θ, ϕ) from other directions specified by (θ', ϕ'), and is defined by

$$J_f = \frac{1}{4\pi} \int_{4\pi} L_f(\theta', \phi')p(\cos \Theta) \, d\Omega' \qquad (3.65)$$

where $d\Omega' = \sin \theta' \, d\theta' \, d\phi'$ is an element of solid angle, and the integration is performed over all directions (i.e. over 4π steradians). $p(\cos \Theta)$ is the *phase function* of the scattering, and describes the angular distribution of the scattered radiation in terms of the angle Θ through which the radiation has been deflected. This angle is given by

$$\cos \Theta = \cos \theta \cos \theta' + \sin \theta \sin \theta' \cos(\phi - \phi') \qquad (3.66)$$

Further modifications to equation (3.64) are possible. The only one we will consider now is the effect of black-body emission. If the radiation is in thermal equilibrium with the medium through which it is propagating, we need to add a term for emission to equation (3.64):

$$\frac{dL_f(\theta, \phi)}{dz} = -(\gamma_a + \gamma_s)L_f(\theta, \phi) + \gamma_s J_f + \gamma_a B_f \qquad (3.67)$$

where B_f is the spectral radiance of black-body radiation at the appropriate temperature T, given by equation (2.30):

$$B_f = \frac{2hf^3}{c^2} \frac{1}{\exp(hf/kT) - 1} \qquad (3.68)$$

Various special cases of equation (3.67) find important applications in different aspects of remote sensing. We will consider first the case in which only absorption is significant, so that we can set γ_s and B_f to zero. This gives the equation

$$\frac{dL_f}{dz} = -\gamma_a L_f$$

which is equivalent to equation (3.60). Note that we do not need to specify the direction (θ, ϕ), because there is no scattering so the direction does not change. If γ_a varies with position z, the solution of this equation can be written as

$$L_f(z) = L_f(0) \exp(-\tau(z)) \qquad (3.69)$$

where

$$\tau(z) = \int_0^z \gamma_a(z') \, dz' \qquad (3.70)$$

is the integral of the absorption coefficient with respect to distance from 0 to z. This quantity τ is called the optical thickness of the path from 0 to z.

Next, we will add thermal emission back into our model, but will continue to ignore scattering. The radiative transfer equation in this case is

$$\frac{dL_f}{dz} = \gamma_a(B_f - L_f) \qquad (3.71)$$

If the frequency is low enough for the Rayleigh–Jeans approximation to be valid, this can be rewritten as

$$\frac{dT_b}{dz} = \gamma_a(T - T_b)$$

where T_b is the brightness temperature of the radiation and T is the physical temperature of the medium through which it is propagating. For constant T, the solution to this equation is

$$T_b(z) = T_b(0) \exp(-\tau(z)) + T(1 - \exp(-\tau(z))) \qquad (3.72)$$

Equation (3.72), which has an obvious application to the correction of microwave radiometer measurements for the effect of atmospheric absorption, has an intuitive reasonableness. If radiation travels through only a small amount of absorbing material, so that τ is small, the brightness temperature is hardly changed; but if the optical thickness is large, the brightness temperature of the emerging radiation is equal to the physical temperature of the medium.

Finally, we will consider the solution of equation (3.71) in the case where B_f is not constant because the temperature is not constant. This could be, for example, a model of atmospheric temperature sounding, where the brightness temperature of upward-travelling radiation is measured at a point above the bulk of the Earth's atmosphere and used to make deductions about the temperature distribution within the atmosphere. It is much simpler in this case to work with the optical thickness τ rather than the position z, and we can note from equation (3.70) that there is a monotonic relationship between τ and z. On this basis, the solution of equation (3.71) can be written as

$$L_f(\tau) = L_f(0) \exp(-\tau) + \int_0^\tau B_f(\tau') \exp(\tau' - \tau) \, d\tau' \qquad (3.73)$$

or, if the Rayleigh–Jeans approximation is valid, as

$$T_b(\tau) = T_b(0) \exp(-\tau) + \int_0^\tau T(\tau') \exp(\tau' - \tau) \, d\tau' \qquad (3.74)$$

As before, an intuitive understanding of these equations is possible. For simplicity, we will consider equation (3.74), and assume that it applies to radiation propagating upwards from the Earth's surface (where z and τ are zero), through the atmosphere, for some total distance z_{max} (see figure 3.20). The brightness temperature of the radiation emitted from the Earth's surface is $T_b(0)$, and the total optical thickness of the path is τ, so the first term in equation (3.74) corresponds to equation (3.69) and just represents the absorption of the surface radiation. The integrand considers some position z such that the optical thickness between this point and $z = 0$ is τ'. The optical thickness between this point and $z = z_{max}$ is $(\tau - \tau')$, so we expect the contribution from this point to be reduced by a factor of $\exp(-(\tau - \tau'))$. The optical thickness $(\tau - \tau')$ from the point in the atmosphere from which the radiation originates, to the point at which it is measured, is often called the *optical depth* of the former point.

3.4.2 *Absorption and scattering by macroscopic particles*

In section 3.4.1, we introduced the concepts of the absorption and scattering coefficients. These were defined as bulk properties of the medium through which the radiation is propagating, but they arise from interactions between the radiation and the particles of which the medium is composed. In this section, we will examine the factors that relate the absorption and scattering coefficients to physical properties of the particles.[5] We begin by assuming that the particles are much smaller than the wavelength of the radiation, so that the electric field at any instant can be assumed to be constant across the particle. However, we will defer a discussion of scattering and absorption by individual

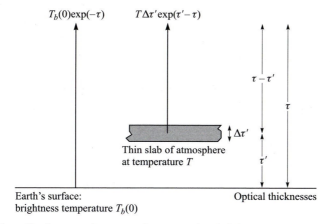

Figure 3.20. Contributions to the upward-propagating brightness temperature of the atmosphere. Heights above the Earth's surface are measured using the optical thickness τ.

---

[5] A much fuller treatment of the material presented in this section is given by Van de Hulst (1957, 1981). Useful material is also provided by Schanda (1986).

molecules to chapter 4, in which we consider propagation of radiation through the atmosphere.

A useful concept in understanding absorption and scattering by particles is the polarisability, which we have already met in section 3.1.1 and which will again be denoted by the symbol α. This is defined such that the dipole moment p induced in the particle when it is placed in an electric field E is given by αE. The direction of the dipole moment is the same as that of the electric field vector, but in an oscillating electric field, such as we will have in the case of electromagnetic radiation, the dipole moment might not be in phase with the electric field. For this reason, the polarisability α will in general be complex.

First, we will consider the absorption of electromagnetic energy by the particle. If the induced dipole moment p is varying with time, the instantaneous rate of power dissipation is given by $E\,dp/dt$. Putting

$$E = E_0 \exp(i\omega t)$$

for the electric field in the vicinity of the particle, we see that the dipole moment must be given by

$$p = \alpha E_0 \exp(i\omega t)$$

However, we recall that the physical meaning of these complex exponential expressions is contained in their real parts. We can assume that E_0 is real without any loss of generality, so we can rewrite our expression for E as

$$E = E_0 \cos(\omega t)$$

and obtain the following expression for dp/dt:

$$\frac{dp}{dt} = -\mathrm{Re}(\alpha)\omega E_0 \sin(\omega t) - \mathrm{Im}(\alpha)\omega E_0 \cos(\omega t)$$

The time-average of the dissipated power can then be obtained by multiplying these two expressions together, and averaging the result over time. This gives the mean dissipated power as

$$\langle P \rangle = -\frac{\omega \mathrm{Im}(\alpha)}{2} E_0^2$$

At this point, we can introduce the *absorption cross-section*, which we denote by the symbol σ_a. It is defined such that, if electromagnetic radiation of flux density F is incident on the particle, the mean absorbed power is given by

$$\langle P \rangle = \sigma_a F$$

Using equations (2.4), (2.7) and (2.8), we find that the absorption cross-section is given by

$$\sigma_a = -\frac{k}{\varepsilon_0} \mathrm{Im}(\alpha) \tag{3.75}$$

where k is the wavenumber of the radiation. It is clear that σ_a has the dimensions of an area, and this area has a physical interpretation: it is as though the particle absorbs all of the radiation incident within this cross-sectional area.

We can also note from equation (3.75) that the imaginary part of the polarisability cannot be negative.

For a small spherical particle of radius a and refractive index n (which may be complex), the polarisability is given by

$$\alpha = 4\pi\varepsilon_0 a^3 \left(\frac{n^2 - 1}{n^2 + 2}\right) \tag{3.76}$$

so the ratio of the absorption cross-section to the geometrical cross-section (πa^2) is

$$-\frac{8\pi a}{\lambda}\, \text{Im}\left(\frac{n^2 - 1}{n^2 + 2}\right)$$

Because of our assumption that the radius a of the particle is much smaller than the wavelength λ, this expression will almost always be much smaller than 1. Thus, in the small-particle limit, the particle will absorb much *less* power than the power carried by a cross-section of the radiation equal to the geometrical area of the particle.

Next, we consider scattering by our small ($\ll \lambda$) particle. Classical electrodynamics shows that an oscillating electric dipole whose strength is described by

$$p = p_0 \sin(\omega t)$$

radiates a mean power of

$$\langle P \rangle = \frac{\mu_0 \omega^4 p_0^2}{12\pi c}$$

so the time-average of the power reradiated by a particle of polarisability α in an electromagnetic wave whose electric field vector is given by

$$E = E_0 \exp(i\omega t)$$

must be

$$\langle P \rangle = \frac{\mu_0 \omega^4 E_0^2}{12\pi c} |\alpha|^2$$

We can define the *scattering cross-section* σ_s in a manner similar to the absorption cross-section, such that the scattered power is $\sigma_s F$ when the particle is illuminated by radiation of flux density F. Again using equations (2.4), (2.7) and (2.8), we obtain the result

$$\sigma_s = \frac{k^4}{6\pi\varepsilon_0^2} |\alpha|^2 \tag{3.77}$$

As for the absorption cross-section, the scattering cross-section is much smaller than the geometrical cross-section of the particle, provided that the particle is small compared with the wavelength of the radiation.

We noted in section 3.4.1 that, in considering the scattering of radiation, we need to know how much energy is scattered in different directions. Equation (3.65) introduced the phase function as a means of specifying this. For the small-particle scattering described by equation (3.77), the phase function is given by

$$p(\cos \Theta) = \frac{3}{2}(1 - \cos^2 \Theta) \tag{3.78}$$

(Note that the p in this expression is *not* the electric dipole moment!) Equation (3.78) shows that the scattering is maximum in the forward ($\Theta = 0$) and backward ($\Theta = \pi$) directions, and zero for scattering through $\pi/2$. The factor of $3/2$ ensures that the expression is correctly normalised, so that the integral over all directions is equal to 4π. For comparison, the phase function for an isotropic scatterer would be 1 in all directions.

Equations (3.75), (3.77) and (3.78) apply only to particles that are very small compared with the wavelength of the radiation. In this case, the scattering is often referred to as Rayleigh scattering. For larger particles, the situation becomes more complicated. If the particles are spherical with radius a, the problem can sometimes be analysed in terms of the dimensionless variable

$$x = \frac{2\pi a}{\lambda}$$

Rayleigh scattering then corresponds to the case when x is very small. The term Mie scattering is used to describe the situation for larger values of x. The extinction (attenuation) and scattering cross-sections can be expressed as

$$\frac{\sigma_e}{\pi a^2} = \frac{2}{x^2} \sum_{l=1}^{\infty} (2l + 1)(|a_l^2| + |b_l^2|) \tag{3.79}$$

$$\frac{\sigma_s}{\pi a^2} = \frac{2}{x^2} \sum_{l=1}^{\infty} (2l + 1)\mathrm{Re}(a_l + b_l) \tag{3.80}$$

where a_l and b_l, which are the *Mie coefficients*, depend on the complex refractive index and on the value of x. Equations (3.79) and (3.80) can be written as power series in x, in which case the first few terms are

$$\frac{\sigma_e}{\pi a^2} = -\mathrm{Im}\left\{ 4x \frac{n^2 - 1}{n^2 + 2} + \frac{4}{15}x^3 \left(\frac{n^2 - 1}{n^2 + 2}\right)^2 \frac{n^4 + 27n^2 + 38}{2n^2 + 3} \right.$$

$$\left. + x^4 \mathrm{Re}\left\{ \frac{8}{3}\left(\frac{n^2 - 1}{n^2 + 2}\right)^2 \right\} + \cdots \tag{3.81}$$

and

$$\frac{\sigma_s}{\pi a^2} = \frac{8}{3}\left|\frac{n^2 - 1}{n^2 + 2}\right|^2 x^4 + \frac{16}{45}\left|\frac{(n^2 - 1)^2(n^2 - 2)}{(n^2 + 2)^3}\right|x^6 + \frac{32}{27}\left|\frac{n^2 - 1}{n^2 + 2}\right|^3 x^7 + \cdots \tag{3.82}$$

We can recognise the first term in equation (3.81) as the result we have already derived for the absorption cross-section of a very small sphere, and the first term in equation (3.82) as the Rayleigh scattering cross-section. A simple iterative procedure for calculating the Mie coefficients, suitable for computing, has been presented by Deirmendjian (1969, pp. 14–15) and restated by Ulaby et al. (1981, volume 1, pp. 290–291).

Figure 3.21 shows the dependence of the ratios $(\sigma_e/\pi a^2)$ and $(\sigma_s/\pi a^2)$ calculated as a function of x for a refractive index $n = 5.67 - 2.88i$. For small values of x (the Rayleigh scattering region), the scattering cross-section is proportional to x^4, and the absorption cross-section is proportional to x. Since the absorption cross-section is much larger than the scattering cross-section, the absorption cross-section is also proportional to x. The figure also shows that, when x increases beyond about 1, the cross-sections pass through a maximum value that is of the order of the geometric cross-section, and then tend towards a constant value, usually with some oscillations. The grey lines in figure 3.21 show the asymptotic behaviour at small values of x. For the scattering cross-section, only the Rayleigh formula is shown since this provides an adequate approximation for values of x up to about 1. For the attenuation cross-section, the grey lines show the full expression of equation (3.81) (curve) and just its first term (straight line).

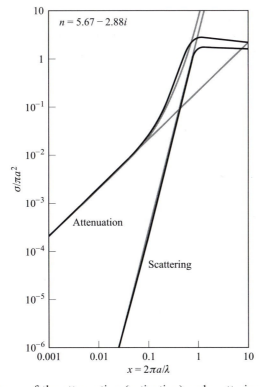

Figure 3.21. Dependence of the attenuation (extinction) and scattering cross-sections on the parameter x for a spherical particle of refractive index $n = 5.67 - 2.88i$. The grey curves show the asymptotic behaviour at small values of x.

Finally, we need to examine how the absorption and scattering cross-sections, which are defined for single particles, are related to the absorption and scattering coefficients. We will consider a thin slab of material, having cross-sectional area A and thickness Δz, with radiation of flux density F incident normally upon it. The material contains N particles per unit volume, each of which has absorption cross-section σ_a and scattering cross-section σ_s. It is clear that the slab must contain $NA\,\Delta z$ particles, so the total power that is absorbed by them is given by $F\sigma_a NA\,\Delta z$, provided that Δz is small enough. The flux density is thus reduced by $F\sigma_a N\,\Delta z$, so from equation (3.60) we can see that the absorption coefficient is

$$\gamma_a = N\sigma_a \qquad\qquad (3.83)$$

A similar argument for scattering leads to the result that

$$\gamma_s = N\sigma_s \qquad\qquad (3.84)$$

As an example, let us consider absorption and scattering of 1-cm microwaves by water droplets. We will take two hypothetical but fairly realistic cases: a cumulus cloud, in which the typical droplet radius $a = 2 \times 10^{-5}$ m and the number density $N = 2 \times 10^7\,\mathrm{m}^{-3}$; and a light rainfall, for which $a = 1\,\mathrm{mm}$ and $N = 100\,\mathrm{m}^{-3}$. At a wavelength of 1 cm, the refractive index n of water is $5.67 - 2.88i$, so for the cloud droplets, for which $x = 0.0126$, equation (3.81) gives $\sigma_e = 3.58 \times 10^{-12}\,\mathrm{m}^2$ and equation (3.82) gives $\sigma_s = 7.73 \times 10^{-17}\,\mathrm{m}^2$. (In fact, just taking the first terms of the equations will give these values with sufficient accuracy.) The absorption cross-section is given by $\sigma_a = \sigma_e - \sigma_s$ as $3.58 \times 10^{-12}\,\mathrm{m}^2$. Multiplying these cross-sections by the number density of the droplets, we find that the scattering coefficient for cloud is $1.55 \times 10^{-9}\,\mathrm{m}^{-1}$, and the absorption and attenuation coefficients are $7.16 \times 10^{-4}\,\mathrm{m}^{-1}$ (approximately 0.3 dB/km [6]).

For the rain droplets $x = 0.63$. This is rather too large for equation (3.81) to give an accurate value for the attenuation cross-section, which is $4.78 \times 10^{-6}\,\mathrm{m}^2$, but equation (3.82) gives a reasonable estimate of the scattering cross-section as $1.40 \times 10^{-6}\,\mathrm{m}^2$ (the true value is $1.76 \times 10^{-6}\,\mathrm{m}^2$). The absorption cross-section is thus $3.02 \times 10^{-6}\,\mathrm{m}^2$, and the corresponding coefficients are as follows: scattering, $1.76 \times 10^{-4}\,\mathrm{m}^{-1}$ (0.8 dB km^{-1}); absorption, $3.02 \times 10^{-4}\,\mathrm{m}^{-1}$ (1.3 dBkm^{-1}); and attenuation, $4.78 \times 10^{-4}\,\mathrm{m}^{-1}$ (2.1 dB km^{-1}). Here, scattering, although still smaller than absorption, occurs at a significant level, and this illustrates the possibility of monitoring rainfall from ground-based *rain radars*.

---

[6] If the intensity of radiation (or anything else) is reduced by a factor x, this can equivalently be expressed as an attenuation by $10\log_{10}(x)$ decibels (dB). Thus, for example, an absorption coefficient of $y\,\mathrm{m}^{-1}$ can also be expressed as $10\log_{10}(e)\,y \approx 4.34\,y\,\mathrm{dB\,m}^{-1}$.

3.4.3   *Simple models of volume scattering*

It should be quite clear from the previous sections that volume scattering is a mathematically complicated phenomenon, and that no single model can embrace the range of possibilities that may occur. In this section, therefore, we will describe just one practical model of volume scattering. This was originally derived for the microwave backscatter from a snowpack, and is based on the work of Stiles and Ulaby (1980).

Figure 3.22 shows the situation schematically. Radiation is incident on the snowpack at an angle θ to the normal, and a fraction $T(\theta)$ of the incident power is transmitted across the interface, where

$$T(\theta) = |t(\theta)|^2$$

and $t(\theta)$ is the appropriate Fresnel coefficient (equation (3.31.2) or (3.31.4)). This radiation is refracted towards the normal at an angle θ', calculated from Snell's law (equation (3.30)).

Once inside the snowpack, the radiation undergoes volume scattering. In deriving equation (3.63), we saw that the power reflected out of an infinite volume-scattering medium in a simple one-dimensional model depended only on the ratio $\gamma_s/\gamma_a = \sigma_s/\sigma_a$. Plausibly, we may write the power backscattered from an infinite three-dimensional medium as

$$\frac{\gamma_s \cos \theta'}{2\gamma_e} = \frac{\sigma_s \cos \theta'}{2\sigma_e}$$

where the factor $(\cos \theta')/2$ has been introduced to account for the fraction of the total scattered power that is directed in the backscattering direction. However, the medium is not infinite, but instead has a depth d. The optical thickness of the path from the snow surface to the ground beneath is given by

$$\tau(\theta') = \gamma_e \, d \, \sec \theta'$$

and the fraction of power remaining after traversing this path twice (once in each direction) is $\exp(-2\tau(\theta'))$. If this fraction is zero, the medium is effectively infinitely deep so that volume scattering will contribute fully, whereas if it is 1

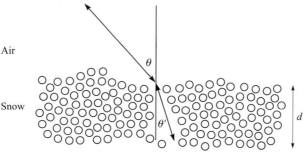

Figure 3.22. Volume scattering of microwave radiation from a snowpack (schematic).

the medium is transparent and volume scattering will not contribute at all. Thus, we may plausibly write

$$\frac{\gamma_s \cos \theta'}{2\gamma_e}(1 - \exp(-2\tau(\theta')))$$

for the fraction of the power entering the snowpack that is volume-scattered back to the surface. This radiation must again be transmitted through the snow–air interface, so our expression for the volume-scattering component of the backscatter coefficient becomes

$$T^2(\theta)\frac{\gamma_s \cos \theta'}{2\gamma_e}(1 - \exp(-2\tau(\theta')))$$

To complete the model, we need also to consider the surface scattering component and the component that is scattered from the snow–ground interface. We can write the latter term as

$$T^2(\theta)\sigma^0_{\text{ground}} \exp(-2\tau(\theta'))$$

to take into account that this radiation must pass through the air–snow interface twice and must also traverse the optical thickness $\tau(\theta')$ twice, so our final formula becomes

$$\sigma^0(\theta) = T^2(\theta)\left(\frac{\sigma_s \cos \theta'}{2\sigma_e}(1 - \exp(-2\tau(\theta'))) + \sigma^0_{\text{ground}} \exp(-2\tau(\theta'))\right)$$
$$+ \sigma^0_{\text{surface}}(\theta) \tag{3.85}$$

It should be emphasised that this is in fact a very simple volume-scattering model that does not take account of, amongst other things, multiple scattering. Nevertheless, it illustrates the principal considerations involved in such models.

3.5 Reflection and emission from real materials

We have now considered the factors that govern the reflection of electromagnetic radiation from solid and liquid surfaces in some detail, and we ought now to try to use these ideas to understand the reflectance properties of some real materials. By extension, we can also consider the emission properties in those parts of the electromagnetic spectrum – namely, the thermal infrared and microwave regions – in which they are important. The treatment in this section is rather brief. A fuller discussion can be found in, for example, Elachi (1987).

3.5.1 *Visible and near-infrared region*

The visible and near-infrared (VIR) region of the electromagnetic spectrum, from $0.4\,\mu$m to about $2\,\mu$m, is still the most important for remote sensing of the Earth's surface. The majority of remote sensing systems (with the exception of those designed to probe the Earth's atmosphere) operate in this region, and the data are intelligible to comparatively unskilled users.

Minor but significant use is still made of broad-band optical data, in which the reflected radiation is integrated over the whole of the visible part of the spectrum. This is, for example, roughly what is achieved by black-and-white (panchromatic) photography, and it is clear that what is measured is the spectrally averaged albedo. Figure 3.23 illustrates typical values of the albedo of various materials.

The lowest albedos are usually shown by water surfaces. The refractive index of water for visible light is roughly 1.33, so for normally incident radiation the Fresnel coefficient is given by equations (3.33.1) or (3.33.2) as 0.14 for the amplitude, and hence 0.02 for the intensity. The low albedo of pure water can therefore be explained in terms of the fact that the refractive index of water is not very different from 1. For 'real', naturally occurring, water, the albedo can be somewhat higher as a result of scattering by suspended particulate matter, or lower because of absorption. The refractive index of pure ice is similar to that of water, but naturally occurring ice generally contains a significant number density of trapped air bubbles that give rise to volume scattering, and hence a higher albedo.

The albedos of clouds (which are composed of small water droplets or ice crystals) and snow (a mixture of ice crystals, air and, in the case of wet snow, liquid water) are dominated by volume-scattering effects. We showed in section 3.4 that the extent of volume reflectance is governed by the ratio γ_s/γ_a. If we again consider a cumulus cloud in which the droplet radius $a = 2 \times 10^{-5}$ m, we see that the particle size parameter $x = 2\pi a/\lambda$ must be very large (several hundred) at optical wavelengths, so we expect the scattering cross-section to be of the order of πa^2. Taking the number density of the droplets to be $N = 2 \times 10^7 \, \mathrm{m}^{-3}$, we therefore find the scattering coefficient to be of the order of $0.02 \, \mathrm{m}^{-1}$. How can we estimate the absorption coefficient? Figure 3.1 shows that the absorption coefficient of pure water is of the order of

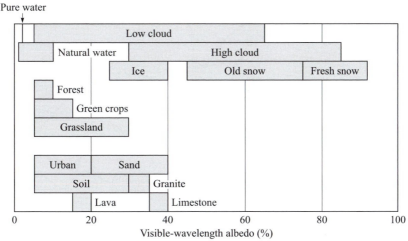

Figure 3.23. Typical values of albedo integrated over the visible waveband for normally incident radiation. (Mostly after Schanda, 1986.)

$0.01\,\mathrm{m}^{-1}$. However, one cubic metre of cloud contains only $4\pi a^3 N/3 = 7 \times 10^{-7}\,\mathrm{m}^3$ of water, so the average absorption coefficient of the cloud is roughly $10^{-8}\,\mathrm{m}^{-1}$. Thus, $\gamma_s/\gamma_a \approx 4 \times 10^6$, and we expect virtually all the radiation that enters the cloud to be scattered out of it, provided that it is optically thick. Since the attenuation length is of the order of $50\,\mathrm{m}$, this condition is easily satisfied. A similar argument applies to snow.

Minerals, soils and the materials from which the visible parts of urban areas are composed show albedos typically in the range 5–40%. The dominant factor in these cases is the refractive index, since the scope for volume scattering is very limited. It is clear that the refractive index must be somewhat larger than that of water, typically between 1.6 and 4.5. If the materials are wet, the refractive index contrast between air and the material is reduced, and the reflectance is also reduced. Soils with a high content of organic material, which strongly absorbs light, also show low reflectance.

Finally, we consider the case of vegetation. Figure 3.23 shows quite low values of albedo for these materials in the optical region. This is again a consequence of absorption, by light-absorbing pigments (principally *chlorophyll* in the case of green vegetation), and is unsurprising since plants derive energy by absorbing light. However, figure 3.23 gives a very incomplete picture of the reflectance properties of vegetation in particular. To understand these better, we should examine the variation of reflectance with wavelength throughout the VIR region.

Figure 3.24 illustrates schematically the spectral reflectance (which we use as a general term to denote certain but unspecified conditions of illumination and viewing geometry) of various materials, in the wavelength range from 0.5 to $2.5\,\mu\mathrm{m}$. The data are simplified, and fine detail has been omitted. Nevertheless, it is clear that in many cases the shape of such a curve (often called the *spectral*

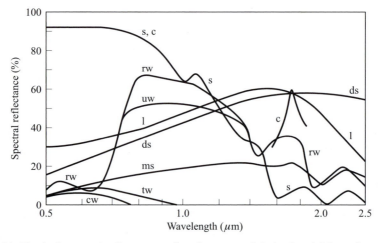

Figure 3.24. Typical spectral reflectances of various materials in the visible and near-infrared regions (schematic). s: snow; c: cloud; rw: ripe wheat; uw: unripe wheat; l: limestone; ds: dry soil; ms: moist soil; tw: turbid water; cw: clear water.

signature) is characteristic of the material, and indeed we are familiar with the idea that the colour of an object often gives a major clue to identifying it. The manner in which remotely sensed images with spectral content can be analysed to identify the probable distribution of materials represented within it will be discussed in chapter 11.

Let us comment on each of the spectral signatures of figure 3.24 in turn. In the case of water, there is little spectral structure, just a decline in reflectance at longer wavelengths as a result of increasing absorption. Limestone also exhibits little spectral structure (at the resolution of the figure), and dry loam broadly follows the same behaviour. However, moist loam exhibits a spectral reflectance that (a) is lower than that of dry loam, as we expect because of the reduced contrast in refractive index, and (b) shows a number of oscillations in the infrared part of the spectrum. These are due to *absorption lines* (absorption maximum near particular wavelengths) of water molecules. Much of this same absorption-line structure can also be seen in the spectral signature of snow.

Finally, we turn to the very characteristic spectral signature of vegetation (see also Curran, 1985), illustrated in figure 3.24 by the example of wheat. As we have remarked already, the low reflectance in the optical region is largely governed by the presence of pigments. The most important of these is chlorophyll, which has absorption maxima at 0.45 and $0.65\,\mu\text{m}$ (respectively, blue and red), and consequently gives rise to a local maximum in the spectral reflectance at about $0.55\,\mu\text{m}$. This is the explanation for the green colour of much vegetable matter. The other important botanical pigments are carotene and xanthophyll (which give orange–yellow reflectance spectra) and the anthocyanins (red–violet). These latter pigments are dominant in the autumn, when the chlorophyll decomposes in many species, and give rise to the spectacular colours of autumn leaves (see Justice et al., 1985).

The high reflectance of vegetation in the near-infrared, roughly between 0.7 and $1.3\,\mu\text{m}$, is a volume-scattering effect. It is principally caused by multiple internal reflections of the radiation from hydrated cell walls in the *mesophyll* of the leaves, with little absorption. The transition from low reflectance in the red part of the spectrum to the much higher reflectance in the near-infrared is rather sharp, and is often termed the '*red edge*'. Above about $1.3\,\mu\text{m}$, the absorption of infrared radiation by water becomes significant (again we see the oscillatory structure that we noted for moist loam and for snow) and the reflectance is reduced. If the vegetation is stressed or diseased, the cell structure is less well developed or is damaged and the high reflectance between 0.7 and $1.3\,\mu\text{m}$ is reduced. Stressed vegetation may also have a lower chlorophyll content than healthy vegetation, which will tend to increase the reflectance in the red part of the spectrum, so that the steepness of the red edge will be reduced. This phenomenon forms the basis of a number of techniques for assessing the amount and health of the vegetation present in a remotely sensed image, and will be discussed in more detail in chapter 11.

3.5.2   *Emissivities in the thermal infrared region*

The thermal infrared region was defined in chapter 2 as the range of wavelengths from 3 to $15\,\mu$m, and we saw in section 2.6 that objects at normal terrestrial temperatures radiate maximally in this part of the electromagnetic spectrum. The emissivity of a body (loosely speaking, its efficiency of emitting black-body radiation) at a given wavelength is negatively related to the reflectivity at the same wavelength, such that when the reflectivity is zero the emissivity is 1 and vice versa. For most naturally occurring materials, with the exception of metals and some minerals, the refractive index in the thermal infrared region is close to 1, which means that the reflectivity is low and the emissivity high. Figure 3.25 shows the emissivities of various materials. It should be noted that the emissivities shown in the figure are somewhat schematic, since the surface roughness of a material will also influence its emissiv-

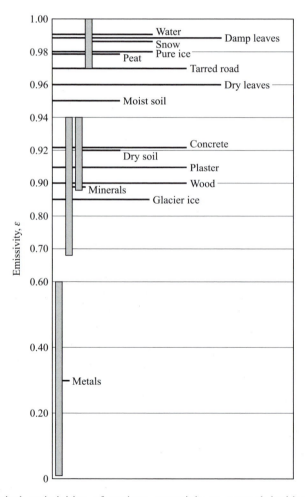

Figure 3.25. Typical emissivities of various materials at normal incidence in the range 8–12 μm. Note the change of scale at $\varepsilon = 0.90$.

ity. For example, a granite rock has a normal emissivity of about 0.89, but if the same material is highly polished, its emissivity will fall to about 0.80. In fact, the emissivity can be written generally in terms of the bidirectional reflectance distribution function (BRDF) as

$$\varepsilon_p = 1 - \int (R_{pp} + R_{qp}) \cos \theta_1 \, d\Omega_1 \qquad (3.86)$$

where ε_p denotes the emissivity for *p*-polarised radiation and R_{qp} is the BRDF for radiation that is incident in the *q*-polarised state and reflected in the *p*-polarised state. θ_1 is the angle between the reflected radiation and the surface normal, and $d\Omega_1 = \sin \theta_1 \, d\theta_1 \, d\phi_1$ where ϕ_1 is the azimuth angle of the reflected radiation. The integration is carried out over 2π steradians; that is, $\theta_1 = 0$ to $\pi/2$ and $\phi_1 = 0$ to 2π. Emissivities can thus be calculated using equation (3.86) and the discussion of BRDFs presented in section 3.3.

3.5.3   Emissivities in the microwave region

Thermally generated radiation can also be detected in the microwave region of the electromagnetic spectrum; that is, at wavelengths between 1 mm and 1 m (frequencies between 0.3 and 300 GHz). As with thermal infrared emission, the physical parameter that determines the quantity of radiation that is emitted at a given temperature is the emissivity, and the principles that determine the emissivity are the same. In practical terms, however, the situation is more complicated than for thermal infrared emission. First, the range of wavelengths over which microwave observations can be made is much larger than for infrared observations. The latter are normally made in a single, rather broad, waveband from (typically) 8 to 12 μm, or perhaps in two narrower wavebands within this range, whereas microwave observations are routinely made at a number of frequencies spanning the range from (typically) 4 to 40 GHz – a factor of 10. It is therefore necessary to consider the variation of emissivity with frequency. Secondly, microwave observations are often made at angles away from the surface normal, so it is important to consider the dependence of the emissivity on the incidence direction.[7] Finally, the emissivities are often significantly different for different polarisation states, so that the dependence of polarisation must also be considered. These factors greatly increase the difficulty of providing a simple characterisation of the microwave emissivities of 'typical' materials. We will therefore content ourselves with illustrating the main features. A much fuller discussion is presented by Ulaby et al. (1982, volume 2).

For a homogeneous material (i.e. one in which volume-scattering effects are unimportant) with a smooth surface, the emissivity is given by $1 - |r|^2$, where r

---

[7] The widespread use of the term 'incidence direction' is potentially misleading, since we are discussing emitted, and not reflected, radiation, and perhaps a term such as 'outgoing direction' or 'viewing direction' would be more logical.

is the Fresnel reflection coefficient appropriate to the direction and polarisation of the radiation (this follows from equation (3.86)). Figure 3.26 illustrates this type of behaviour. The figure has been calculated using the Fresnel coefficients for a material with a dielectric constant of $18.5 - 31.3i$, which is appropriate for calm sea water at a temperature of $20°C$ and a frequency of $35\,GHz$. It can be noted from the figure that the emissivity for V- (vertically) polarised radiation increases to a maximum near 1 at about $80°$. This is the phenomenon of the Brewster angle that we noted in section 3.2. The reason that the emissivity does not quite reach a value of 1 is because the dielectric constant is not real but complex. (We may also note in passing that the dielectric constant, and hence emissivity, of a water surface depends on temperature.)

Continuing with our example of sea water, we can consider the effect of varying the salinity. Increasing the salinity will increase the electrical conductivity of the water, and hence the Fresnel reflection coefficients, which will in turn lower the emissivity. However, we saw in figure 3.2 that the dielectric constant of sea water differs significantly from that of pure water only at frequencies below about 5 GHz, so we should not expect any significant dependence of emissivity on salinity above this frequency. We may also consider the effect of a rough surface, such as would be produced by wind action. This will lower the reflectance, and hence raise the emissivity, for observing directions away from the normal. Off-nadir observations of the microwave emission from a sea surface therefore have the potential to measure the sea state. The effect is greater at higher frequencies, which is consistent with the scattering models that we discussed in section 3.3.4, and is greater for horizontally than for vertically polarised radiation.

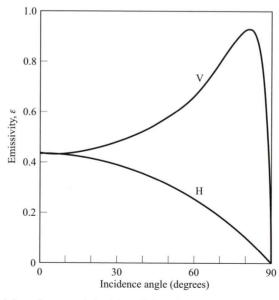

Figure 3.26. Emissivity of a material with a dielectric constant $18.5 - 31.3i$ as a function of incidence angle, for vertically and horizontally polarised radiation.

The microwave emissivity of a bare soil surface is dominated by the surface roughness and by the moisture content of the soil. The dielectric constant of water in the microwave region is much higher than that of soil (typically 3), so increasing moisture content will increase the reflectance and hence decrease the emissivity. Typical emissivities for soil surfaces lie in the range 0.5 to 0.95.

Volume-scattering effects are negligible in soils unless they are extremely dry. However, these effects can dominate in determining the emissivity properties of more open structures such as dry snowpacks and vegetation canopies. In such cases, the emissivity is given by

$$1 - \frac{\gamma_s}{\gamma_e}$$

for an optically thick medium in which multiple scatter is insignificant. Vegetation canopies have microwave emissivities typically in the range 0.85 to 0.99. A deep, dry snowpack has an emissivity of about 0.6. If the medium is not optically thick, the measured emission will include a contribution from the surface below, for example from the soil surface below a vegetation canopy. For a dry snowpack, the optical thickness at normal incidence is proportional to the total mass of ice per unit area of the pack, and this effect can therefore be used to estimate the snow mass. For wet snow, volume-scattering effects are insignificant and the emissivity is much higher, typically 0.95.

3.5.4   Microwave backscattering coefficients

From a practical point of view, it is as difficult to present characteristic data on the microwave backscattering properties of 'typical' materials as for their microwave emissivities. Again, the number of combinations of different observing parameters (frequency, polarisation and incidence angle) is very large, and the influence of natural variations in the physical properties of the scattering materials must also be taken into account. However, by way of example, figure 3.27 (which is mainly based on material presented by Long, 1983) illustrates the angular dependence of the backscattering coefficient σ^0 of a few representative materials at X-band. A much fuller discussion is presented by Ulaby et al. (1982, volume 2).

The backscatter from a concrete road surface is almost entirely due to surface scattering processes, and the low values of σ^0 at angles away from normal are consistent with a smooth surface, as expected. Surface scattering also dominates the backscatter from wet snow, the sea (in which case, the influence of wind speed on surface roughness is apparent), and urban areas. The dependence of the backscatter from a sea surface on the wind speed means that wind speed over the sea can be inferred from appropriate backscatter measurements (obviously, one would avoid incidence angles close to $10°$ where the sensitivity is very small). In fact, there is also a small dependence on wind *direction* (the data in figure 3.27c have been plotted for the case when the radar look direction

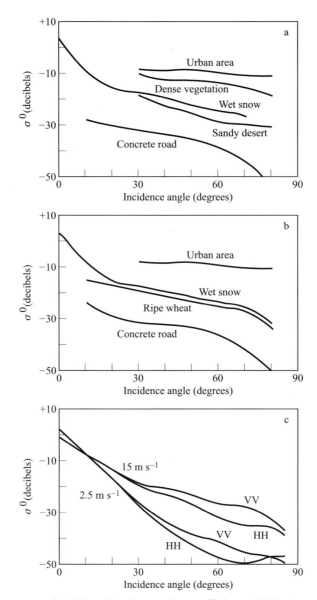

Figure 3.27. Representative X-band backscattering coefficients. (a) Various materials at HH-polarisation; (b) various materials at VV-polarisation; (c) sea surfaces at HH and VV polarisations. The curves have been labelled with the corresponding wind speed.

is upwind), which can be used to infer the wind velocity. The microwave backscattering properies of ocean surfaces are discussed further in section 9.4.1.

The high backscatter from urban areas, not strongly dependent on incidence angle, arises mainly from the presence of large numbers of planar surfaces, oriented at right angles to one another, in such areas. Figure 3.28 illustrates one of the mechanisms by which this occurs. If two reflecting planes are arranged perpendicularly (e.g. one plane could be the wall of a building and the other a road surface), incident radiation that strikes the concave region

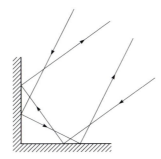

Figure 3.28. Radiation is reflected by two perpendicular planes so that it returns along the incidence direction. This is an important contributor to the strong microwave backscatter from urban areas.

between the planes from any direction in the plane that contains the normals to both surfaces will be reflected back along its incidence direction. In fact, if *three* planar surfaces meet at right angles, radiation from *any* direction that strikes the concave region between them will be scattered back along its incidence direction. Thus, a region that contains a large number of such 'inside corners' will give strong specular scattering from most incidence directions.

Dry deserts, vegetation canopies and dry snowpacks are examples of media from which volume-scattering effects are important. In general, the backscattering in such cases will show a weaker dependence on incidence angle than is observed in cases where surface scattering dominates. A dry desert can be assumed to be optically thick (physical depth much larger than the attenuation length). However, for dry snow and vegetation canopies, this assumption may not be true. For example, radiation incident close to vertically on an agricultural crop is likely to be scattered significantly from the soil below. In the case of dry snow, the assumption of optical thickness is often not valid since the attenuation depth may be many metres. Thus, a dry snow cover a few tens of centimetres deep may be practically invisible to microwave radiation.

The data presented in figure 3.27 show only the co-polarised (like-polarised) backscattering coefficients: namely HH and VV. In general, the cross-polarised components (HV and VH) are extremely small unless multiple volume-scattering occurs.

PROBLEMS

1. Show that the real and imaginary parts of the refractive index of a non-magnetic material are given by

$$m = \sqrt{\frac{\sqrt{\varepsilon'^2 + \varepsilon''^2} + \varepsilon'}{2}}$$

$$\kappa = \sqrt{\frac{\sqrt{\varepsilon'^2 + \varepsilon''^2} - \varepsilon'}{2}}$$

2. A commonly used approximation to the absorption length is

$$l_a \approx \frac{\sqrt{\varepsilon'}}{2\pi\varepsilon''}\lambda_0$$

where λ_0 is the free-space wavelength. Show that this approximation is accurate to within 1% if $\varepsilon''/\varepsilon' < 0.28$.

3. Sea water has an electrical conductivity σ of typically about 4 S m^{-1}. The real part of its dielectric constant at radio frequencies of 100 MHz and below is 88.2. By assuming that the imaginary part is given by equation (3.23), find the absorption length of electromagnetic waves at 100 MHz and 100 kHz.

4. Randomly polarised radiation at a wavelength of 3 cm is incident on a plane water surface at an angle of 83° to the normal. Calculate the Stokes vector, and hence polarisation state, of the reflected radiation. The dielectric constant of water at 3 cm is $63.1 - 32.1i$.

5. (For mathematical enthusiasts.) Show that the diffuse albedo for specular reflection of perpendicularly polarised radiation is

$$\frac{3n^2 - 2n - 1}{3(n+1)^2}$$

and confirm that this expression has the correct limiting forms as the refractive index n tends to 1 and to infinity.

6. A rough surface has a BRDF proportional to $\cos(\theta_0)\cos(\theta_1)$, where θ_0 and θ_1 are, respectively, the angles of the incident and scattered radiation measured from the surface normal. The BRDF has no azimuthal dependence. Show that, if the albedo for normally incident radiation is 1, the diffuse albedo of the surface is 2/3.

7. Consider scattering from a surface with r.m.s. height variation $= \lambda/2$, where λ is the wavelength. Show that the stationary phase model can describe this scattering provided that the incidence angle is less than about 60° and $m < 0.60$, whereas the scalar model requires $m < 0.25$, where m is the r.m.s. surface slope variation.

8. Prove equation (3.63).

9. A typical pack of freshly fallen snow has a density of 100 kg m^{-3} and consists of ice crystals which can be modelled as spheres of radius 0.5 mm. Show that if the snowpack is sufficiently deep it would be expected to have a visible-band albedo close to 1, and estimate the depth necessary for this to occur. The absorption coefficient for light in ice can be taken as 0.01 m^{-1}, and the density of pure ice is 920 kg m^{-3}.

4

Interaction of electromagnetic radiation with the Earth's atmosphere

In chapter 3, we discussed principally the interaction of electromagnetic radiation with the surface and bulk of the material being sensed. However, the radiation also has to make at least one journey through at least part of the Earth's atmosphere, and two such journeys in the case of systems that detect reflected radiation, whether artificial or naturally occurring. Each time radiation passes through the atmosphere it is attenuated to some extent. In addition, as we have already seen in section 3.1.2 and figure 3.4, the atmosphere has a refractive index that differs from unity so that radiation travels through it at a speed different from the vacuum speed of $299\,792\,458$ m s^{-1}. These phenomena must be considered if the results of a remotely sensed measurement are to be corrected for the effects of atmospheric propagation, or if they are to be used to infer the properties of the atmosphere itself. We have already considered them in general terms in discussing the radiative transfer equation (section 3.4). In this chapter, we will relate them more directly to the constituents of the atmosphere.

4.1 Composition and structure of the gaseous atmosphere

At sea level, the principal constituents of the dry atmosphere are molecules of nitrogen (about 78% by volume), oxygen (21%) and the inert gas argon (1%). There is also a significant but variable (typically 0.1% to 3%) amount of water vapour, often specified by the *relative humidity H*. This is defined by equation (4.1):

$$H = \frac{p_{\text{water}}}{p_{\text{sat}}(T)} \tag{4.1}$$

as a fraction between zero and 1 (or more commonly as a percentage), where p_{water} is the *partial pressure* of the water vapour, which can be defined as the product of the total atmospheric pressure with the volume fraction of water vapour, and $p_{\text{sat}}(T)$ is the saturated vapour pressure of water at temperature T. Figure 4.1 shows the variation of the saturated vapour pressure of water with temperature. As an example, at $20°C$ $p_{\text{sat}} = 2.34$ kPa, so if the total atmo-

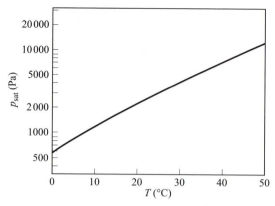

Figure 4.1. Temperature dependence of the saturated vapour pressure of water.

spheric pressure is 100 kPa and the relative humidity is 80%, the volume fraction of water vapour is 1.9%.

In addition to the gases already mentioned, the atmosphere contains a regionally variable quantity of carbon dioxide (currently about 0.035% by volume) and traces, measured in parts per million, of many other gases. Table 4.1 summarises the normal gaseous composition of the atmosphere.

Atmospheric pressure and density diminish with height above the Earth's surface. This is because the molecules, acted upon by gravity, tend to sink to

Table 4.1. Gaseous constituents of the Earth's atmosphere

The third column shows the fraction by volume of the gas at sea level, and the fourth column shows the total mass of gas found in a column through the entire atmosphere

Gas	Chemical formula	Volume fraction	Total mass kg m^{-2}
Nitrogen	N_2	0.7808	7797
Oxygen	O_2	0.2095	2389
Argon	Ar	9.34×10^{-3}	133
Carbon dioxide	CO_2	3.5×10^{-4}	5.6
Neon	Ne	1.8×10^{-5}	0.13
Helium	He	5.2×10^{-6}	7.5×10^{-3}
Methane	CH_4	1.8×10^{-6}	1.0×10^{-2}
Krypton	Kr	1.1×10^{-6}	3.4×10^{-2}
Carbon monoxide	CO	$0.06 - 1 \times 10^{-6}$	$0.06 - 1 \times 10^{-2}$
Sulphur dioxide	SO_2	1.0×10^{-6}	2.9×10^{-2}
Hydrogen	H_2	5.0×10^{-7}	4.0×10^{-4}
Ozone	O_3	$0.01 - 1 \times 10^{-6}$	5.4×10^{-3}
Nitrous oxide	N_2O	2.7×10^{-7}	4.0×10^{-3}
Xenon	Xe	9.0×10^{-8}	4.0×10^{-3}
Nitric oxide	NO_2	$0.05 - 2 \times 10^{-8}$	$0.02 - 4 \times 10^{-4}$
Total dry atmosphere		1	1.032×10^4
Water vapour	H_2O	$0.001 - 0.028$	$6.5 - 180$

the surface but are prevented from doing so fully by thermal excitation. The distribution of density with height is thus governed by the Boltzmann distribution and is approximately exponential. There are, however, significant variations from this approximate dependence and it is conventional to divide the atmosphere into several layers. The lowest of these is the *troposphere* (approximately 0–11 km above the Earth's surface), in which the temperature decreases with height; this is overlain by the *stratosphere* (11–50 km), in which the temperature is roughly constant up to about 35 km, then increases with height; the *mesosphere* (50–80 km); and the *thermosphere* (above about 80 km). The height ranges just specified are for typical conditions at temperate latitudes: there is considerable seasonal and latitudinal variation. Figure 4.2 shows the variation of temperature, pressure and density of the standard (mid-latitude) atmosphere with height.

The absolute temperature T, pressure p and density ρ of the atmosphere can be assumed to be related to one another by

$$\frac{\rho T}{p} = \frac{M_m}{R} \tag{4.2}$$

where M_m is the mass of one mole of atmospheric gas and R is the gas constant (8.314 J K^{-1}). This equation is based on the assumption that the atmosphere behaves as an ideal gas. For heights up to about 100 km, M_m has a more or less constant value of about 0.02896 kg.

The atmospheric pressure p at a height z is a measure of the mass of air, and hence the number of molecules, above z. This follows because the Earth's gravitational field strength g may be assumed to be constant over the range of heights for which p is significant, so that

$$p(z) = M_m g \int_z^\infty N(z')\, dz' \tag{4.3}$$

where $N(z')$ is the molar concentration (number of moles per unit volume) at height z'. For example, figure 4.2b shows that the pressure falls to half its sea-level value at about 5.5 km, and to 1% at about 31 km, so we may state that half of the atmosphere, by mass, is found below 5.5 km and 99% below 31 km. A spaceborne remote sensing system thus looks through all of the atmosphere, while an airborne system typically looks through only a quarter of it. The variation of pressure with height can be modelled rather crudely as an exponential,

$$p(z) = p_0 \exp\left(-\frac{z}{z_0}\right) \tag{4.4}$$

equivalent to assuming that the graph of figure 4.2b is a straight line, which will be convenient for considering the behaviour of atmospheric sounding systems. For the troposphere, a value of z_0 (the *scale height*) of roughly 7.5 km is appropriate.

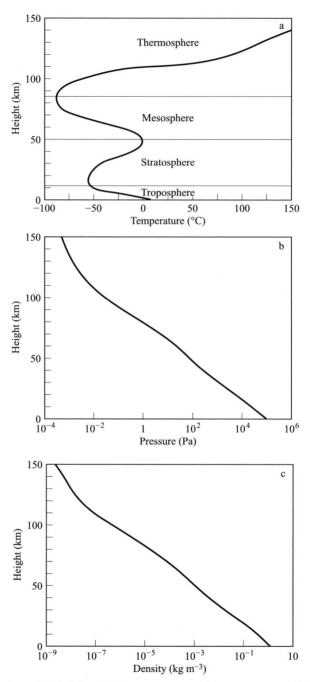

Figure 4.2. Variation with height of (a) temperature, (b) pressure and (c) density of the standard atmosphere.

Graphs similar to figure 4.2b can also be drawn for the partial pressures of the individual molecular species listed in table 4.1 that compose the atmosphere. Provided that we do not consider altitudes much above 100 km, where the heavier molecules are less favoured, and provided that the gas is 'well mixed', the graph will follow the shape of figure 4.2b fairly accurately. Most of the atmospheric gases are well mixed in this sense, with the notable exception of ozone, which, as well as being found in the troposphere, is also generated (by the action of solar ultraviolet radiation) in the stratosphere.

4.2   Molecular absorption and scattering

A detailed understanding of the processes by which molecules absorb and scatter electromagnetic radiation requires considerable knowledge of quantum mechanics, which is beyond the scope of this book.[1] We may state, however, that the energy of an individual molecule cannot be varied continuously, but must be one of a number, in principle infinite, of discrete values called *energy levels*. If a molecule absorbs electromagnetic radiation, it must be promoted from one energy level to another, and hence only certain values of the energy increase ΔE are allowed. Planck's law states that the frequency f of the electromagnetic radiation is given by

$$\Delta E = hf \tag{4.5.1}$$

where h is Planck's constant, although it is often more convenient to write this equation in terms of the angular frequency ω:

$$\Delta E = \frac{h}{2\pi}\omega = \hbar\omega \tag{4.5.2}$$

Thus, we expect that molecules will absorb 'selectively', at particular frequencies, which are usually called *absorption lines*.

4.2.1   *Mechanisms of molecular absorption*

There are three main mechanisms by which molecules can absorb electromagnetic radiation. The first of these, requiring the largest amounts of energy, involves the promotion of electrons to higher energy levels. These are termed *electronic transitions*. Calculation of the energy levels for any but the simplest of molecules is an extremely difficult task, so we will illustrate the idea with reference to the hydrogen atom. In this case, the electronic energy levels are given by

$$E_n = -\frac{me^4}{32\pi^2\varepsilon_0^2\hbar^2}\frac{1}{n^2}$$

---

[1] Schanda (1986) gives a considerably more detailed account.

where m is the electron mass (strictly, the electron's reduced mass, which, in the hydrogen atom, is 99.95% of the electron mass) and n is a *quantum number* that can take only positive integer values. Substituting the values of the constants into the formula, we find that

$$E_n = -\frac{2.177 \times 10^{-18}\,J}{n^2}$$

although it is often more convenient to use the *electron-volt* (symbol eV) as the unit of energy, where $1\,eV \approx 1.602 \times 10^{-19}\,J$, so that the formula becomes

$$E_n = -\frac{13.59\,eV}{n^2}.$$

In its *ground state* (the configuration with lowest energy), hydrogen has $n = 1$. The smallest increase in energy therefore corresponds to the transition from $n = 1$ to $n = 2$, which requires an increase in energy of 10.2 eV. This is typical of the energies required for electronic transitions, and from equation (4.5.1) we see that the frequency of the electromagnetic radiation needed to cause such transitions will therefore be of the order of 10^{15} Hz.[2] The corresponding wavelength is thus a few tenths of a micrometre, so that we expect to find the absorption lines due to electronic transitions in the ultraviolet and visible regions of the electromagnetic spectrum.

The second mechanism of molecular absorption that we shall consider is vibration. The molecular bond between atoms behaves more or less like a spring. To model this, we will consider a diatomic molecule consisting of two atoms, with masses m_1 and m_2, connected by a spring with force constant (defined as dF/dx, where F is the tension and x is the extension) k, as shown in figure 4.3. Classical physics gives the natural angular frequency of this system as

$$\omega_0 = \sqrt{\frac{k(m_1 + m_2)}{m_1 m_2}} \tag{4.6}$$

Figure 4.3. Classical model of molecular vibration. Atoms of masses m_1 and m_2 are connected by a spring of force constant k.

---

[2] Note that in molecular spectroscopy, it is common to specify the frequency of a transition using its *wavenumber*, normally defined as $1/\lambda$, where λ is the wavelength. For example, the wavenumber of the $n = 1 \to 2$ transition for atomic hydrogen is about $82\,000\,cm^{-1}$. Note that the spectroscopist's definition of wavenumber differs from the physicist's definition (equation (2.6)) by a factor of 2π.

and quantum mechanics gives the energy levels as

$$E_v = \left(v + \frac{1}{2}\right)\hbar\omega_0 \tag{4.7}$$

where v is a quantum number that can take any non-negative integer value. This quantum number can change only by ± 1, so in fact the only possible amount of energy that can be absorbed is $\Delta E = \hbar\omega_0$, giving an absorption line at the resonant frequency $f = \omega_0/2\pi$. Because the force constant k is of the order of 1000 N m^{-1}, the resonant frequency is typically between 10^{13} and 10^{14} Hz, corresponding to wavelengths generally in the thermal infrared region.

The last absorption mechanism we shall discuss is rotation. We will consider a simple diatomic molecule consisting of two atoms, with masses m_1 and m_2, separated by a fixed distance d (figure 4.4). Classically, this system can rotate about the centre of mass of the two atoms. The moment of inertia of the system is given by

$$I = \frac{m_1 m_2}{m_1 + m_2} d^2 \tag{4.8}$$

and, according to quantum mechanics, the energy of such a state is given by

$$E_J = \frac{J(J+1)\hbar^2}{2I} \tag{4.9}$$

where J is a quantum number that can take any non-negative integer value. When electromagnetic radiation is absorbed, J must increase by 1. Calculating ΔE from equation (4.9), for the transition from quantum number J to $J+1$, and substituting into equation (4.5), we find that the frequency of a rotation absorption line is given by

$$f = \frac{(J+1)h}{4\pi^2 I}$$

As an example, we can consider the carbon monoxide molecule. This has $m_1 = 2.66 \times 10^{-26}$ kg, $m_2 = 1.99 \times 10^{-26}$ kg and $d = 1.13 \times 10^{-10}$ m, giving it a moment of inertia $I = 1.45 \times 10^{-46}$ kg m^2. We thus calculate that the $J = 0 \to 1$ transition will occur at a frequency of 116 GHz, which is in the microwave region. In general, we expect to find the rotational absorption lines in the microwave or far-infrared regions of the electromagnetic spectrum, with frequencies typically between 10^{10} and 10^{12} Hz.

Although we have now outlined the most important mechanisms governing molecular absorption lines, there are further complications to be considered. *Combinations* of mechanisms can operate at the same time. For example, the

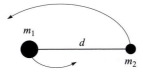

Figure 4.4. Classical model of molecular rotation. Atoms of masses m_1 and m_2, separated by distance d, rotate about their common centre of mass.

energy level of a molecule can be described by both a rotational quantum number J and a vibrational quantum number v, and both of these may change in a transition. This gives rise to a more complicated vibrational–rotational spectrum in which a vibrational absorption line has fine structure super-imposed on it as a result of different rotational transitions. We should also note that not all possible transitions can in fact be excited by electromagnetic radiation. For example, the hydrogen molecule H_2 has a symmetric distribution of electric charge,which means that, classically speaking, an electric field cannot exert a force on it. We should therefore not expect molecular hydrogen to absorb electromagnetic radiation by vibrational or rotational transitions.

We have implicitly suggested that a molecular transition occurs at a single frequency, so that the absorption line in the spectrum has a width of zero. In fact, all lines are broadened to some extent. The Heisenberg uncertainty prin-ciple imposes a minimum line width, although this is negligible compared with other sources of line-broadening. In proportion to the frequency of the line, the effect is largest for electronic transitions, and even for these it is only of the order of 1 part in 10^8. Much more significant is the effect of thermal motion of the gas. The line width Δf due to this effect, which is usually called *Doppler broadening*, is given by

$$\frac{\Delta f}{f} = \sqrt{\frac{RT}{M_m c^2}} \qquad (4.10)$$

where f is the frequency of the line, R is the gas constant, T is the absolute temperature, M_m is the mass of one mole of the gas, and c is the speed of light. Equation (4.10) shows that increasing the temperature will broaden the line, and that heavier molecules will exhibit narrower lines than lighter molecules. The fractional broadening due to this effect is typically 1 part in 10^6.

Another important mechanism is *pressure broadening*, also called *collision broadening*. The molecules of the gas collide with one another and with other molecules in the atmosphere, and these collisions disturb the state of the mole-cule. The pressure broadening can be written approximately as

$$\Delta f \approx \frac{\sigma N_A p}{\sqrt{M_m RT}} \qquad (4.11)$$

where p is the gas pressure, σ is related to the collision cross-section for the molecules, and is of the order of 10^{-19} m^2, and N_A is the Avogadro number.

We can illustrate these formulae with two examples. First, we consider the 0.76 μm absorption line of O_2 at an altitude of 50 km, near the top of the stratosphere. The temperature at this altitude can be taken as 271 K, so equa-tion (4.10) gives the Doppler broadening $\Delta f/f$ as 9×10^{-7}. The pressure is about 80 Pa, so equation (4.11) gives the pressure broadening as $\Delta f \approx 6 \times 10^5$ Hz, and hence $\Delta f/f \approx 1.5 \times 10^{-9}$. In this case Doppler broadening dominates, and the line width can be specified as 3.5×10^8 Hz or about 0.01 cm^{-1}. Our second example is the 22.2 GHz absorption line of H_2O at sea level. Taking the temperature as 288 K, we find from equation (4.10) that the Doppler broadening is $\Delta f/f = 1.2 \times 10^{-6}$, and taking the pressure as 10^5 Pa, equation (4.11) gives the pressure broadening as $\Delta f/f = 4.1 \times 10^2$. In this case the pressure broadening dominates, and the line width is approximately 0.9 GHz (0.03 cm^{-1}).

Table 4.2 summarises the main absorption lines in the Earth's atmosphere.

4.2.2   Molecular scattering

If we model a molecule very simply as an electrically conducting sphere of radius *a,* much smaller than the wavelength λ of the radiation, it will have a polarisability

$$\alpha = 4\pi\varepsilon_0 a^3$$

(this follows from equation (3.76) by setting the refractive index $n = \infty$), and hence a scattering cross-section, given by equation (3.77), of

$$\sigma_s = \frac{128\pi^5 a^6}{3\lambda^4} \tag{4.12}$$

Table 4.2. Principal molecular absorption lines in the Earth's atmosphere

Wavelength (μm)	Visible–infrared region Molecule	Wavelength (μm)	Molecule
0.26	O_3	3.9	N_2O
0.60	O_3	4.3	CO_2
0.69	O_2	4.5	N_2O
0.72	H_2O	4.8	O_3
0.76	O_2	4.9	CO_2
0.82	H_2O	6.0	H_2O
0.93	H_2O	6.6	H_2O
1.12	H_2O	7.7	N_2O
1.25	O_2	7.7	CH_4
1.37	H_2O	9.4	CO_2
1.85	H_2O	9.6	O_3
1.95	CO_2	10.4	CO_2
2.0	CO_2	13.7	O_3
2.1	CO_2	14.3	O_3
2.6	H_2O	15	CO_2
2.7	CO_2		

Frequency (GHz)	Microwave region Molecule	Frequency (GHz)	Molecule
22.2	H_2O	119	O_2
60	O_2	183	H_2O

This rather crude model describes Rayleigh scattering by individual molecules. We can estimate the region of the electromagnetic spectrum in which it is important as follows.

From equation (4.3) we know that the molar concentration (number of moles per unit volume) of the molecules in the atmosphere, integrated through the whole atmosphere, is given by

$$\frac{p_0}{M_m g}$$

where p_0 is the sea-level pressure, M_m the molar mass of the molecules and g the gravitational field strength. Thus, we can estimate the optical thickness of the atmosphere due to molecular Rayleigh scattering as

$$\tau_s \approx \frac{N_A p_0}{M_m g} \frac{128\pi^5 a^6}{3\lambda^4}$$

where N_A is Avogadro's number. Assuming that this scattering is significant when $\tau_s > 1$, and substituting the values $p_0 = 10^5$ Pa, $M_m = 0.029$ kg, $a = 10^{-10}$ m, $N_A = 6 \times 10^{23}$, $g = 10$ m s^{-2}, we find that λ must be less than about 0.25 μm. Thus, molecular Rayleigh scattering is significant mainly in the ultraviolet region of the electromagnetic spectrum, although it is also notice-able in the visible region. The λ^{-4} dependence of equation (4.12) implies that blue light will be very much more strongly scattered than red light, and this is the reason why the clear sky appears blue, since visible skylight is mainly due to molecular scattering. It is also the principal explanation of red skies in the region of the rising and setting sun, since here the sunlight has travelled through a very long atmospheric path so that significant amounts of blue light have been scattered away from the forward direction. The strong atmo-spheric scattering of blue light, and especially of ultraviolet radiation, is the reason that these wavelengths are normally less important in remote sensing than are the longer wavelengths.

Other molecular scattering mechanisms also occur. In particular, scattering cross-sections are very much larger than implied by equation (4.12) at frequen-cies close to absorption lines. This phenomenon is known as *resonant absorp-tion*. Molecular fluorescence effects can also cause large scattering cross-sections.

Figure 4.5 attempts to summarise the practical implications of sections 4.2.1 and 4.2.2, by illustrating the optical thickness due to the attenuation (i.e. scattering and absorption) of electromagnetic radiation propagating vertically through the entire atmosphere. The figure is rather schematic, since it does not resolve fine spectral details and assumes 'standard' atmospheric conditions. (An alternative presentation of the information in figure 4.5 is given in figure 1.1.) The comparative transparency of the optical region and of the microwave region below the 60 GHz oxygen absorption line is clearly apparent, as is the complexity of the infrared region as a result of the large number of molecular transitions. The comparatively transparent regions define the so-called *atmo-*

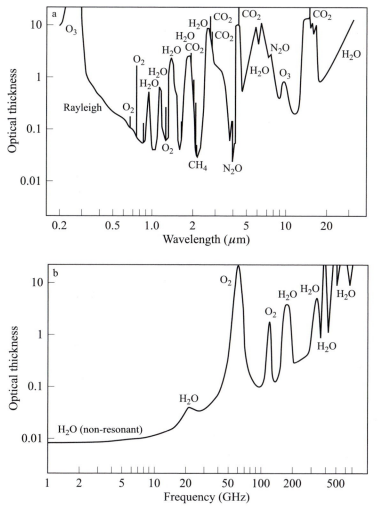

Figure 4.5. Total zenith optical thickness of the standard atmosphere: (a) ultraviolet, optical and infrared region; (b) microwave region.

spheric windows. Roughly speaking, there are two main windows: one in the optical-infrared region and the other in the microwave region. These can, however, be usefully subdivided by the presence of absorption lines.

Figure 4.5 shows the optical thickness of the atmosphere for a vertically propagating ray. If the ray travels obliquely through the atmosphere, however, it is clear that it will encounter a larger number of molecules, so we would expect the optical thickness to be increased. We can derive a very simple model of this phenomenon by assuming that the Earth is flat, and considering a ray that makes an angle θ to the horizontal. The optical thickness is given by

$$\tau = \int_0^\infty \gamma(x) \, dx$$

where $\gamma(x)$ is the attenuation coefficient at distance x along the ray. We know that x is related to the height z in the atmosphere by

$$x = \frac{z}{\sin\,\theta}$$

so we can rewrite our expression for the optical thickness as

$$\tau = \frac{1}{\sin\,\theta} \int_0^\infty \gamma(z)\,dz$$

In other words, the optical thickness is increased by a factor of $1/\sin(\theta)$ with respect to its value for a vertical ray. This very simple formula is sufficiently accurate for θ greater than about $15°$; that is, for factors up to about 4. For smaller angles, it is necessary to take the Earth's curvature into account. (This is obvious, because the $1/\sin(\theta)$ formula implies that the optical thickness is infinite for a horizontal ray, whereas we know that such a ray will, because of the Earth's curvature, emerge from the atmosphere and hence propagate within it for a finite, and not infinite, distance.) The required correction depends on the distribution of the attenuation coefficient with height, which we can model as an exponential decay

$$\gamma(z) = \gamma(0)\,\exp\left(-\frac{z}{z_0}\right)$$

where z_0 is the scale height. Note that this is analogous to the simple model of the distribution of atmospheric pressure with height that we introduced in equation (4.4), and we expect that for most atmospheric species the appropriate value of z_0 will be about 8 km.

Figure 4.6 shows the factor by which the optical thickness for oblique propagation is increased relative to vertical propagation, for three values of the scale height z_0. Figure 4.6a shows the range of θ from 0 to $15°$, and it can be seen that the effect of the scale height is significant in this range. Figure 4.6b shows the range from $15°$ to $90°$. Here, the simple $1/\sin(\theta)$ formula is accurate, so only one curve is shown.

4.3 Microscopic particles in the atmosphere: aerosols

In the preceding sections we have considered the gaseous composition of the atmosphere. However, it also contains solid and liquid components that can have a significant effect on the propagation of electromagnetic radiation. These components generally exhibit much greater spatial and temporal variability than the gaseous components, so they are rather harder to characterise. In this section we discuss aerosols, and in the following section the larger ice and water particles that constitute fog, cloud, rain and snow.

Aerosols (atmospheric *haze*) are suspensions of very small solid particles or liquid droplets, with radii typically in the range 10 nm to 10 μm. They can be regarded as being suspended in the atmosphere since, because of their small

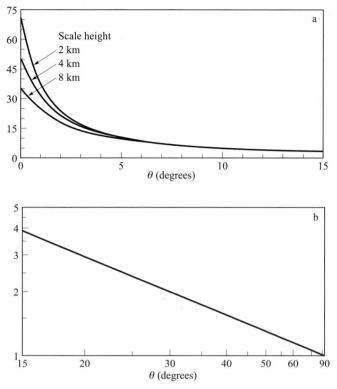

Figure 4.6. The factor by which the optical thickness of the atmosphere is increased for oblique propagation relative to vertical propagation. θ is the angle to the horizontal. (a) The range 0–15°, showing the effect of the scale height of the attenuating medium; (b) the range 15–90°, in which the influence of the scale height is negligible. Figuure 4.6b has been plotted with logarithmic scales to emphasise the $1/\sin(\theta)$ dependence.

size, the speed at which they fall under gravity is very small. For example, a particle of radius 1 μm will fall at something like 10^{-4} m s^{-1}, which means that under conditions of absolutely still air it would take of the order of 100 days to fall through 1 km.

Most aerosols are found in the atmospheric *boundary layer*, the lowest kilometre or so of the atmosphere in which transport processes are dominated by wind turbulence and by atmospheric convection.[3] This indicates that the aerosols are largely generated from the Earth's surface, for example in liberating and then carrying aloft solid microscopic dust particles from the land surface and water droplets from the ocean surface. Consequently, the type and size distributions of aerosols are strongly dependent on the local meteorological conditions and the local nature of the Earth's surface – whether urban, rural, oceanic, volcanic, and so on. For visible-wavelength radiation (say a wavelength of 0.55 μm), the attenuation coefficient at sea level typically ranges

[3] Aerosols are, however, also found in the stratosphere.

between 0.05 and 0.5 km^{-1}, and the total (vertical) optical thickness of a tropospheric aerosol is typically 0.1 to 1. If we compare this range of values with the gaseous optical thickness at wavelength 0.55 μm of about 0.2, as shown in figure 4.5a, we can see that the aerosol component of the atmosphere is radiatively significant.

The attenuation coefficient of an aerosol at optical and infrared wavelengths is dominated by scattering rather than absorption. The dependence of the attenuation coefficient on wavelength is usually represented approximately by the *Ångström relation*

$$\gamma = \gamma_0 \lambda^{-n} \tag{4.13}$$

where γ_0 is a constant and n is called the *Ångström exponent*. For particles that are very small compared with the wavelength, the Rayleigh scattering formula implies that $n = 4$, and for very large particles, the fact that the scattering cross-section is of the order of the geometrical cross-section implies that $n = 0$. For aerosols, the Ångström exponent n is usually between 0.2 and 2. Maritime aerosols usually exhibit the largest particle sizes (the modal droplet radius is around 0.2 μm) and the smallest values of n. For most other aerosols, the value of n is around 1, although at high altitudes (stratospheric and higher-altitude tropospheric aerosols) the very small particle sizes give $n \approx 2$.

4.4   Larger particles: fog, cloud, rain and snow

At any one time, about half of the Earth's surface will be covered by cloud. Visible-wavelength sensors, and to some extent those operating in other regions of the electromagnetic spectrum, will be limited by the presence of significant amounts of cloud cover, and this can be a serious problem in the case of spaceborne sensors that revisit a particular location comparatively infrequently (see chapter 10). For example, it has been estimated that a Landsat satellite, which revisits a particular location once every 16 days, will obtain a cloud-free scene of a particular location in Britain only once per year, and a scene with 1 okta of cloud (an okta is one eighth of the sky obscured by cloud) only twice per year. The probability of less than 10% cloud cover in a single Landsat observation over the continental USA is about 0.05 to 0.4, and the number of observations needed to obtain a probability of 75% of less than 10% cloud cover is between 5 and 100 (Goetz, 1979). In high-latitude regions, the problems of cloud cover can be significantly worse (Marshall et al., 1994).

Precipitation can also have a significant effect on the propagation of electromagnetic radiation. In the case of rainfall, the rain rate is the dominant factor since this largely controls both the size distribution of the droplets and their number density.

The attenuation of electromagnetic radiation by these meteorological phenomena can be considered in both a positive and a negative light. For observations intended to characterise the Earth's surface, it is an inconvenience since it must be corrected for, or may even render the observation impossible. On the other hand, it may be possible to derive useful information about the meteorological phenomenon itself from the effect on radiation, for example in microwave rain radars.

Fog and low-altitude (up to about 3000 m) cloud consists of water droplets, but these are very much larger than the droplets in aerosols, having modal radii from 10 to 50 μm. At visible and infrared wavelengths, scattering is again dominant. We can model the scattering fairly crudely by assuming that all the droplets have the same radius a and that their scattering cross-section is just the geometrical area πa^2. If the number density of the droplets (the number of droplets per unit volume) is N, the scattering coefficient is therefore given by

$$\gamma_s \approx \pi a^2 N$$

The mass density of liquid water in the fog or cloud is given by

$$\rho = \frac{4}{3}\pi a^3 N \rho_w \tag{4.14}$$

where ρ_w is the density of water, and we have assumed that the droplets are spherical, so we see that we can write the scattering coefficient as

$$\gamma_s \approx \frac{3\rho}{4a\rho_w} \tag{4.15}$$

Since the droplet radius is not strongly dependent on the mass density ρ, this equation implies that the scattering coefficient is roughly proportional to the mass density of liquid water in the cloud. This ranges from about 10^{-4} kg m^{-3} for fog and thin cloud, to about 10^{-2} kg m^{-3} for the densest clouds, so we should expect scattering coefficients in the region of 1 to 100 km^{-1}. Cloud layers are typically of the order of 1 km thick, so all but the thinnest clouds (and fog layers, which are typically 50 m thick) are optically opaque, with optical thicknesses in the range 1–100.

Similar remarks apply to the higher-altitude clouds that are composed of ice crystals. The crystals are of a similar size and density to the water droplets, and the calculation on which equation (4.14) is based is too crude to incorporate the fact that the ice crystals are not spherical.

We saw in section 3.4.2 that absorption dominates hugely over scattering for cloud droplets at microwave frequencies. Putting together equations (3.75) and (3.76), we see that the absorption cross-section of a small sphere of radius a and dielectric constant $\varepsilon' - i\varepsilon''$ is given by

$$\sigma_a = \frac{12\pi a^3 k \varepsilon''}{(\varepsilon' + 2)^2 + \varepsilon''^2}$$

where k is the wavenumber of the radiation. The dielectric constant of water in the microwave region can be described quite accurately using the Debye equation (3.20), and if we substitute this equation into our expression for σ_a we obtain

$$\sigma_a = \frac{12\pi a^3 k(1 + \omega^2\tau^2)\omega\tau\varepsilon_p}{((\varepsilon_\infty + 2)(1 + \omega^2\tau^2) + \varepsilon_p)^2 + \omega^2\tau^2\varepsilon_p^2}$$

for the absorption cross-section at angular frequency ω. Making the substitutions $\omega = ck = 2\pi f$ where f is the frequency, and approximating to the low-frequency limit $\omega\tau \ll 1$, we find

$$\sigma_a \approx \frac{48\pi^3\tau\epsilon_p}{(\varepsilon_\infty + \varepsilon_p + 2)^2 c} a^3 f^2 \tag{4.16}$$

Although this expression appears complicated, the right-hand side is just a constant multiplied by $a^3 f^2$. Multiplying this by the number density N of the water droplets to obtain the absorption coefficient, and making use of equation (4.14), we see that this simplified (low-frequency) model predicts that the absorption coefficient should be proportional to ρf^2, where ρ is the mass density of liquid water in the cloud. In fact, a somewhat more accurate approximation over the usual range of microwave frequencies is

$$\gamma_a \approx 0.6 \left(\frac{\rho}{\text{kg m}^{-3}}\right)\left(\frac{f}{\text{GHz}}\right)^{1.9} \text{dB km}^{-1} \tag{4.17}$$

Thus, we see that even a thick layer of dense cloud is comparatively transparent to microwave radiation, introducing of the order of 1 dB attenuation at 50 GHz. At frequencies below about 15 GHz, absorption by cloud is clearly negligible.

The water droplets in rain are roughly 100 times larger than those in clouds, being of the order of 1 mm in radius. At visible and infrared wavelengths, scattering is again dominant over absorption, and since the particle size is very much larger than the wavelength, the scattering cross-section of a droplet is of the order of the geometric cross-section. The angular distribution of the scattered radiation is, however, quite complicated – this fact is obvious from the existence of various rainbow phenomena.

The radii of the droplets in a particular rain shower are distributed over a rather large range. We can define a droplet size distribution $N(a)$, such that $N(a)\,da$ is the number of droplets per unit volume having radii between a and $a + da$, and various empirical forms of this distribution have been defined (e.g. Laws and Parsons, 1943; Marshall and Palmer, 1948; Joss and Gori, 1978). Assuming that the scattering cross-section of an individual drop is given by πa^2, the scattering coefficient is then

$$\gamma_s = \int_0^\infty \pi a^2 N(a)\,da \tag{4.18}$$

and the mass density of liquid water in the rain is

$$\rho = \int_0^\infty \frac{4}{3}\pi a^3 \rho_w N(a) \, da \qquad (4.19)$$

where ρ_w is the density of water. The drop size distribution is mainly governed by the *rain rate*, usually specified in millimetres per hour. Table 4.3 illustrates the values calculated from equations (4.18) and (4.19) for rain rates of 1 mm h^{-1} (a light rainfall) and 100 mm h^{-1} (a tropical downpour), using the drop size distribution given by Joss and Gori (1978). If we assume that the scattering coefficient is proportional to the rain rate R raised to some constant power, the data of table 4.3 can be interpolated by equation (4.20):

$$\gamma_s = 4.9 \times 10^{-5} \left(\frac{R}{\text{mm h}^{-1}} \right)^{0.73} m^{-1} \qquad (4.20)$$

Thus, we might expect the scattering coefficient in a heavy rain shower ($R = 25$ mm h^{-1}) to be of the order of 5×10^{-4} m^{-1} (about 2 dB km^{-1}).

The interactions between microwave radiation and rain are not so easy to calculate, because the droplets are similar in size to the wavelength of the radiation. The dimensionless parameter $x = 2\pi a/\lambda$ that determines whether the Rayleigh or Mie scattering formulae should be used can vary over a wide range as a result of the rather broad distribution of droplet sizes. Figure 4.7 shows the approximate dependence of the scattering and absorption coefficients on frequency for rain rates of 1 mm h^{-1} and 100 mm h^{-1}, from which it can be seen that (a) both scattering and absorption increase with rain rate, (b) absorption dominates over scattering except at high frequencies and rain rates, and (c) both absorption and scattering can be significant at frequencies above about 10 GHz.

4.5   The ionosphere

The ionosphere is an ionised layer above the Earth's atmosphere, extending from about 70 km to a few hundred kilometres above the surface. The ionisation is produced by extreme ultraviolet and X-radiation from the Sun, and it can have a significant effect on the propagation of radio-frequency electromagnetic radiation.

We saw in chapter 3 that the dielectric constant of a plasma is given by

Table 4.3. Calculated scattering coefficient and mass density
of liquid water for two different rain rates

Rain rate (mm h^{-1})	γ_s (m^{-1})	ρ (kg m^{-3})
1	4.9×10^{-5}	3.3×10^{-5}
100	1.4×10^{-3}	2.2×10^{-3}

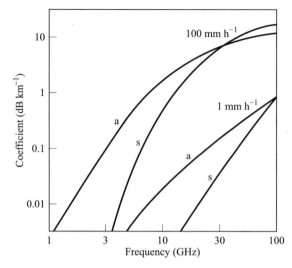

Figure 4.7. Typical microwave absorption (*a*) and scattering (*s*) coefficients for rainfall.

$$\varepsilon_r = 1 - \frac{Ne^2}{\varepsilon_0 m_e \omega^2} \qquad (3.24)$$

where N is the number density of the electrons, e is the charge and m_e the mass of an electron, and ω is the angular frequency of the radiation. The plasma frequency is

$$\omega_p = \sqrt{\frac{Ne^2}{\varepsilon_0 m_e}} \qquad (3.25)$$

and the dielectric constant is positive or negative according as ω is greater or less than ω_p. The maximum value of the electron density N in the ionosphere is of the order of $10^{12}\,\mathrm{m}^{-3}$, implying that ω_p is about $6 \times 10^7\,\mathrm{s}^{-1}$ ($f_p \approx 9\,\mathrm{MHz}$). Thus, at microwave frequencies (and above) the dielectric constant is positive and very slightly less than 1.

We can use the binomial expansion to approximate the square root of equation (3.24) and hence obtain the refractive index of a plasma for the case when $\omega \gg \omega_p$:

$$n = \sqrt{\varepsilon_r} \approx 1 - \frac{Ne^2}{2\varepsilon_0 m_e \omega^2} \qquad (4.21)$$

This is purely real, so there is no absorption of radiation. The phase velocity v of electromagnetic waves is given by equations (3.5) and (3.6) as

$$v = \frac{c}{n}$$

where c is the speed of light *in vacuo*, and in this case it is clearly greater than c. This seems paradoxical, since it appears to contradict Einstein's postulate that the speed of light represents an upper speed limit, until we recall that this speed limit applies to the propagation of *information* and that, as discussed in section 3.1.3, the information in a wave is propagated at the group velocity and not the phase velocity.

Combining equations (3.5), (3.6) and (4.21), we may write the dispersion relation for radiation propagating in a plasma at a frequency very much higher than the plasma frequency as

$$\omega = \frac{ck}{1 - \dfrac{Ne}{2\varepsilon_0 m_e \omega^2}}$$

so we can evaluate the group velocity from equation (3.27):

$$v_g = \frac{d\omega}{dk} = \frac{c}{1 + \dfrac{Ne^2}{2\varepsilon_0 m_e \omega^2}} \tag{4.22}$$

This is clearly *less* than c, as expected. Furthermore, we can use equation (4.22) to calculate the time t taken for a pulse of radiation to travel through a finite region (for example, all) of the ionosphere. Since

$$t = \int \frac{dz}{v_g}$$

where z measures propagation distance, we obtain

$$t = \frac{z}{c} + \frac{e^2}{2\varepsilon_0 m_e \omega^2 c} \int N \, dz \tag{4.23}$$

The right-hand side of this expression can be interpreted as follows: the first term is just the time taken for light to travel the distance z *in vacuo*, and the second consists of a frequency-dependent constant multiplied by the integrated number-density of electrons along the path through the ionosphere.

As an example of the use of equation (4.23), consider the propagation of a pulse of microwaves at a frequency of 10 GHz vertically through the entire ionosphere. The value of $\int N \, dz$ in this case is typically 3×10^{17} m^{-2}, so the second term on the right-hand side of equation (4.23) has a value of 4.0×10^{-10} s. Thus, the pulse is delayed by 0.40 ns, or 0.12 m, compared with a pulse travelling the same distance through free space. A pulse at 5 GHz would be delayed by four times this amount. We will consider these delays again when we discuss radar altimeters in chapter 8 and the Global Positioning System (GPS) in appendix 1.

The electron density in the ionosphere is very variable, both temporally and spatially. The ionisation is significantly greater on the Earth's sunlit side than on the night side, it is strongly affected by variations in solar activity, and its spatial distribution is correlated with both altitude and geomagnetic latitude. Figure 4.8 illustrates typical mid-latitude electron densities as functions of altitude for day and night. As we have seen in deriving equation (4.23), the the integrated electron density $\int N dz$ through the whole of the ionosphere is an

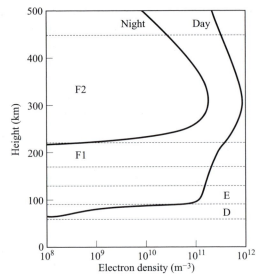

Figure 4.8. Typical electron densities in the ionosphere. The figure also shows the approximate positions of the layers into which the ionosphere is conventionally divided.

important quantity in considering the propagation time of microwave pulses. This quantity is often called the *total electron content* (TEC). It has a typical daytime value of $3 \times 10^{17} \, \text{m}^{-2}$, and is usually about ten times less at night.

We remarked earlier that at frequencies below the plasma frequency, the dielectric constant of the ionosphere is negative. This implies that the refractive index is purely imaginary, and hence that radiation will be absorbed. The ionosphere is thus increasingly opaque as the frequency decreases below about 10 MHz, and this places a lower frequency limit on spaceborne remote sensing. (It does not apply to airborne techniques, of course, since these do not involve looking through the ionosphere.) However, we can note in passing that the opacity of the ionosphere at sufficiently low radio frequencies does have a beneficial effect, since it allows HF ('short wave') radio signals to propagate for long distances round the Earth's surface by bouncing between the surface and the ionosphere.

4.6 Atmospheric turbulence

One further and potentially important influence that the atmosphere can have on the propagation of electromagnetic radiation is as a result of atmospheric turbulence. This is always present to a greater or lesser extent in the lower atmosphere, and causes variations in the density, and hence refractive index, of the air. The phase of an electromagnetic wave is corrupted by these variations, and this adversely affects the behaviour of an imaging system.

The most useful way to describe the effects of this kind of phase fluctuation, which is of course statistical rather than deterministic, is by specifying the *structure function*. This is usually defined as the variance of the phase difference between two points which would, in the absence of such effects, be in phase. It is therefore measured in radians squared, and is usually a function of the wavelength of the radiation as well as of the separation between the two points. The time-scale over which the phase variance is measured is also often important.

So far as the effect on the resolution of an imaging system is concerned, an approximate idea can be obtained by replacing the turbulent medium by a notional aperture whose size is equal to the separation at which the structure function reaches a value of 1 radian squared. For visible light travelling through the whole of the Earth's atmosphere, this separation is typically 0.2 m, corresponding to an angular resolution (calculated from the diffraction formula, equation (2.42)) of about 3×10^{-6} radians or about 1 second of arc. This scattering angle, which we will denote by $\Delta\theta$, is the limiting angular resolution that can be achieved by an *upward*-looking observation (such as an astronomical observation) through the whole atmosphere. However, for a *downward*-looking observation we also need to consider the effective height at which the scattering occurs. We will take this to be the scale-height of the atmosphere (of the order of 8 km), which we will denote by H, and we will assume that the observation is made from a height much greater than H. In our very simplified model, we will assume that all the scattering takes place at the height H.

Figure 4.9a shows a ray AS propagating obliquely through the atmosphere at an angle $\Delta\theta$ to the vertical. At S, it is scattered into a cone of directions from SE, which is clearly vertical, to SD. Figure 4.9b similarly shows the scattering of the ray BS, where S is the same point as in figure 4.9a. We can see, therefore, that a downward-looking observation from directly above the point S will collect radiation originating from any point between A and B. The distance between these points is approximately $2H\,\Delta\theta$, so this is the limiting *linear* resolution that is achievable. Substituting the values for H and $\Delta\theta$, we find that this limiting resolution is of the order of 5 cm.

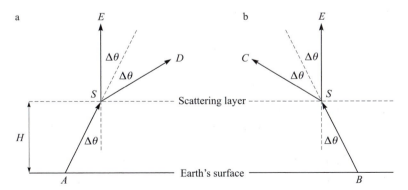

Figure 4.9. Schematic representation of the scattering of light travelling through the atmosphere. (a) The ray AS is scattered into the cone of directions between SE and SD; (b) the ray BS is scattered into the cone of directions between SC and SE.

Turbulence in the lower atmosphere causes similar effects at radio frequencies. However, for observations from satellites at radio frequencies, the *ionosphere* poses a potentially more serious problem. It is not really possible to quote a typical value for the ionospheric structure function because of the great variability alluded to earlier, but we may note that the phase variance will be proportional to λ^2 (because of the plasma dispersion relation, discussed in section 4.5), and that it will be greater near the geomagnetic poles and equator, and during the daytime.

PROBLEMS

1. Show that Doppler broadening will dominate over pressure broadening of a spectral line for a gas at temperature T and pressure p provided that the wavelength of the spectral line is less than $kT/p\sigma$, where σ is the collision cross-section defined in equation (4.11).

2. The attenuation coefficient of a typical tropospheric aerosol is $0.1 \, \text{km}^{-1}$ at sea level, and the total optical thickness of the aerosol for a vertical path through the atmosphere is 0.2. By assuming that the density, and hence the attenuation coefficient, of the aerosol obeys a negative exponential distribution with height (i.e. similar to equation (4.4)), calculate the scale height of the aerosol layer.

3. The meteorological visibility in fog is defined as the distance that gives an optical thickness of 4. If a typical fog has a mass density of $10^{-3} \, \text{kg}$ water per cubic metre and gives a visibility of $100 \, \text{m}$, estimate the size of the water droplets.

4. Use equation (4.16) to show that, at low microwave frequencies, the absorption coefficient of a cloud of water droplets is given approximately by

$$0.5\left(\frac{\rho}{\text{kg}\,\text{m}^{-3}}\right)\left(\frac{f}{\text{GHz}}\right)^2 \text{dB km}^{-1}$$

(i.e. independent of the droplet size), where ρ is the mass of water in kilograms per cubic metre of cloud. Assume that, for water at microwave frequencies, $\varepsilon_p = 75.9$, $\varepsilon_\infty = 4.5$ and $\tau = 9.2$ ps. Hence, by assuming that this relationship holds throughout the microwave region, discuss the statement that clouds are transparent to microwave radiation. Consider only absorption; that is, ignore any scattering from the droplets. Cloud water content ranges typically from $10^{-6} \, \text{kg m}^{-3}$ for haze to $10^{-2} \, \text{kg m}^{-3}$ for cumulonimbus.

5. Use the data of table 4.3 to estimate the effective raindrop radius, number density and sedimentation rate for rain rates of 1 and 100 mm h^{-1}, assuming that all the drops are spherical and have the same radius.

5

Photographic systems

5.1 Introduction

Aerial photography, as we remarked in chapter 1, represents the earliest modern form of remote sensing system. Despite the fact that many newer remote sensing techniques have emerged since the first aerial photograph was taken in 1858, aerial photography still finds many important applications, and there are many books that discuss it in more detail than will be possible in this chapter. The interested reader is referred, for example, to chapters 2 to 5 of Avery and Berlin (1992). Aerial photography is familiar and well understood, and is a good point from which to begin our discussion of types of imaging system. In particular, it provides a convenient opportunity to introduce some of the imaging concepts that will be useful in discussing some less familiar systems in later chapters.

Photography responds to the visible- and near-infrared parts of the electro-magnetic spectrum. It is, in the context of remote sensing, a passive technique, in that it detects existing radiation (reflected sun- and skylight), and an imaging technique, in that it forms a two-dimensional representation of the radiance of the target area. In this chapter, we shall consider the construction, function and performance of photographic film, especially its use in obtaining quantitative information about the geometry of objects. The chapter then discusses the effects of atmospheric propagation, and concludes by describing the character-istics of some real instruments and giving a brief account of the applications of the technique.

5.2 Photographic film

Figure 5.1 shows schematically the construction of monochrome (black-and-white) photographic film. An emulsion[1] containing crystals of a silver halide (bromide, chloride or iodide) is held in suspension in a transparent layer of gelatin, supported on a polyester base. The gelatin layer is typically about 10 μm thick, and the crystals, or grains, of silver halide are typically 0.1 to 5 μm in size.

---

[1] In fact, not strictly an emulsion since the term normally refers to a suspension of one *liquid* in another.

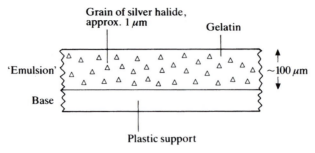

Figure 5.1. Schematic diagram showing the construction of a typical panchromatic film.

As we shall see in this and later chapters, the fundamental principle involved in detecting electromagnetic radiation is to observe changes in the energy of electrons. In the case of photographic film, the interaction that takes place is that an electron is transferred from a halide ion to a silver ion, leaving an atom of silver and an atom of the halogen. For example, in the case of silver bromide, the reaction can be represented as

$$Ag^+Br^- \xrightarrow{hf} Ag + Br$$

where '*hf*' denotes the absorption of a photon. Normally, this reaction reverses itself spontaneously, but if several silver atoms are produced in close proximity to one another, the clump of atoms is stable, and is known as a *development centre*. A film in which some of the halide grains contain development centres is said to contain a *latent image*. After its exposure to light, and hence the creation of a latent image, the film is *developed* by treating it with a chemical reducing agent (developer). This is carefully chosen so that it is capable of reducing Ag^+ ions to Ag atoms only in the presence of a development centre, which acts as a nucleus for the growing silver crystal. The remainder of the Ag^+ ions are unchanged by the action of the reducing agent, and are later washed away. The end result of this process is that those areas of the film that were exposed to light are converted to solid silver, which is opaque, while the remaining parts of the film are transparent. This is called a *negative*, because if it is viewed by passing light through it, areas that received light during the exposure stage will appear dark, and those that did not will appear light.

We can see from this brief description of the theory (the Gurney–Mott theory) of the photographic process that an exposure to light sufficient to cause only a few atoms of silver to be formed in a given grain of halide will allow the entire grain, containing perhaps 10^{10} ions, to silver. The process of photographic detection thus involves an amplification by a factor of roughly 10^9. A fuller account of the Gurney–Mott theory is presented by Omar (1975).

5.2.1 *Film types*

5.2.1.1 *Black-and-white film*
This is the simplest type of film, and is still in wide use.

The photochemical reaction on which the photographic process depends was discussed in section 5.2. Since it depends on the oxidation of the halide ion, the energy required for this process will govern the maximum wavelength (minimum photon energy) to which the film can respond. For the halide ions in the crystalline state, these maximum wavelengths lie in the range 0.4–0.5 μm, so photographic film should respond only to blue, violet and ultraviolet radiation (and in fact to shorter wavelengths, as far as X-rays. It is the opacity of the optical glass and the material of the film itself that prevent these shorter wavelengths from causing problems). In fact, the range of sensitivity is extended by the use of *sensitising dyes*. Panchromatic (i.e. 'all colours') black-and-white film normally has an upper sensitivity limit of about 0.70 or 0.72 μm. The lower limit is in practice set by the opacity of the glass lenses used in the system, at about 0.35 μm, although, if an extended response into the ultraviolet is required, quartz lenses can be used to give a cutoff at about 0.30 μm. Figure 5.2 illustrates the spectral sensitivity of typical panchromatic black-and-white film.

5.2.1.2   Infrared film

The sensitivity range of panchromatic film can be extended further into the infrared region by the use of appropriate sensitising dyes, to give an upper cutoff wavelength of about 0.9 μm. The spectral sensitivity of such a film is also illustrated in figure 5.2, from which it can be seen that the sensitivity to visible wavelengths is notably non-uniform, with a peak sensitivity to blue light and a rather low sensitivity to yellow and green. In practice, however, this kind of film is often used in conjunction with a filter that eliminates wavelengths below about 0.7 μm, thus giving a uniform response between about 0.7 and 0.9 μm. This procedure is often referred to as *true infrared photography*.

5.2.1.3   Colour film

The human eye can distinguish about 200 shades of grey, but about 20 000 tints and shades of colour. Thus, far more information can be deduced from a

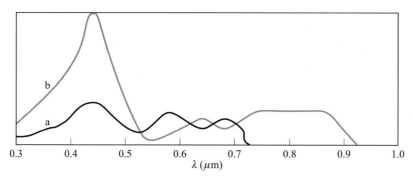

Figure 5.2. Typical spectral sensitivity of (a) black-and-white and (b) infrared panchromatic films, as a function of wavelength λ.

colour image than from a black-and-white image, even if the means of doing so is merely visual examination.

Colour films are constructed with three layers of emulsion instead of just one. By the incorporation of suitable dyes, the uppermost layer is made sensitive to blue light, the middle one to blue and green light, and the lowest layer to red and blue light. A yellow filter (i.e. a filter that removes blue light) is interposed between the upper and middle layers, to prevent blue light from exposing all three layers. The process of developing colour film is a complicated one: the yellow filter layer must be removed (by bleaching) and coloured dyes must be formed in the activated regions of the three emulsion layers. Two main types of colour film exist: normal (negative) film and reversal (positive) film. In normal film, the activated regions of the upper, middle and lower layers form dyes that are, respectively, yellow, magenta and cyan in transmission. The result of this is that when the developed film is viewed in transmitted light the colours are reversed, so that black is exchanged for white, red for cyan, and so on. In reversal film, the same dyes are used but the processing (which is more complicated) ensures that the dyes are formed in the unactivated parts of the layers, rather than the activated parts. Reversal film is more commonly used for aerial photography than normal negative film. The two processes are illustrated in figure 5.3, and figure 5.4 shows the typical spectral sensitivity of the three emulsion layers.

5.2.1.4   *False-colour infrared film*

False-colour infrared (FCIR) film has a similar construction to colour infrared film. The three emulsion layers are sensitive to near-infrared (0.7–0.9 m), green and red light, and the yellow filter is placed in front of all three layers. After exposure and development to produce a positive transparency, the parts of the

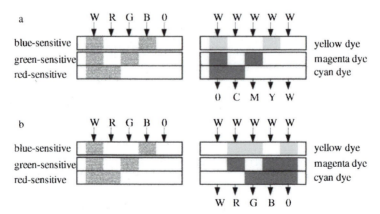

Figure 5.3. (a) Normal (negative) colour film; (b) reversal (positive) colour film. The diagrams on the left show the film after exposure to light, the shaded areas indicating the activated regions of the emulsion layers. The diagrams on the right show the regions in which yellow, magenta and cyan dyes are formed after development. The symbols for colour are W = white, R = red, G = green, B = blue, C = cyan, M = magenta, Y = yellow, 0 = no light = black. (See the plate section for colour version of figure 5.3.)

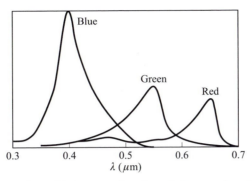

Figure 5.4. Typical spectral sensitivity, as a function of the wavelength λ, of the three emulsion layers in a colour film.

film that were exposed to infrared radiation transmit red light, those that were exposed to red radiation transmit green light, and those that were exposed to green radiation transmit blue light, as shown in figure 5.5. This is termed the *colour shift*.

False-colour infrared film was originally developed during the Second World War to assist in the location of military equipment hidden by camouflage netting. Real vegetation has a high reflectance in the near-infrared region of the spectrum (as discussed in section 3.5.1), and hence appears bright red in FCIR film, whereas material painted to simulate the visible appearance of vegetation has a low reflectance in the infrared region, and appears blue in the FCIR film. Since the 1940s, many civilian uses have been developed for FCIR film, most of which are derived from its sensitivity to vegetation.

5.3   Performance of photographic film: speed, contrast and spatial resolution

The response of a photographic film to incident radiation is characterised most simply by three parameters: the speed, contrast and spatial resolution. The *speed* of a film refers to the length of time for which it must be exposed to light of a given irradiance, in order to achieve a significant change in its opacity after processing. The speed is usually quantified by an ASA (American Standards Association), ISO (International Standards Organisation) or DIN

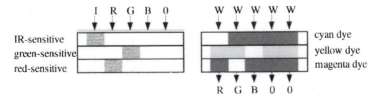

Figure 5.5. Reversal FCIR film, showing (left) activated regions after exposure, (right) the regions in which dye is formed after development. The symbols for the 'colours' are as for figure 5.3, with the addition of I for infrared radiation. (See the plate section for colour version of figure 5.5.)

(Deutsche Industrie Norm) number, with larger numbers corresponding to faster films and hence shorter exposure times, although for films intended specifically for aerial photography the AFS (aerial film speed) index is used.

The speed of a film describes its response to light of a single intensity. The *contrast*, on the other hand, describes the effect of changing the irradiance (or the exposure time). If a small change produces a large change in the opacity of the processed film, the film is said to have a high contrast, and conversely.

The radiometric response of a film can be summarised graphically by the *characteristic curve* (or Hurter–Driffield curve). This is a graph of the optical density D of the processed film against the logarithm of the exposure X to which it has been subjected. The optical density is defined by

$$D = - \log_{10} T$$

where T is the intensity transmission ratio of the film, so that, for example, a film that transmits 1% of the light incident upon it has an optical density of 2. The exposure X is defined as the product of the irradiance E at the film with the exposure time Δt, so that its units are W m^{-2} s. However, two important points should be made about exposure. The first is that its use implies that it is only the product $E \, \Delta t$ that is significant, for example that simultaneously doubling E and halving Δt will not change the appearance of the processed film. This assumption is known as *reciprocity*, and under extreme conditions (Δt very large or very small) it breaks down, a phenomenon termed *reciprocity failure*.

The second point that should be made regarding exposure is that *photometric units* are normally used in place of the radiometric units (W m^{-2} for irradiance and exitance, etc.) that were introduced in section 2.5. Photometric units are weighted with respect to the nominal spectral sensitivity of the human eye, using the function $V(\lambda)$ shown in figure 5.6. The photometric quantity corresponding to irradiance is called the *illuminance*, and it is defined by equation (5.1):

$$E_v = K \int_0^\infty E_\lambda V(\lambda) \, d\lambda \qquad (5.1)$$

In this equation, E_v (the 'v' stands for 'visible') is the illuminance, E_λ is the spectral irradiance, and K is a constant with the value 680 lumens per watt. All of the radiometric quantities defined in section 2.5 have their photometric equivalents, defined analogously to equation (5.1). Table 5.1 summarises the correspondences between these quantities and the names of their units.

A typical characteristic curve is shown in figure 5.7. The horizontal (asymptotic) parts of the curve are labelled D_{fog} and D_{sat}. The fog level, D_{fog}, is caused by slight opacity of the film base and gelatin layer, and by the fact that a small proportion of the grains that have not been exposed will nevertheless be developed. This may have a number of causes, including failure of the development process to distinguish between grains with and without development centres, and the generation of development centres by stray gamma-rays or X-rays. The

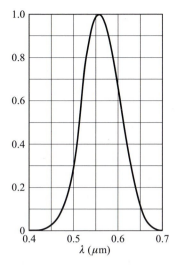

Figure 5.6. The function $V(\lambda)$ that defines the nominal sensitivity of the light-adapted human eye, and hence the photometric units.

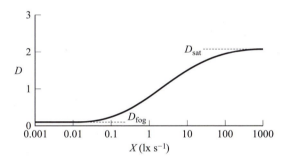

Figure 5.7. The characteristic curve of a typical medium-speed film.

Table 5.1. Corresponding radiometric and photometric quantities

Radiometric quantity	Unit	Photometric quantity	Unit
Radiant power	watt (W)	Luminous flux	lumen (lm)
Radiant intensity	W sr^{-1}	Luminous intensity	candela (cd) = lm sr^{-1}
Radiance	W m^{-2} sr^{-1}	Luminance	cd m^{-2}
Irradiance	W m^{-2}	Illuminance	lux = lm m^{-2}
Radiant exitance	W m^{-2}	Luminous exitance	lux = lm m^{-2}

saturation level, D_{sat}, corresponds to the development of all the grains in the film. It is clear that if a film is exposed and processed so as to approach either of these values of D, very little information is available about the exposure X.

The speed of the film is indicated by the value of X at which the curve of figure 5.7 begins to rise. The slope γ of the useful, central, part of the curve is a measure of the contrast. A film with high contrast will have a large value of γ.

Most films have $\gamma \approx 2$. Note that a large value of γ will reduce the input *dynamic range* of the film, that is, the range of values over which X may vary while still producing a useful change in the optical density D.

We may go some way towards understanding the shape of the characteristic curve by using a very simple model. If the grains present a cross-sectional area A to the incident radiation, and are suspended in the gelatin such that there are N grains per unit area of the film, we would expect (assuming that a developed grain is completely opaque) that

$$D_{\text{sat}} = -\log_{10}(1 - AN)$$

If the film is exposed to an irradiance E for a time Δt, the average number of photons striking a particular grain is

$$\frac{AE\,\Delta t}{hf} = \frac{AX}{hf}$$

where h is Planck's constant and f is the frequency of the radiation. We thus expect the probability of a given grain containing a development centre, and hence the opacity of the developed film, to be some increasing function of AX. If the grain size A is reduced, a larger exposure X will be needed to achieve the same opacity, and the film will be slower. If the grains are not all of the same size, but instead have a wide distribution of sizes, the input dynamic range will be increased and thus the film will have a low contrast. It is important to note, however, that the processing applied to the exposed film can also have a pronounced effect on the shape of the characteristic curve, so that the latter cannot truly be said to be a property of the film alone.

Spatial resolution is, roughly speaking, the ability of a remote sensing system to distinguish an extended object from a point. It is one of the most important parameters describing the performance of a system, and we shall return to it again in subsequent chapters. For photographic systems, the resolution is normally expressed in line-pairs (lp) per unit length. That is to say, a bar pattern resembling figure 5.8 is photographed, and said to be resolved if the bar pattern is recognisably reproduced on the negative. The spatial resolution is then the greatest number of these bars, per unit length, that can be resolved *on the negative*.

Figure 5.8. Typical object used to determine the spatial resolution of a photographic system in terms of line-pairs per unit length.

The spatial resolution of a film depends on the grain size, which, as we have remarked, is typically 1 to 10 μm. Higher resolutions will require smaller grains and, as we have seen, this will result in slower film speeds. The highest spatial resolutions available for aerial photographic films are typically 200 lp mm^{-1}, with corresponding low film speeds of about 10 AFS units. Such films are used for reconnaissance. At the other extreme, a fast mapping film might have a spatial resolution of about 20 lp mm^{-1} and a film speed of 1000 AFS units.

5.4   Photographic optics

In this section, we shall consider briefly the optics of photographic systems. It is assumed that the reader is already familiar with the simple theory of image formation by lenses; if not, any elementary textbook on optics can be consulted.

Let us consider first a system with a single lens of focal length f (figure 5.9). The object distance u and image distance v are related by

$$\frac{1}{u} + \frac{1}{v} = \frac{1}{f} \tag{5.2}$$

and it is clear that an object of height x subtending an angle θ will produce an image of height $v \tan \theta$. As the object distance u becomes very much larger than the focal length f, the image distance v tends to f and the image size to fx/u. Clearly, in all practical cases of aerial photography, this will be a justifiable approximation.

Suppose now that we consider an object of uniform exitance, such that the radiance incident at the lens is L, and that the object subtends a small solid angle Ω. The irradiance at the lens is thus ΩL and the total flux intercepted by the lens is

$$\pi \left(\frac{D}{2}\right)^2 \Omega L$$

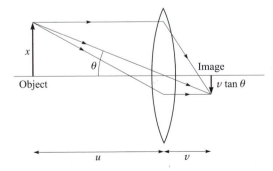

Figure 5.9. Formation of an image by a single converging lens.

where D is the diameter of the lens. Assuming that there are no losses in the lens, all of this flux will be distributed over an area Ωf^2 on the film, giving an irradiance at the film plane of

$$E_{\text{film}} = \frac{\pi D^2 L}{4f^2}$$

Thus, the ratio of the irradiance at the film to the radiance at the lens is given by

$$\frac{E_{\text{film}}}{L} = \frac{\pi}{4}\left(\frac{D}{f}\right)^2 \tag{5.3}$$

that is, it is determined by the ratio f/D, which is called the *f/number* of the lens. The smaller[2] the *f*/number, the larger the lens and the brighter the image. Lenses can be constructed with *f*/numbers as small as about 1, although most lenses used in aerial mapping have *f*/numbers in the range 5 to 10. The *f*/number of a lens can be increased by 'stopping down': that is, by reducing the diameter of an aperture placed just behind the lens.

Most aerial photographic systems in fact use *compound lenses*, which have the advantage of giving increased focal length without making the lens assembly physically larger. Figure 5.10 illustrates a compound lens using two converging lenses, with focal lengths f_1 and f_2 respectively. We shall again assume that the object is located at infinity (which in practice means at a distance very much greater than f_1), so that the first lens (the *objective lens*) forms an image of it in its focal plane. By comparing figure 5.10 with figure 5.9, it is clear that the combined effect of the two lenses is equivalent to that of a single lens of focal length $f_1 v_2/u_2$; that is, the focal length of the first lens multiplied by the magnification of the second. All of the remarks we have made about the single-lens system remain valid so long as we substitute this effective focal length for the simple focal length f.

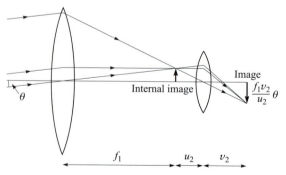

Figure 5.10. Formation of an image by a compound lens. The object is at infinity and subtends a small angle θ.

---

[2] The inverse relationship between the $f/$ number and the image brightness arises because the $f/$ number is defined as f/D, not D/f.

In fact, an even greater saving of space is made if the second lens of the compound system is of the diverging, rather than the converging, type. This is the usual construction of telephoto lenses.

5.4.1   Scale, coverage and ground resolution

The scale of a map and, by extension, of an aerial photograph is the number less than unity that expresses the ratio of the size of the representation of an object on the map to the size of the real object. 'Large scale' is usually taken to mean greater than $1/50\,000$ and 'small scale' less than $1/500\,000$.

Figure 5.11 shows schematically the geometry of a vertical aerial photograph; namely, one in which the optical axis of the camera is directed vertically, normal to the ground surface. We are again assuming that the distance H is much larger than the focal length f, so that the distance from the lens to the film plane can be taken as f. It is clear from simple geometrical considerations that the scale of the image formed at the film plane (i.e. the scale of the negative) is given by

$$s = \frac{f}{H} \qquad (5.4)$$

although, of course, prints having larger scales can be prepared by photographic enlargement from the negative. It is also clear that, if the negative has a width w, the width of the corresponding region on the ground – the coverage of the aerial photograph – will be given by

$$\frac{w}{s} = \frac{wH}{f} \qquad (5.5)$$

The spatial resolution at the ground is also determined by the imaging geometry, and by the resolution of the film. In section 5.3 we introduced the film resolution in terms of the maximum number of line-pairs per unit length

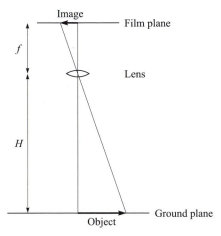

Figure 5.11. Schematic illustration of the formation of a vertical aerial photograph.

resolvable on the negative. If we denote this resolution by l, which has the dimensions of 1/length (so that, for example, a resolution of 100 lp/mm^{-1} corresponds to $l = 10^5$ m^{-1}), the corresponding resolution at the ground must be given by l/s. In other words, a target of the form of figure 5.8 will be resolvable provided the spacing between the bars is at least H/lf. This is conventionally regarded as being equivalent to the ability to resolve two *points* separated by a distance r_g, where

$$r_g = \frac{H}{2lf} \tag{5.6}$$

Comparison of equations (5.5) and (5.6) shows that, other things being equal, higher spatial resolution (smaller r_g) requires a larger focal length f and will hence give a smaller coverage.

The use of line-pairs per unit length as a measure of film resolution is a comparatively crude measure. A more informative representation is given by the *modulation transfer function* (MTF). This describes the ability of the film to record sinusoidal variations in intensity, as a function of their spatial frequency. Because of the fact, mentioned but not proved in section 2.3, that any 'reasonable' function can be constructed from a (possibly infinite) set of sinusoids of different amplitude, phase and frequency, this approach contains in principle all the information about the spatial response of the film.

Figure 5.12 shows a sinusoidal variation of intensity with position. Apart from its phase, which we shall ignore, this function is characterised by its spatial frequency q (cycles per unit length) and its modulation m, defined as

$$m = \frac{I_{\max} - I_{\min}}{I_{\max} + I_{\min}} \tag{5.7}$$

The MTF of a photographic system is defined as the ratio of the output modulation, namely that produced on the film, to the input modulation, namely that on the target. It is a function of the spatial frequency q, and for photographic systems this is conventionally defined as the spatial frequency in the film plane. Figure 5.13 shows typical MTFs for coarse and fine-grained films.

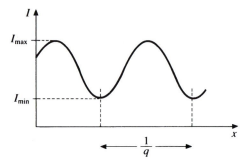

Figure 5.12. A sinusoidal variation of intensity I with position x. This form of variation is characterised by its modulation $m = (I_{\max} - I_{\min})/(I_{\max} + I_{\min})$, and its spatial frequency q.

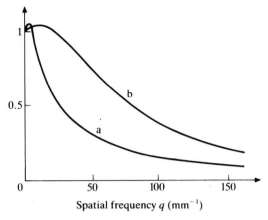

Figure 5.13. Modulation transfer functions for typical photographic films: (a) low resolution; (b) high resolution.

In fact, the spatial resolution of a photographic system is a combination of the performances of the film and of the optics. We saw in chapter 2 that diffraction at an aperture of diameter D broadens plane-parallel radiation into a cone of angle $\approx \lambda/D$ radians, where λ is the wavelength, and this also sets a limit on the resolution. For example, a system operating at a nominal wavelength of 0.5 μm with an objective lens of diameter 1 cm will be diffraction-limited to an angular resolution of about 5×10^{-5} radians. A point-like object will be imaged as a blurred spot on the negative. If the focal length of the lens is 150 mm, the radius of this spot will be about 7.5 μm. Thus, if the resolution of the film is greater than about 70 lp mm^{-1}, it is the lens, rather than the film, that will determine the system resolution. The combined effect of the different components of the photographic system on the spatial resolution can conveniently be described using MTFs. The MTF can be defined for each component, and these are then multiplied to calculate the MTF for the system.

Note that all of our considerations of ground scale, resolution and coverage have been derived for the case of *vertical* aerial photography. *Oblique* photography (figures 5.14 and 5.15) gives much greater coverage, but at the expense

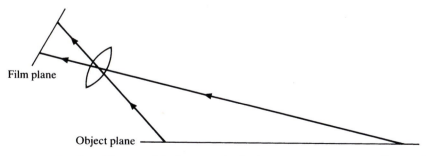

Figure 5.14. The oblique aerial photograph obtains greater coverage at the expense of variable scale and resolution.

Figure 5.15. An oblique photograph, showing wide coverage and variable scale. The image is of Cambridge, England. (Cambridge University Collection: Copyright reserved.)

of variable resolution and scale. For this reason, it is generally unsuitable for quantitative analysis.

5.5 Photogrammetry and stereogrammetry

One of the most important group of applications for which aerial photographs are used is based on measuring the geometrical properties of the image. These properties are especially simple if it can be assumed that the camera optics are free from distortion, in which case they can be derived from the fact that rays will be undeviated in angle if they pass through the optical centre of a lens. (This is the assumption that has been used in drawing figures 5.9 to 5.11 and 5.14. It is very well justified for mapping cameras, less well so for reconnaissance systems.)

It is convenient to describe the geometry of a vertical aerial photograph using a Cartesian coordinate system, as shown in figure 5.16. We suppose that the camera's optical axis is directed along the $-z$-axis, such that the

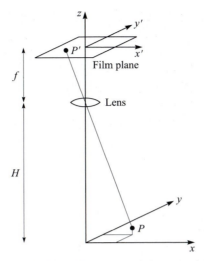

Figure 5.16. Geometry of a vertical aerial photograph in Cartesian coordinates. An object at P is imaged at P'.

optical centre of the lens is located at $(0, 0, H)$. The point P, with coordinates (x, y, h), is imaged at the point P' on the film, where the coordinates of P' are $(x', y', H + f)$.[3]

By simple geometry, we obtain the results

$$x' = -\frac{fx}{H - h} \tag{5.8}$$

and

$$y' = -\frac{fy}{H - h} \tag{5.9}$$

Setting $h = 0$ in these equations shows that the image of a horizontal surface has a constant scale of f/H, as we have already seen through equation (5.4). This fact is the basis of simple photogrammetry, in which the lengths, areas and orientations of shapes on horizontal ground can be determined from a photograph, as illustrated by figure 5.17.

5.5.1   *Relief displacement*

Equations (5.8) and (5.9) show that an aerial photograph also contains some information about the heights of objects. For example, suppose that a vertical object such as a tower has its base at coordinates $(x, 0, 0)$ and its top at $(x, 0, h)$. Equation (5.8) then shows that the base will be imaged at $(-fx/H, 0)$ and the top at $(-fx/[H - h], 0)$, so that the two are separated by a height-dependent amount. This is the phenomenon of *relief displacement*, illustrated schemati-

---

[3] We are again assuming that the height H is very much larger than the focal length f. Strictly speaking, f should be replaced by the image distance v throughout this section.

Figure 5.17. A vertical aerial photograph of horizontal terrain (a village in Cambridgeshire, England). A 500-m grid and a simple map of the road network have been overlaid to show the correspondence between photograph and map coordinates. (Cambridge University Collection: Copyright reserved.)

cally in figure 5.18. Qualitatively, objects appear to lean away from the *principal point* of the photograph (the point at $x' = y' = 0$), as can be seen in figure 5.19.

The distance h' is the projection of the height h onto the film plane, and is called the relief displacement. It is given by

$$h' = \frac{fx}{H-h} - \frac{fx}{H} = \frac{hfx}{H(H-h)} = \frac{hx'}{H-h} \tag{5.10}$$

which is approximately hx'/H if $H \gg h$. Thus, the height of a vertical object can be determined from a single vertical aerial photograph, provided we know

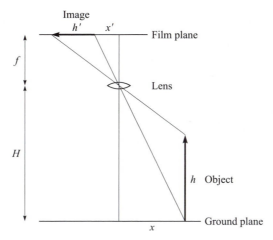

Figure 5.18. Relief displacement. The top and bottom of a vertical object are imaged at different points on the film.

the height from which the photograph was obtained, and provided the object is not located at the principal point.

If we differentiate equation (5.10) with respect to h, substitute equation (5.8), and make the approximation that $H \gg h$, we find that

$$\frac{\partial h'}{\partial h} \approx \frac{x'}{H}$$

Thus, if the smallest change in h' that can be resolved on the film is $\Delta h'$, the smallest corresponding change in height is given by

$$\Delta h = \frac{H \, \Delta h'}{x'}$$

This shows that the accuracy with which heights can be determined will improve with distance from the principal point. Taking x' to have the largest possible value, namely $w/2$ (where w is the width of the negative), gives the optimum height resolution as

$$\Delta h \approx \frac{2H \, \Delta h'}{w} \tag{5.11}$$

For a typical mapping system, $w = 230 \, \text{mm}$ and $\Delta h'$ is of the order of 0.1 mm. This implies that $\Delta h \approx H/1000$. This is a common rule of thumb.

5.5.2 Stereophotography

The method of relief displacement, relying on the measurement of h', depends for its success on two conditions being met. The first is that the object should not be located at the principal point, and the second is that both the top and bottom of the object should be visible in the photograph. While this second condition may be met for some buildings, for example, it will often not apply in the case of topographic features. In such cases, however, the height information can be retrieved provided that a *pair* of photographs, from different locations, is available. This is the technique of *stereophotography*, and the procedure for determining the topography is called *stereogrammetry*. Figure 5.19 shows a pair of stereophotographs.

To show formally that topographic information can be retrieved from a stereopair, we return to the derivation of equations (5.8) and (5.9). The camera lens in that case was located at the coordinates $(0, 0, H)$. Now, however, we will suppose that a second photograph is acquired from the coordinates $(B, 0, H)$, that is, from the same height but from a point separated by distance B, termed the *baseline*, from the first. If we denote the coordinates of the image point P' in this second photograph by (x'_2, y'_2), where these are measured relative to the principal point of the second photograph, it follows that the point at (x, y, h) will be imaged at

$$x'_2 = -\frac{f(x - B)}{H - h} \tag{5.12}$$

$$y'_2 = -\frac{fy}{H - h} \tag{5.13}$$

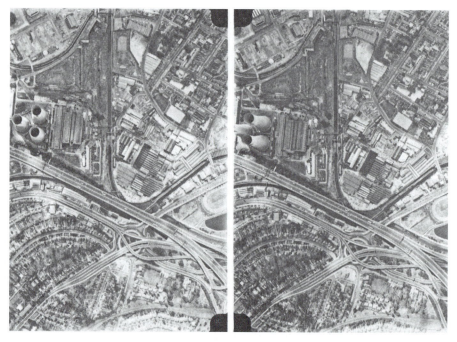

Figure 5.19. A pair of stereophotographs. Each shows the phenomenon of relief displacement (e.g. the cooling towers, left centre), and by viewing the pair together an impression of three-dimensional relief can be obtained. The images were recorded from a height of approximately 900 m, and have a coverage of about 1.3 × 0.9 km. They show a motorway interchange in Birmingham, England. (Cambridge University Collection: Copyright reserved.)

Thus, for any point that is imaged in both photographs, we can determine three image coordinates, namely the values of x', y' and x_2' (y_2' does not provide any more information since it is the same as y'), and this is sufficient to allow us to determine the three object coordinates x, y and h. Specifically, by solving equations (5.8), (5.9) and (5.12), we obtain

$$x = \frac{x'B}{x' - x_2'} \tag{5.14.1}$$

$$y = \frac{y'B}{x' - x_2'} \tag{5.14.2}$$

$$h = H + \frac{fB}{x' - x_2'} \tag{5.14.3}$$

Comparison of equations (5.8) and (5.12) shows that the difference in the relief displacement between the two photographs is $Bf/(H - h)$. Thus, we expect the accuracy achievable in determining the height to increase as the baseline B is increased. However, this will also have the effect of reducing the overlap between the two photographs, namely the area common to both, as shown in figure 5.20. From the figure, it is clear that the width of the overlap is given by

$$c = \frac{wH}{f} - B \tag{5.15}$$

where H is the height, f the focal length, B the baseline and w the film width. A reasonable compromise between the requirements to maximise both c and the height accuracy is to set

$$B \approx 0.4 \frac{wH}{f}$$

giving an overlap width

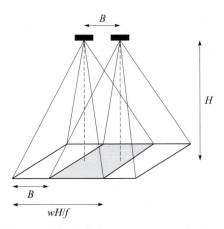

Figure 5.20. Overlap area (shaded) between two vertical aerial photographs.

$$c \approx 0.6 \frac{wH}{f}$$

that is, 60% of the width of the region imaged by a single photograph. In this case, the accuracy Δh with which heights can be determined is given by

$$\Delta h \approx \frac{H \, \Delta h'}{0.4w}$$

where $\Delta h'$ is the accuracy with which differences in relief displacement can be measured from the negatives. Taking typical values of $w = 230$ mm, $f = 150$ mm and $\Delta h' = 0.1$ mm as in section 5.5.1, and we find that a typical value of the *base–height ratio* B/H will be about 0.6, giving an overlap width c of approximately $0.9H$ and a height accuracy of about $H/900$, similar to the result we obtained for relief displacement in section 5.5.1.

The actual analysis of heights from a pair of stereophotographs may be made by measuring from prints, or by scanning them to convert them to digital form and then analysing in a computer. Various optical–mechanical instruments are available to assist the process of deriving heights from stereophotographs. However, a common method of analysing stereophotographs where quantitative height information is not required is to view them through an instrument called a stereoscope that presents one image to each eye. The human brain is familiar with the interpretation of the slightly different perspectives seen by the two eyes; indeed, this is the normal functioning of binocular vision. The object of the stereoscope, then, is to fool the brain into reconstructing a sensation of three-dimensionality.

Most people's eyes are about 7 or 8 cm apart, and focus comfortably on objects held about 50 cm distant. Because of this, stereophotographs obtained with this base–height ratio (approximately 1/6) will appear to have the correct perspective when interpreted by the brain. Conversely, if the base–height ratio is greater than 1/6, the reconstructed image will appear to have a proportional vertical exaggeration. This vertical exaggeration is merely an artefact of the way the brain interprets parallax, but it is often useful since it enhances subtle variations in relief.

5.6 Atmospheric propagation

In chapter 4, we discussed in general the effects of the atmosphere on electromagnetic radiation propagating through it. For a photographic system, the most important of these effects are likely to be the limiting spatial resolution imposed by atmospheric turbulence (although this will only be significant for high-magnification systems of high intrinsic resolution), and the reduction in image contrast as a result of atmospheric scattering.

The contrast[4] C of a scene, or of some part of it, is commonly defined as

---

[4] 'Contrast' here does not have the same meaning as in section 5.3. To avoid confusion, we refer explicitly to 'scene contrast' and 'film contrast'.

$$C = \frac{L_{\mathrm{max}} - L_{\mathrm{min}}}{L_{\mathrm{max}} + L_{\mathrm{min}}} \tag{5.16}$$

where L_{max} and L_{min} are, respectively, the maximum and minimum luminances (or, in radiometric terms, radiances) of the scene. Other definitions of scene contrast are also used, for example $L_{\mathrm{max}}/L_{\mathrm{min}}$, but the definition of equation (5.16) is convenient as it varies between zero (corresponding to completely uniform luminance) to 1 (corresponding to a minimum luminance of zero). Absorption of radiation *by* the atmosphere will reduce the luminances, whereas the corresponding reradiation *from* the atmosphere will increase them.

Let us write E_s for the illuminance at the Earth's surface due to skylight (and also to direct sunlight, if it is a clear day), E_a for the (upwards) illuminance of the atmosphere itself, T for the intensity transmittance of the atmosphere and r for the reflectivity (defined in section 3.3.1) of the scene. The contrast of the scene, measured just above it so that the effects of the atmosphere can be ignored, is then

$$C = \frac{r_{\mathrm{max}} - r_{\mathrm{min}}}{r_{\mathrm{max}} + r_{\mathrm{min}}}$$

The luminous exitance immediately above a part of the scene having reflectivity r is just rE_s. However, above the atmosphere this value will be reduced by a factor of T as a result of atmospheric attenuation, and increased by the upwelling skylight contribution E_a, giving a total upwelling illuminance of $rE_sT + E_a$. Thus, the scene contrast detected above the atmosphere becomes

$$C' = \frac{r_{\mathrm{max}} - r_{\mathrm{min}}}{r_{\mathrm{max}} + r_{\mathrm{min}} + \dfrac{2E_a}{E_sT}} \tag{5.17}$$

The scene contrast is therefore reduced, and it may be necessary to use film with a high contrast to compensate for the correspondingly small range of illuminance present at the camera.

Figure 5.21 shows typical sunlight and atmospheric illuminances as a function of the solar elevation angle. The figure shows the sunlight, skylight and total illuminances for downwelling radiation at the Earth's surface, and the upwelling illuminance at altitudes above approximately 8000 m. As an example, let us consider the scene contrast between two objects with reflectivities of 0.1 and 0.2. The intrinsic scene contrast is clearly 0.33. For a solar elevation of 30°, figure 5.21 indicates that $E_s \approx 45$ klx and $E_a \approx 2.5$ klx, so if we take the atmospheric transmittance $T = 0.8$, we find that the scene contrast C' above the atmosphere will be reduced to about 0.23. When the solar elevation is 5°, $E_s \approx 4.5$ klx and $E_a \approx 800$ lx, so the scene contrast C' will be reduced to only 0.13.

Equation (5.17) shows the importance of the atmospheric transmission T in determining the scene contrast above the atmosphere. When T is small, the contrast will be reduced. We have seen in section 4.2.2 that the optical thickness of the atmosphere increases as the wavelength decreases through the

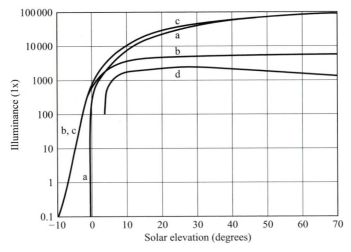

Figure 5.21. (a) Illuminance at the Earth's surface due to direct sunlight; (b) illuminance at the Earth's surface due to skylight; (c) total illuminance at the Earth's surface (a + b); (d) upwelling illuminance of the atmosphere.

visible band, so we expect the scene contrast to be greater for red light than for blue light. In section 4.3 we discussed the influence of haze, and noted that the optical thickness of the tropospheric aerosol layer can vary between about 0.1 and 1, giving values of T between 0.9 (clear sky) and 0.4 (hazy).

5.7   Some instruments

In this section, we describe some actual photographic systems, to illustrate the principles discussed earlier in the chapter. The first example we shall consider is an *airborne mapping camera*. There is a wide variety of such cameras, so rather than select the product of a particular manufacturer we will describe a typical instrument. This might have a film format of 230 mm × 230 mm, and operate at a focal length of 150 mm, achieving a spatial resolution of 50 lp mm^{-1} and very small distortion. Let us suppose that such a camera is operated from a height of 3000 m (approximately the greatest flying altitude possible without pressurisation of the aircraft). Equation (5.5) shows that the ground coverage achieved will be 4.6 km × 4.6 km, and equation (5.6) shows that the ground resolution will be roughly 0.2 m. Such a resolution implies that the photographs would be suitable for mapping at scales up to about 1/500.[5] The height accuracy achievable from an altitude of 3000 m will be given by the approximate rule $H/1000$ as about 3 m. Figure 5.17 shows a typical aerial photograph from a mapping camera.

Airborne photographic cameras generally hold very large rolls of film, capable of recording typically 500 to 1000 images. Since flights are of compara-

---

[5] The 'rule of thumb' here is that the maximum scale of map that can be made from an image with a ground resolution r_g is roughly 0.5 mm/r_g.

tively short duration, these can be rapidly returned to a laboratory for processing. *Space photography* obviously poses a greater problem in this regard. One
solution is to carry the camera on a short-duration manned mission. This was
the approach adopted for the *Metric Camera*, which was flown on board
Spacelab-1, a short mission of the space shuttle flown in winter 1983 at an
altitude of 250 km. The Metric Camera had a focal length of 305 mm, a film
format of 230 mm × 230 mm (it was modified from a standard aerial mapping
camera) and an effective spatial resolution of about 35 lp mm^{-1}. Its coverage
was thus approximately 190 km × 190 km, with a ground resolution of about
12 m, suitable for mapping at scales up to 1/50 000. Plate 1 shows part of an
image obtained by the Metric Camera.

The Metric Camera, and the *Large Format Camera* flown on space shuttle
missions in 1981 and 1984, were carried on short-duration missions that provided limited continuity of data. Other photographic instruments have, however, been flown on manned spacecraft, notably the Russian space station *Mir*.
Amongst other instruments, this carries a camera designated *KFA-1000*, which
has a film format of 300 mm × 300 mm and a focal length of 1000 mm. From a
nominal altitude of 350 km, this achieves a coverage of approximately
100 km × 100 km and a ground resolution of 20 m.

Finally, we should mention the Corona, Argon and Lanyard American
military reconnaissance satellites flown during the 1960s and early 1970s.
These were short-duration unmanned satellites in extremely low-altitude
orbits, chosen to give very high ground resolution. They carried cameras code-
named 'Keyhole'. Canisters of exposed film were ejected from the satellites and
retrieved by high-altitude aircraft. As an example, Keyhole 4B was carried on a
series of Corona satellites between 1967 and 1972, at a nominal minimum
altitude of 150 km. The film format was 55 mm × 760 mm and the focal length
610 mm, giving a coverage of 14 km × 190 km, and the effective spatial resolution was 160 lp mm^{-1}, giving a ground resolution of about 1 m. Since 1995
much of this imagery (over 800 000 images) has been declassified (Campbell,
1996; MacDonald, 1995). Figure 5.22 shows an enlarged portion of a Keyhole
image of central Moscow.

5.8   Applications of aerial and space photography

The applications of aerial photography are in general well known, and (with
the exception of the use of infrared film) the correspondence between a photographic image and the perception provided by the human eye and brain is
sufficiently great that most applications in any case have an intuitive feel
about them. The main advantages of photography as a remote sensing technique are that it is controllable and comparatively inexpensive, and that photographic optical systems can be made with sufficient precision and lack of
distortion that quantitative spatial information can be obtained from the
images. In this way, aerial photography has found very widespread application
in mapping and surveying, for example in geology and hydrology, in terrain

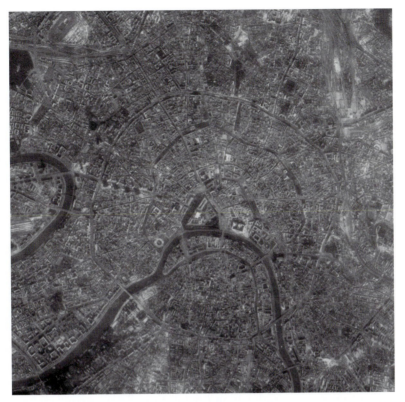

Figure 5.22. Extract of a Keyhole photograph of central Moscow, recorded on 22 November 1967. The extract shows an area 8 km × 8 km.

analysis, archaeology, field mapping, regional planning, the study of crop types and diseases, and so on. Colour film, though more expensive and complicated to process, is widely used, especially in cases where vegetation is studied (e.g. in agriculture, forestry and ecology) but also in geomorphology, hydrology and oceanography.

Black-and-white near-infrared film has proved especially useful in studying soil moisture and erosion, and in archaeological surveying. False-colour infra-red film has found applications in the classification of urban areas, in monitoring soil moisture, in disaster assessment, and especially in vegetation mapping and monitoring. Plate 2 shows a typical FCIR photograph used to study vegetation distribution.

PROBLEMS

1. Estimate the illuminance at the Earth's surface due to solar illumination when the Sun is 45° above the horizon. Use equation (5.1), but approximate the function $V(\lambda)$ shown in figure 5.6 by a function that is 1 for λ between 0.51 μm and 0.61 μm, and zero everywhere else. Compare your answer with figure 5.21.

2. A typical 35-mm camera with a standard lens and normal outdoor film can be assumed to have the following parameters: focal length $f = 50\,\text{mm}$; spatial resolution $l = 40\,\text{lp mm}^{-1}$; film format $w = 25\,\text{mm} \times 35\,\text{mm}$. Estimate the performance of this system when used to obtain vertical aerial photographs from an altitude of 5000 m.

3. A vertical aerial photograph reveals a tall building. The foot of one corner of the building has (x, y) coordinates (30.5, 62.0) (both measured in millimetres from the lower left-hand corner of the negative), and the top of the same edge appears at (19.0, 58.0). The corresponding coordinates for an adjacent edge of the building are (30.5, 73.0) and (19.0, 71.5). Given that the camera's focal length was 88 mm and the aircraft's altitude 212 m, find (a) the coordinates of the negative's principal point (directly below the camera), (b) the height, and (c) the width, of the building.

6

Electro-optical systems

6.1 Introduction

In chapter 5 we discussed photographic systems, and although these provide a familiar model for many of the concepts to be addressed in this and subsequent chapters, they nevertheless stand somewhat apart from the types of system to be discussed in chapters 6 to 9. In the case of photographic systems, the radiation is detected through a photochemical process, whereas in the systems we shall now consider it is converted into an electronic signal that can be detected, amplified and subsequently further processed electronically. This clearly has many advantages, not least of which is the comparative simplicity with which the data may be transmitted as a modulated radio signal, recorded digitally and processed in a computer.

In this chapter, we shall consider electro-optical systems, interpreted fairly broadly to include the visible, near-infrared and thermal infrared regions of the electromagnetic spectrum. The reason for this is a pragmatic one, since many instruments combine a response in the visible and near-infrared (VIR) region with a response in the thermal infrared (TIR) region, and much of the technology is common to both. Within this broad definition we shall distinguish imaging systems, designed to form a two-dimensional representation of the two-dimensional distribution of radiance across the target, and systems used for profiling the contents of the atmosphere. It is clear that an imaging system operating in the VIR region has much in common with aerial photography, and systems of this type are in very wide use from both airborne and spaceborne platforms. We shall therefore begin our discussion with these systems.

6.2 VIR imaging systems

6.2.1 Detectors

We saw in chapter 5 that the detection process in photography involves a photochemical reaction between the incident radiation and ions of silver halide. In electro-optical systems, the radiation falls on a suitable detector,

which generates an electrical signal dependent upon the intensity of the radiation. In this section, we consider the main types of detector, although a discussion of the numerical values of their sensitivities is beyond our scope. Chen (1985) provides a fuller discussion.

One of the earliest types of detector, still in fairly wide use for the VIR region, is the *photomultiplier*. This is a vacuum-tube device, shown schematically in figure 6.1. It consists of an evacuated glass vessel, containing a number of electrodes that are maintained at different electric potentials. A photon falls on the electrode at the most negative potential (the *photocathode*), where it causes the ejection of an electron by the *photoelectric effect*. This electron is accelerated towards an intermediate electrode (*dynode*) at a more positive potential, and the increased kinetic energy that it acquires causes it to eject more than one electron from the dynode. This process is repeated several times, the number of electrons increasing at each dynode, until the flow of electrons reaching the most positive electrode (the *anode*) represents a measurable current A. The size of this current depends on the intensity of the incident radiation.

The minimum photon energy that can be detected by such a device is the *work function* of the material from which the photocathode is made. The work function W is defined as the energy difference between an electron *in vacuo* and an electron within the material, namely the energy required to eject the latter. For metals, values of W lie typically in the range 2 to 5 eV, so the maximum wavelength to which such devices are sensitive is about 0.6 μm. Mixtures of alkali metals, however, can have significantly smaller work functions, and sensitivity up to about 1 μm can be obtained.

The photomultiplier is a very sensitive device with a rapid response time of the order of 1 ns. Its main disadvantages are its mechanical fragility, its comparatively large size, and the fact that it needs a high operating potential of about 1 kV.

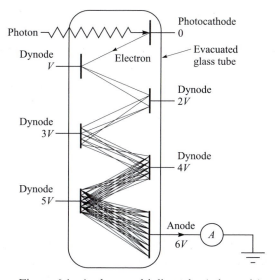

Figure 6.1. A photomultiplier tube (schematic).

Radiation in the near-infrared region is usually detected using a *photodiode*. This is a semiconductor junction device, usually indium antimonide (InSb) or lead sulphide (PbS), in which an incident photon generates a current or voltage across the junction. The signal is proportional to the light intensity. The theory underlying the operation of photodiodes is quite complicated, and is outlined below.

A semiconductor diode consists of two abutting pieces of semiconducting material. One piece has been *doped* with a trace of impurity that gives rise to an excess of electrons, the other with an impurity giving a deficit of electrons. These are referred to as n-type and p-type material, respectively, since the effective charge carriers are, respectively, negatively (electrons) and positively ('holes', i.e. absences of electrons) charged. At the junction between the two materials, holes from the p-type material diffuse into the n-type material, where they combine with the free electrons. A corresponding effect also occurs in the opposite direction, with electrons diffusing into the n-type material, and this gives rise to a *depletion region* of very low conductivity, typically about 1 μm in width. Because there is now an excess of positive charge in the n-type material and an excess of negative charge in the p-type material, there is an electric field from the n-type to the p-type across the depletion region, and this inhibits the further diffusion of charges.

If an external electric field is applied from the p-type to the n-type material (*forward bias*), the internal field is overcome to some extent and the depletion region is made narrower. A current flows, and its magnitude increases approximately exponentially with the applied voltage difference. If, however, the external field is applied in the opposite sense (*reverse bias*), the depletion region is made wider and a very much smaller current is able to pass.

If now the diode, with no external bias, is subjected to incident electromagnetic radiation, a photon may be able to create an electron-hole pair in the p-type material. If the electron diffuses into the depletion region, it will be accelerated by the internal field into the n-type material, and the work done will appear as a voltage, proportional to the intensity of the radiation, across the diode. This is called the *photovoltaic mode* of operation. Similarly, of course, if the electron-hole pair is created in the n-type material, the positively charged hole will be accelerated into the p-type material. If, on the other hand, the diode is reverse-biassed, increasing the width of the depletion region, the external field maintains the field within the depletion region and the charge-carriers will generate a current. This is called the *photoconductive mode* of operation, and it gives a much faster response (of the order of 1 ns) compared with the photovoltaic mode (about 1 μs). The reason for this is that the response time is determined by the capacitance of the depletion region, which is inversely proportional to its width. The photoconductive operation of a photodiode is illustrated in figure 6.2.

The maximum photon wavelength (i.e. minimum energy) that can be detected by a photodiode is determined by the energy required to create an electron-hole pair. This is termed the *band-gap* of the semiconductor.

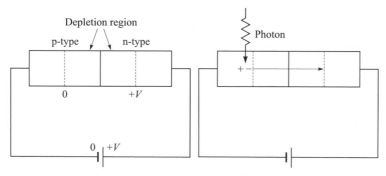

Figure 6.2. Operation of a photodiode (schematic). The p-type region contains free positive charges and the n-type region free negative charges, except in the depletion region. An incident photon creates an extra pair of charge-carriers in the p-type material, and the negative charge is accelerated by the potential difference V into the n-type material, thus generating a current. A current in the same direction will be produced if the photon creates an electron-hole pair in the n-type material, in which case a positive charge-carrier is accelerated into the p-type material.

Semiconductors such as germanium have relatively large band-gaps, giving sensitivity up to only $1.7\,\mu$m, but PbS responds up to about $3\,\mu$m and InSb to about $5\,\mu$m.

Photomultipliers and photodiodes are both single-element detectors, which means that, in order to use them to produce images, they must either be combined in large arrays of detectors, or scanned over the target. Clearly, the construction of an array of detectors is more practical for the photo-diode (which can be made very small) than for the photomultiplier, but another possibility is represented by the *charge-coupled device* (CCD).[1] This is another semiconductor device, consisting of many (of the order of 1000) identical elements in a linear array. Each of these elements is light-sensitive, developing and then storing free charges as a result of illumina-tion. By suitable manipulation of the voltages applied to the elements of the array, these charges can be moved from one element to the next, until they are finally 'read out' at the end of the array. In other words, the device operates as a shift-register.

It is possible to construct CCDs in both one-dimensional (linear) and two-dimensional (planar) forms. The two-dimensional CCD has two major advantages. The first follows from the fact that the controlling voltages used to shift the charges from one element to the next are applied only to the edges of the array and not to each individual element. Thus, an array consisting of $n \times m$ elements needs only $(n + m)$ connections, so that arrays of the order of 1000×1000 elements are feasible. Each element is of the order of $10\,\mu$m wide, so the whole array is only a centimetre or so in width. The second advantage is that, since the array does not need to be scanned in order to form an image, each element can 'look at' (i.e. collect photons from)

[1] These are now increasingly familiar through their use in digital and video cameras.

the target for a longer time than is the case for a scanned system, which increases the potential sensitivity.

An earlier approach to two-dimensional detection is represented by the *vidicon*, which is a type of television camera. Figure 6.3 shows schematically the design of a simple vidicon, consisting of a lens and shutter, and a sheet of photoconductive material that is coated, on the side nearer the lens, with a transparent conductor. The other side of the photoconductive sheet is illuminated by an electron beam that can be deflected electrostatically, in a manner similar to that used in a cathode-ray oscilloscope, to impinge upon any part of the sheet.

The mode of operation is as follows. First, with the shutter closed, the electron beam is used to coat the back of the photoconductive plate with electrons. The electron beam is then switched off, and the shutter is opened and closed, leaving a distribution of charge on the sheet corresponding to the distribution of light intensity. Finally, the back of the sheet is scanned by the electron beam, resulting in an electric current (the read-out) from those parts that were illuminated.

6.2.2   Imaging

Having reviewed the principal methods by which VIR radiation is detected, we now consider the means by which images, namely two-dimensional representations of the scene radiance, can be built up.

A two-dimensional detector such as a planar-array CCD or a vidicon generates a two-dimensional image. All that is necessary is to ensure that the detector views the scene for long enough to acquire and process sufficient photons from it. However, we should note that, if the platform carrying the detector is in motion relative to the target, it may be necessary to compensate for this motion in order to avoid 'motion blurring' of the

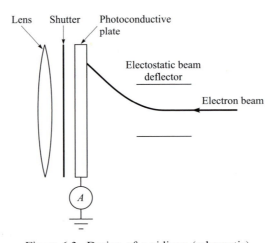

Figure 6.3. Design of a vidicon (schematic).

image. This mode of operation can conveniently be thought of in terms of the detector 'staring' at a scene, then moving on to stare at the next scene, and so on. For this reason, it is sometimes called *step-stare imaging*, and is illustrated in figure 6.4. However, if the detector is one- or zero-dimensional (a linear array or a single detector), some form of mechanical scanning is required. In the case of a linear array of detectors, this scanning can be achieved using the motion of the platform. A strip of rezels, oriented perpendicularly to the direction of motion, is imaged instantaneously, and the adjacent strip of rezels is imaged when the platform has moved a distance equal to the width of the rezels. In a pleasing domestic analogy, this mode of imaging is often called *push-broom imaging*, and it is illustrated in figure 6.5. This type of imaging is used, for example, by the HRV (high-resolution visible) sensor carried on the SPOT satellites.

In the case of a single (zero-dimensional) detector, it is clear that some form of mechanical scanning will also be required, since the forward motion of the platform can provide scanning in only one direction. The usual way of achieving this mechanical scanning is through a rotating or oscillating mirror, which scans the *instantaneous field of view* (IFOV) from side to side, in a direction approximately perpendicular to the forward motion. An instrument operating on this principle is usually called a *line scanner*, and the mode of operation is often called *whisk-broom imaging*. It is in widespread use for airborne and spaceborne instruments, for example the TM (Thematic Mapper) instrument carried on recent Landsat satellites. It is illustrated in figure 6.6.

It will be apparent that care must be taken to match the speed of the side-to-side scanning to that provided by the forward motion. If the former is too small, or the latter too large, some strips of the surface will not be imaged. The ideal relationship between the two is governed by the width Δx of the IFOV measured in the direction of platform motion. If the platform is not to advance by a distance of more than Δx during the time ΔT taken for one line to be scanned, the scan time must satisfy

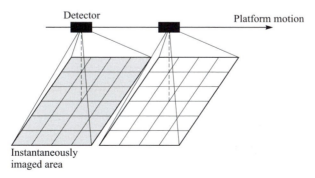

Figure 6.4. 'Step-stare' imaging by a two-dimensional detector. The elements within the instantaneously imaged area are called resolution elements or *rezels*; the corresponding elements in the image are called picture elements or *pixels*. In practice, the number of rezels or pixels is of the order of 1000×1000.

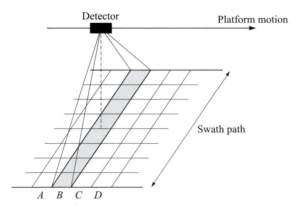

Figure 6.5. 'Push-broom' imaging by a one-dimensional detector (linear array). The detector is currently imaging the shaded strip *B* of rezels; strip *A* was imaged previously, and strips *C* and *D* will be imaged when the platform has moved forward a sufficient distance.

$$\Delta T \leq \frac{\Delta x}{v} \tag{6.1}$$

where v is the platform velocity. The Landsat TM sensor, for example, has an effective IFOV width $\Delta x = 30$ m, and an equivalent ground speed $v = 6.46 \times 10^3$ m s^{-1}, so equation (6.1) implies that the scan time must be at most 4.6 ms. In fact, the sensor scans 16 lines simultaneously in a time of 71.4 ms, giving an effective scan time for each line of 4.5 ms.

If the platform that carries the sensor is *not* in motion relative to the Earth's surface, then of course it is necessary to provide mechanical scanning of a single detector in *two* directions. This is the case for geostationary imagers (geostationary orbits are discussed in chapter 10). Scanning in the east–west direction is normally achieved by spinning the satellite about its north–south axis (this also helps to stabilise the orientation of the axis), while north–south scanning is effected using a mirror within the instrument that rotates in a series

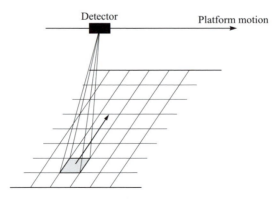

Figure 6.6 'Whisk-broom' imaging by a line scanner. The instantaneous field of view (shaded) is scanned sideways by a mirror within the instrument. Forward scanning is achieved by the forward motion of the platform.

of steps. This mode of operation, illustrated in figure 6.7, is usually called *spin-scan imaging*.

The number of steps made by the scanning mirror determines the number of east–west scan lines. For a typical geostationary imager this is of the order of 2000–5000, giving scan lines a few kilometres wide at the Earth's surface. The angular velocity of the satellite's spin then determines the time taken to acquire a full image. Higher angular velocities will give shorter imaging times, but also mean that the detector views a given region of the Earth's surface for a shorter time and hence provides a less sensitive measurement. Typically, the angular velocity is of the order of 100 r.p.m., so the time required to acquire a complete image is of the order of 20–50 min.

6.2.3   *Spatial resolution*

We showed in section 2.7 that the angular resolution of an imaging system is in general limited by diffraction effects to $\approx \lambda/D$, where λ is the wavelength of the detected radiation and D is the diameter of the objective lens, mirror, or in general the first obstruction encountered by the incoming radiation. Provided that this angular resolution is small (i.e. that the objective lens is many wavelengths in diameter), the corresponding linear resolution at the Earth's surface can be written approximately as $H\lambda/D$, where H is the distance from the sensor to the surface. However, the design of the instrument may degrade the spatial resolution to a value significantly poorer than this. We have already seen an example of this in section 5.4.1, where it was stated that the spatial resolution of a photographic system is determined by both the camera (which includes a contribution from diffraction) and by the film. In the case of VIR imaging systems, the spatial response of the film must be replaced by the spatial response of the detector, or array of detectors, but the principle is the same. Since a detector has a finite size, and since the signal derived from it is, in effect, an average of the radiation intensity over the entire detector, it is clear that the spatial resolution cannot be better than the size of the detector *projected through the system's optics onto the Earth's surface*. This is the concept of the

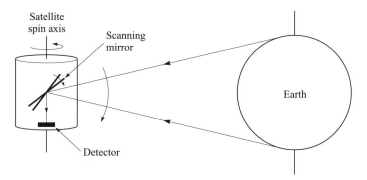

Figure 6.7. 'Spin-scan' imaging by a geostationary satellite.

rezel introduced in the previous section. For a simple optical system charac-terised by a focal length f and a detector size a, the size of the rezel will be Ha/f.

Roughly speaking, then, we may state that the spatial resolution of the system will be the larger of the two terms $H\lambda/D$ and Ha/f. If the former term is larger, the data are *oversampled*, so that adjacent pixels in the image will be strongly correlated and the image will contain less information than the number of pixels would imply. If the latter term is larger, the spatial resolution of the system could be improved simply, for example by increasing the focal length. Thus, a design is in some sense optimal if the two terms are approxi-mately equal, and this is in fact normally the case.[2] A further complication may arise if the scan time ΔT is short enough that adjacent scan lines on the ground actually overlap one another significantly. In this case, the rezel size will be smaller than the IFOV, and again the data are oversampled. In any case, the spatial resolution will be proportional to the height H of the sensor above the Earth's surface. Combining this observation with equation (6.1), we can see that the height and speed of the platform, and the scan time ΔT, cannot be varied independently. This imposes operational limitations for airborne appli-cations, and design considerations for spaceborne applications (for which the relationship between v and H is fixed). We shall refer to these limitations again in chapter 10.

The foregoing discussion has outlined some of the issues relating to the spatial resolution of VIR imaging systems. It should be noted that spatial resolution is in fact a rather elusive concept (see, for example, the discussion by Forshaw et al. (1983)). For the present, we note that although the concepts of spatial resolution, rezel size and IFOV are all different, in practice they are of similar magnitude.

6.2.4 *Spectral resolution*

Most VIR instruments designed for observing the Earth's surface are capable of discriminating between different wavelength components in the incident radiation. The number of spectral bands provided by such instruments varies from a few to a few hundred. Where there are only a few bands, with compara-tively large bandwidths in the range 10 to 100 nm, these are normally defined by filters. Higher spectral resolutions (i.e. smaller bandwidths), to approxi-mately 0.1 nm, are generally obtained by using prisms or diffraction gratings to disperse the spectrum of the incoming radiation onto an array of detectors. Alternatively, a single detector can be scanned mechanically across the spec-trum.

Dispersion by a prism is illustrated in figure 6.8. Radiation undergoes a deviation when it enters and leaves the prism, the deviations being given by Snell's law (equation (3.30)). The refractive index of glass varies somewhat with

---

[2] Perhaps not too surprisingly, this is also true of the human eye.

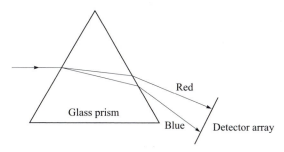

Figure 6.8. Dispersion of broad-band radiation by a prism.

wavelength, so different wavelengths undergo different deviations. The angular dispersion is not large: for a typical arrangement, the angle between the emerging blue (say $\lambda = 0.40\,\mu$m) and infrared ($\lambda = 1.0\,\mu$m) radiation is of the order of 5° to 10°.

Figure 6.9 illustrates dispersion by a diffraction grating. The grating itself consists of a large number of parallel lines ruled on a transparent (e.g. glass) or reflective (e.g. polished metal) sheet. The lines are regularly spaced at some small separation d. A beam of plane-parallel radiation of wavelength λ, striking such a grating at normal incidence, is diffracted into a plane-parallel beam that makes an angle θ to the normal, where the value of θ is given by

$$\sin\theta = \frac{n\lambda}{d} \tag{6.2}$$

and n is an integer.[3] Thus, if broad-band radiation is incident on the grating, it will be dispersed in angle, producing spectra corresponding to different values of n, called the *order* of the spectrum. We should note that the value $n = 0$ gives

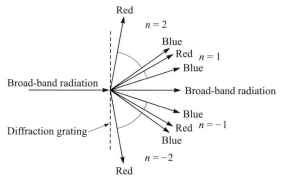

Figure 6.9. Dispersion of broad-band radiation by a diffraction grating. The values of n are the orders of the spectra.

---

[3] This result can of course be proved using the diffraction theory derived in section 2.7. The required amplitude transmittance function is

$$f(y) = \sum_{m=-\infty}^{\infty} \delta(y - md)$$

where δ is the Dirac delta-function introduced in section 2.3

$\theta = 0$ for all wavelengths, so there is a 'zero-order spectrum' of undeviated radiation that has the same spectral content as the incident radiation. This is of course useless from the point of view of spectral discrimination.

Equation (6.2) shows that the angular dispersion of a given range of wavelengths can be increased by making the line spacing d smaller. Values of d as small as 1.3 μm are possible, giving an angular dispersion of about 15° for the first-order spectrum between 0.4 and 0.7 μm. This is about three times better than can be achieved with a glass prism. The angular dispersion is greater for the higher-order spectra, but we should note that, for these higher orders, the maximum usable range of wavelengths will be reduced. If the range of wavelengths is too large, different orders of different wavelengths can overlap. Thus, if a diffraction-grating spectrometer is designed for especially high spectral resolution, it is generally necessary to filter the radiation first.

6.2.5   *Major applications of VIR images*

Visible and near-infrared imagery, normally with quantitative or even calibrated radiance values, has found enormous application in remote sensing. Part of its popularity may be ascribed to the greater ease with which such data may be interpreted, since the wavelength range corresponds largely to the sensitivity range of the human eye, but to a great extent also many important processes modulate the image radiance in this range of wavelengths. Since a VIR image is similar to an aerial photograph, many applications of the former technique can be carried over to VIR imagery. In this section, we cannot discuss any of these applications at great length, but we can at least illustrate something of their range. A good idea of the current applications of VIR and other images can be obtained by scanning the last few years of the main remote sensing journals, listed in the references.

The most obvious set of applications is probably to land-surface mapping, for example for cartography, the delineation of water bodies (including floods), volcanoes, snow and ice, coastal, agricultural, forestry (including forest fires) and urban areas, geological and geomorphological characteristics, and so on. The development of above-ground vegetation biomass (the total organic matter per unit area of the Earth's surface) can be monitored, and this provides the ability to observe changes in the extent and health of vegetation, for example for the commercial estimation of crop and forestry yields, the prediction of crop failure and drought, and monitoring of the impacts of pollution and climate change. Clearly, the scale at which such mapping can be carried out is dependent upon the spatial resolution of the imagery. However, since the ability to distinguish one type of land cover from another is normally reliant on the ability to distinguish their reflectance spectra, the scale is also governed by the spectral resolution and the sensitivity of the data. In some cases, notably vegetation, water bodies and snow, the reflectance spectra are sufficiently characteristic that their delineation can be carried out with minimal ancillary information (see section 3.5.1). For example, a pixel that shows low reflectance in the visible part of the spectrum, especially for red light, and a high reflectance in the near-infrared region, is almost certain to contain a high proportion of green-leaved vegetation. In most cases, however, the association of a particular land-cover type with the measured properties of a pixel is more difficult. This procedure is known as image classification, and is discussed in chapter 11.

Water bodies can also be studied using VIR imagery. The depth of a shallow water body (river, coastal waters or lake) can be estimated by comparing the reflectance in two or more spectral bands. This method relies on the variation with wavelength of the attenuation coefficient, so that the relative contributions from the subaqueous surface and the water itself will differ at different wavelengths, although because of uncertainties in the attenuation coefficient and the subaqueous reflection coefficient the method generally requires calibration. The best spectral bands to use are green and near-infrared, since these give the largest difference in attenuation coefficient. Ocean colour measurements, in which sensitive measurements of the ocean reflectance are made in a number of spectral bands, can be used to identify different bodies of water, for example to distinguish sediment-laden coastal waters from deep ocean waters, and to estimate ocean currents from variations in turbidity. Marine phytoplankton concentrations can be estimated by the increased reflectance, caused by the presence of chlorophyll, in the 0.4–$0.5\,\mu$m band (see e.g. Robinson, 1994). The effect is a small one, so that sensitive instruments and particularly accurate correction for atmospheric propagation effects are needed. Over the open ocean, suspended sediments can be ignored (these are so-called 'type I waters') in estimating chlorophyll concentrations, and the calculation is normally accurate to within a factor of 2, but over shallow waters the suspended sediments, especially those with significant reflection in the yellow band, severely complicate the problem ('type II waters').

Clouds can be delineated and monitored from VIR imagery, measuring the extent of cloud cover, its height and type, as an aid to meteorological investigations. By tracking the motion of clouds, wind speeds can be estimated (e.g. Henderson-Sellers, 1984; Scorer, 1986; Warren and Turner, 1988).

6.2.6   *Atmospheric correction*

For many quantitative applications of VIR imagery, it is desirable to correct the data for the effects of atmospheric propagation.[4] If the data are accurately calibrated, the variable that is measured is the radiance reaching the sensor, but the variable that is wanted is the reflectance (e.g. the BRDF) of the surface. If there were no atmosphere, the calculation would be a straightforward one, requiring only a knowledge of the radiance at the Earth's surface due to solar illumination, and of the geometry of the sun–target–sensor arrangement. However, the presence of the atmosphere significantly complicates the relationship between the solar radiance and the radiance detected at the sensor, as

---

[4] A somewhat fuller discussion of atmospheric correction is given by Cracknell and Hayes (1991).

illustrated schematically in figure 6.10. In this figure, *A* represents direct illumination of the surface by the Sun, but *B* shows that the surface can also be illuminated by radiation that has been scattered from the atmosphere – in other words, by skylight. The rays *C* to *E* show the possible destinations of radiation scattered from the surface. In *C*, it has been scattered back to the surface; in *D* it is transmitted directly to the sensor; and in *E* it also reaches the sensor, although after being scattered from the atmosphere. Finally, ray *F* shows that some sunlight can be scattered directly into the sensor. Thus, rays *A* and *D* represent the simple, 'no atmosphere', situation, and the other rays are the contributions due to atmospheric scattering. These are clearly less significant for a low-altitude airborne observation than for a satellite measurement that views through the entire atmosphere.

Correction of the image data for the effects of atmospheric propagation can be carried out in essentially three ways (e.g. Campbell, 1996, section 10.4). Physically based methods attempt to model the phenomena illustrated in figure 6.10. This is the most rigorous approach, and also the most difficult to apply. The atmospheric scattering and absorption characteristics are calculated by a computer model (the best-known being LOWTRAN 7 (Kneizys et al., 1988) and MODTRAN (Berk et al., 1989)), which requires as input data meteorological, seasonal and geographical variables. In practice, these variables may not all be available with sufficient spatial or temporal resolution, and, in particular, estimation of the contribution of atmospheric aerosols (see section 4.3) is difficult.

The second approach to atmospheric correction of VIR imagery is based on calibration against targets of known reflectance. These targets can be artificially constructed or naturally occurring, but they need to satisfy a number of criteria: (1) their reflectances must be known sufficiently accurately, in the same spectral bands as are used by the imager; (2) the range of reflectances repre-

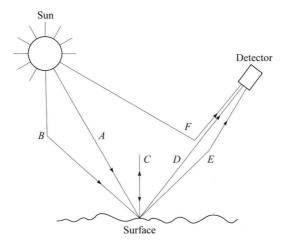

Figure 6.10. Contributions to the radiance measured at a sensor in the presence of atmospheric scattering.

sented by the calibrators must span the range of interest in the scene; (3) each calibrator should cover an area of at least several rezels; (4) the calibrators should be well distributed over the entire scene, so that possible variations of atmospheric conditions from place to place can be assessed and, if necessary, allowed for.

Finally, the simplest, and most widely applied, method of atmospheric correction is based on *dark-pixel subtraction*. For each spectral band present in the image, the minimum-radiance pixel is identified, and this minimum radiance is subtracted from all pixels in the image. This method is quite crude: it assumes that the minimum reflectance in each band is zero, that the atmospheric correction can be modelled adequately as an additive effect, and that the correction does not vary from place to place within the scene. To some extent, visual inspection of an image can determine whether these assumptions are likely to be valid. Zero-reflectance rezels can be provided by shadows and, in the near-infrared region, by water bodies.

6.3   Thermal infrared imagers

6.3.1   Detectors

Detectors for thermal infrared (TIR) radiation can be divided into two classes: quantum detectors and thermal detectors. In a quantum detector, a photon interacts directly with the detector material to change the energy of an electron, whereas in a thermal detector, the radiation is detected as a result of a change in the electrical properties of the material arising from a change in its temperature.

As we remarked in chapter 2, the TIR region of the electromagnetic spectrum occupies the wavelength range between about 3 and 15 μm. The energy of a TIR photon is thus about 0.1–0.4 eV, significantly less than for VIR radiation, and this imposes some difficulty in making suitable photodiodes (quantum detectors). We noted in section 6.2.1 that germanium photodiodes respond to wavelengths up to 1.7 μm, lead sulphide up to about 3 μm and indium antimonide to about 5 μm. More exotic semiconductors, having even smaller band-gaps, can be devised. Two common ones for TIR detection are mercury cadmium telluride ($Hg_{0.2}Cd_{0.8}Te$, also referred to as MCT) and mercury-doped germanium (Ge:Hg), both of which respond to wavelengths up to about 15 μm. It is, however, usually necessary to cool a quantum detector of TIR radiation, especially one designed to operate at the long-wavelength end of the region. This increases the sensitivity of the detector by reducing the number of photons generated by the sensor itself. Cooling is normally provided using liquid nitrogen, which has a maximum temperature of 77 K, or liquid helium (30 K).

In contrast with the quantum detectors, thermal detectors provide very broad spectral range, but at the expense of much lower sensitivity and longer response times. The three main types of thermal detector are thermistor bolometers, thermocouples and pyroelectric devices. The *thermistor bolometer* is a simple device that consists, in essence, of a material (usually carbon, germanium or a mixture of metal oxides) whose resistance varies with temperature. A *thermocouple* uses the *Seebeck effect*, in which a potential difference is generated across a pair of junctions between dissimilar metals when the junctions are held at different temperatures. To amplify the signal, thermocouples are often connected in series as a *thermopile*. A *pyroelectric device* is a crystal that undergoes a redistribution of its internal charges as a result of a change in temperature. Charge separation occurs at the surfaces of the crystal, resulting in a potential difference that can be amplified and detected.

In practice, thermal detectors are little used in remote sensing instruments except where sensitivity is needed at wavelengths significantly longer than about 15 μm.

6.3.2   Imaging

The principles by which thermal infrared imagers form images are similar to those for VIR imagers; that is, both mechanical scanning and arrays of detectors are used (see section 6.2.2). As was mentioned earlier, TIR and VIR functions are often combined in the same instrument.

6.3.3   Spatial resolution

The factors that determine the spatial resolution of a TIR imager are similar to those for a VIR imager, so much of the discussion in section 6.2.3 is applicable. However, we should note that, since the wavelength of TIR radiation is of the order of ten times longer than for VIR radiation, the spatial resolution is generally somewhat coarser. For example, the Thematic Mapper instrument carried on the Landsat-4 and Landsat-5 satellites has six VIR bands, each of which has a rezel size of 30 m, and a TIR band with a rezel size of 120 m.

6.3.4   Spectral resolution and sensitivity

Normally, TIR imaging does not require particularly high spectral resolution. In practice, the wavelength range from 3 to 15 μm is usually split into a small number (often only one or two) channels with a typical width of 1 μm, defined by filters. These channels are generally centred at about 4 μm and about 10 μm, thus avoiding the strong water vapour absorption feature at 6–7 μm (see figure 4.5).

The intrinsic sensitivity of a radiometric observation, at a given wavelength, to black-body radiation at temperature T, can be defined in terms of the Planck radiation law (equation (2.31)) as

$$S = \frac{T}{L_\lambda}\frac{\partial L_\lambda}{\partial T} \qquad (6.3)$$

that is, the ratio of the fractional change in the black-body radiance to the fractional change in the absolute temperature. This function is plotted in figure 6.11 for a temperature of 280 K (a typical terrestrial temperature). As the figure shows, the sensitivity S is about three times greater at a wavelength of $4\,\mu$m than at $10\,\mu$m, and for this reason the 3–5 μm band could be preferred to the 8–14 μm band. However, the figure also shows the spectral radiance L_λ for a black body at a temperature of 5800 K, which is roughly that of sunlight, and it can be seen that this spectral radiance is about 50 times greater at $4\,\mu$m than at 10 μm. Thus, in fact, measurements made in the 3–5 μm band are at risk of 'contamination' from reflected solar radiation, and for this reason this band is less useful than the 8–14 μm band, except for night-time measurements and for measurements of volcanoes.

Since the purpose of a thermal infrared observation is to measure the brightness temperature T_b of the radiation incident upon it, we should consider how the power detected by the instrument varies with T_b. To some extent, this is indicated by the sensitivity S defined in equation (6.3), but we must also consider the effect of the filter used to define the spectral band. For an instrument that collects radiation from a solid angle $\Delta\Omega$ over an area A, the total power P received from black-body radiation of brightness temperature T_b can be written as

$$P = 2hc^2 A\,\Delta\Omega \int\limits_{0}^{\infty} \frac{f(\lambda)\,d\lambda}{\lambda^5\left(\exp\left(\dfrac{hc}{\lambda k T_b}\right) - 1\right)} \tag{6.4}$$

where $f(\lambda)$ is a filtering function that defines the response of the instrument to radiation of wavelength λ. If $f(\lambda) = 1$ for all wavelengths (no filtering), the integral is just

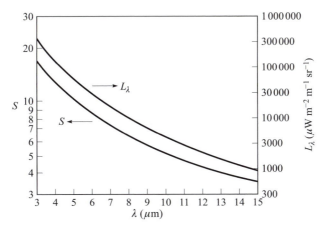

Figure 6.11. The intrinsic sensitivity S (left-hand scale) of black-body radiation at a temperature of 280 K, defined as the ratio of the fractional change in spectral radiance to the fractional change in temperature. The figure also shows the spectral radiance L_λ (right-hand scale) of solar radiation.

$$\frac{1}{15}\left(\frac{\pi k T_b}{hc}\right)^4$$

and the received power varies as T_b^4, in accordance with the laws for black-body radiation. Next we adopt a very simple model for $f(\lambda)$, by supposing that it is equal to 1 for wavelengths between 8 and 14 μm, and zero otherwise. This represents a simple broad-band filter. Figure 6.12 shows the total black-body radiance L, and the radiance L_{8-14} contained between the wavelength limits of the filter,[5] as a function of the brightness temperature between 250 and 300 K. It can be seen that, while L varies as T_b^4, L_{8-14} varies approximately as T_b^5 over this range of brightness temperature. This approximate fifth-power dependence is usual for thermal infrared observations near 300 K, and illustrates the fact that the fraction of the total radiance between the wavelength limits increases roughly in proportion to the brightness temperature over this range.

6.3.5   *Major applications of thermal infrared images*

A calibrated thermal infrared image indicates the brightness temperature of the radiation reaching the sensor. Three factors contribute to this brightness temperature: the physical temperature of the surface being sensed, its emissivity, and atmospheric propagation effects (these are considered in the next section). In most cases, TIR images are used to deduce the surface temperature, and,

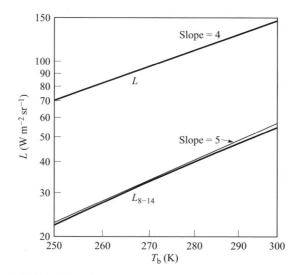

Figure 6.12. Total black-body radiance L, and the radiance L_{8-14} between wavelengths 8 and 14 μm, as functions of the brightness temperature T_b. The thin line shows that L_{8-14} is approximately proportional to T_b^5 in this range.

---

[5] This can be calculated using the methods described in section 2.6.

broadly speaking, we may classify their applications into those in which the surface temperature is governed by man-made sources of heat, and those in which the heating occurs naturally. In the former case, the technique has been used from airborne platforms to determine heat losses from buildings and other structures. It is most useful to perform this kind of survey just before dawn, so that the effects of solar heating will have had the greatest possible time to decay. The technique has also been used, for example, to monitor plumes of hot water generated by power stations.

In the latter case, the principal applications of thermal infrared images are to the identification of clouds and to the measurement of sea-surface temperature and thermal inertia. These are discussed in greater detail below. Other applications include the measurement of soil moisture and water stress, identification of frost hollows and crop types, and so on.

6.3.5.1   Cloud detection

Contamination of a VIR image by the presence of cloud is frequently a problem for the analysis of the Earth's surface, and it is highly desirable to be able to remove from the image those pixels that are affected by cloud. Conversely, one might wish to study the distribution of cloud. In either case, identification of cloudy areas is necessary, and this is most often accomplished using a combination of VIR and TIR imagery.[6] In some cases, this can be performed manually, although automated procedures are clearly desirable.

The brightness temperature of a cloud, when viewed from above, is approximately equal to the *cloud-top temperature*, and hence to the atmospheric temperature at the height of the cloud-top. Thus, clouds tend to be colder than the underlying surface, increasingly so for higher-altitude clouds. A cloud that is optically thick in the VIR band will have a high reflectance, as discussed in section 3.4.1. Thus, the simplest cloud-detection algorithms are designed to look for pixels that are bright (in the VIR band) and cold (in the TIR band). More detailed information on the type of cloud can be obtained by using the brightness-temperature information quantitatively, by examining the texture of the image radiance (see chapter 11), or by measuring the content of liquid water (or solid ice) integrated vertically through the cloud. Figure 6.13 shows an example of cloud detection in VIR and TIR imagery.

6.3.5.2   Sea-surface temperature

The sea-surface temperature (SST) is a quantity of obvious oceanographic, meteorological and climatological importance. In principle, it is straightforward to determine the SST from a thermal infrared observation, since the

---

[6] We should note, however, that cloud detection can also be achieved using only VIR imagery, as suggested by figure 3.24. One common procedure is to examine the ratio of the visible-band and near-infrared reflectances, which is usually close to 1 for cloud surfaces, greater than 1 for water surfaces, and less than 1 for land surfaces. The potential confusion between cloud and snow can be resolved if the instrument also has a band at about 1.6 μm.

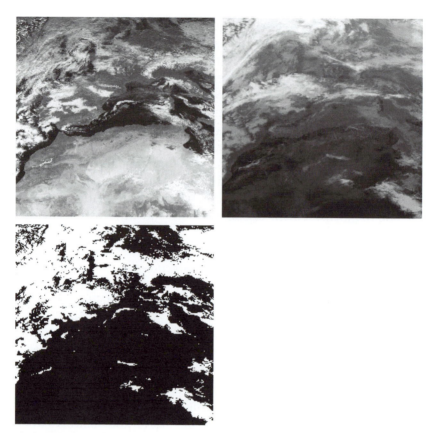

Figure 6.13. Example of simple cloud-detection using visible-wavelength and thermal infrared images. The image at top left is a segment of the visible band of a Meteosat image recorded at 12 noon on 5 January 2000, showing north Africa and most of western Europe. The image at top right is the corresponding segment of thermal infrared data, with low brightness temperatures shown as white and high brightness temperatures as black. At bottom left is a simple cloud image, showing (in white) areas that are bright and cold. (Top images © EUMETSAT and University of Dundee, 2000).

emissivity of sea water is well known (0.993 at a wavelength of $10\,\mu$m – see section 3.5.2). The important problem of determining the brightness temperature of the surface from the brightness temperature measured at the sensor is one of atmospheric correction, discussed in section 6.3.6. This can be a rather large effect, of the order of 5 to 10 K.

There is, however, another potential problem associated with the use of TIR data to determine the sea-surface temperature. The absorption length for TIR radiation in water is very small, typically 0.02 mm or less, which means that the measured brightness temperature is characteristic of only the upper tenth of a millimetre, at most, of the water surface. Oceanographically, the 'surface' is a few centimetres deep, and the physical temperature of the upper tenth of a millimetre can differ from the mean temperature of this layer by an amount of the order of 1 K. This temperature difference, which can be either positive or negative, arises from a combination of evaporative cooling and solar heating.

Figure 6.14 shows an example of sea-surface temperature derived from TIR imagery.

6.3.5.3 *Thermal inertia*

The Earth's surface is subjected to a periodically varying input of heat from the Sun, as a result of the alternation of day and night, and of the seasons of the year. In consequence, the surface temperature fluctuates. The amplitude of the fluctuations in surface temperature depends on a combination of physical properties of the material of which the surface is composed, and this is called the *thermal inertia*; hence, measurement of this amplitude can provide some information about the composition of the Earth's surface. This is the idea behind thermal inertia mapping, and in this section we will develop a simple model of it.

For our model, we will suppose that the material of the Earth's surface is homogeneous and infinitely deep, extending from $z = 0$ at the surface to $z = +\infty$. Heat is propagated by conduction, only in the vertical (z) direction. The thermal behaviour of the material is governed by the heat conduction equation

$$\mathbf{F} = -k\nabla T$$

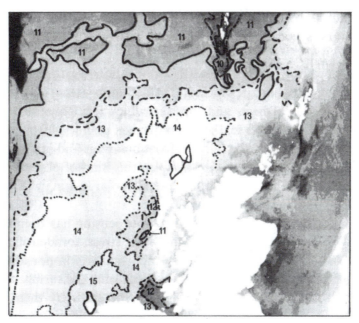

Figure 6.14 Sea-surface temperatures near northern Britain. The contours (in degrees Celsius) have been derived from thermal infrared data obtained on 17 May 1980 by the AVHRR (Advanced Very High Resolution Radiometer). The image has a coverage of about 800 km × 900 km and is reproduced by courtesy of Professor A. P. Cracknell, University of Dundee.

(in which **F** is the vector heat flux, K is the thermal conductivity and T the temperature) and by the heat capacity

$$\nabla \cdot \mathbf{F} = -C\rho\frac{\partial T}{\partial t}$$

where C is the heat capacity per unit mass (often called the *specific heat capacity*), ρ is the density and t is time. Since in fact we are assuming that heat propagates only in the z-direction, these equations can be simplified to

$$F = -K\frac{\partial T}{\partial z} \tag{6.5}$$

and

$$\frac{\partial F}{\partial z} = -C\rho\frac{\partial T}{\partial t} \tag{6.6}$$

If we also assume that the time-dependence of the flux and the temperature is sinusoidal with angular frequency ω, and that the amplitude of the variation in flux at the surface is F_0, it can be shown that the flux F is given by

$$F = F_0 \cos\left(\omega t - z\sqrt{\frac{\omega C\rho}{2K}}\right) \exp\left(-z\sqrt{\frac{\omega C\rho}{2K}}\right) \tag{6.7}$$

and the temperature T is given by

$$T = \frac{F_0}{P\sqrt{\omega}} \cos\left(\omega t - z\sqrt{\frac{\omega C\rho}{2K}} - \frac{\pi}{4}\right) \exp\left(-z\sqrt{\frac{\omega C\rho}{2K}}\right) \tag{6.8}$$

where

$$P = \sqrt{C\rho K} \tag{6.9}$$

is the thermal inertia.

Examining equations (6.8) and (6.9), we see that both the heat flux F and the temperature T vary as exponentially decaying sinusoidal waves. The speed at which the waves travel down into the ground is

$$v = \sqrt{\frac{2K\omega}{C\rho}} \tag{6.10}$$

and the depth over which the amplitude of the waves decreases by a factor of e is

$$z_0 = \sqrt{\frac{2K}{\omega C\rho}} \tag{6.11}$$

We also observe that the ratio of the amplitude of the surface flux variations to the amplitude of the surface temperature variations is $P\sqrt{\omega}$, and that the temperature variations lag the flux variations by $\pi/4$ radians or one eighth of a cycle. Thus, we would expect, on the basis of this simple model, that the daily surface temperature variations should reach a maximum value at about 3 p.m. and a minimum value at about 3 a.m.

In fact, the simple model that we have developed is not directly applicable, because the incident flux does not usually vary sinusoidally, although this particular limitation can be overcome by using Fourier analysis. The calculation of the flux is a complicated one, which includes contributions from direct solar radiation and from the sky, reflection of solar radiation from the surface, emission of infrared radiation from the surface, and geothermal heat flux. It thus depends on the geographical location, time of year, cloud cover, orientation, albedo and emissivity of the surface, amongst other factors.[7] Nevertheless, our simple model gives a good indication of the general trends to be expected. Figure 6.15 shows typical variations of surface temperature for materials of two different thermal inertias. The material with larger thermal inertia exhibits smaller temperature fluctuations, as expected from our model. However, we can note from the figure that the fluctuations are not sinusoidal, and that the maximum temperatures occur at about 2 p.m. instead of 3 p.m., and the minimum temperatures occur just before dawn instead of at 3 a.m.

Figure 6.16 illustrates the typical values of the thermal inertia P and of the *thermal diffusivity* Γ, defined as

$$\Gamma = \frac{K}{C\rho} \tag{6.12}$$

for various materials. Geological materials have thermal inertias ranging from about 400 J m^{-2} s$^{-1/2}$ K^{-1} for light, low-conductivity materials such

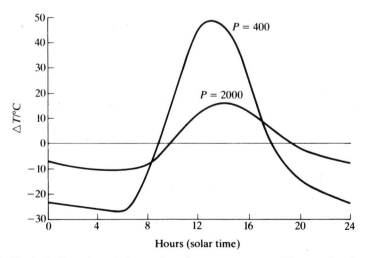

Figure 6.15. Typical diurnal variations of surface temperature. The graphs show ΔT, the surface temperature minus the diurnal mean temperature, plotted against local solar time, for materials with thermal inertias of 400 and 2000 J m^{-2} s$^{-1/2}$ K^{-1}.

---

[7] Elachi (1987) discusses these effects in greater detail.

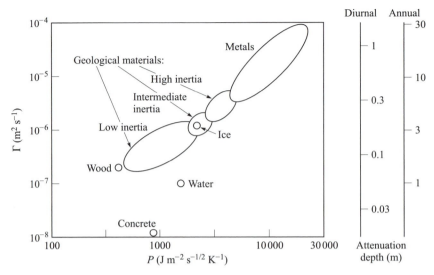

Figure 6.16. Typical values of the thermal diffusivity Γ and the thermal inertia *P* for various materials. The figure also shows the attenuation depth (defined by equation (6.11)) for diurnal and annual fluctuations.

as sand, shale and limestone, to about 4000 J m^{-2} s$^{-1/2}$ K^{-1} for dense, high-conductivity minerals such as quartz. The thermal inertia of metals is typically ten times higher, on account of their higher thermal conductivities and densities, and wood is about ten times lower. There is little contrast between the thermal inertia of water and that of typical minerals. These may nevertheless be distinguished, since a water surface will normally be cooler than a rock surface during the daytime, as a consequence of evaporation, and at night the water surface is correspondingly warmer. A similar effect operates in the case of damp ground, and in this way soil moisture can be assessed, with an accuracy of typically 15%. Dry vegetation may also be distinguished from bare ground at night, since the vegetation and the ground beneath it is physically warmer than the bare ground because of the insulating effect of the vegetation. Again, a converse effect operates in the daytime.

Thermal inertia mapping was carried out from space in 1978–1980, during the HCMM (heat capacity mapping mission). This satellite carried a thermal infrared radiometer (10.5–12.5 μm) called the HRIR (high-resolution infrared radiometer), which had an accuracy of 0.4 K. An example of HRIR imagery is shown in plate 3. The results from this mission indicated that thermal inertia mapping is particularly sensitive to the effects of tectonic disturbance and to lithological boundaries. However, distinguishing one rock type from another is still rather difficult.

Thermal inertia mapping also has an application to archaeological surveying. If one material is buried within another, and the materials have different thermal properties, the flow of heat will be distorted and give rise to a *temperature anomaly* at the surface. This will yield information on the nature of the buried object, but the interpretation is complicated by the fact that the simple one-dimensional problem that we have been describing is no longer applicable, and the thermal diffusion equation must be solved in three dimensions.

6.3.6   *Atmospheric correction of thermal infrared imagery*

As we have already remarked, a thermal infrared radiometer measures the brightness temperature of the radiation reaching the sensor, whereas the quantity that is usually required is the brightness temperature of the radiation leaving the surface. For spaceborne observations, where the full effect of the atmosphere is felt, the difference between these two temperatures can be as large as 10 K, mainly as a result of water vapour, and correction is generally necessary. In this section, we consider the principal techniques for making this correction. More details can be found in, for example, Cracknell and Hayes (1991).

Atmospheric correction of thermal infrared imagery is generally performed by one of three methods. The first of these is physical modelling, for example using the LOWTRAN or MODTRAN models introduced in section 6.2.6. Because the main correction is due to water vapour, which is highly variable both spatially and temporally, physical modelling is rather unsatisfactory unless a detailed characterisation of the atmosphere is available. This can be obtained by local radiosonde observations or by airborne or spaceborne atmospheric sounding (section 6.5).

The second method is the *split-window technique* (e.g. Deschamps and Phulpin, 1980; Závody et al., 1995). In this method, which is in widespread use, the brightness temperatures T_{b1} and T_{b2} are measured in two different but closely spaced spectral bands, for example one centred at 11 μm and one at 12 μm. The brightness temperature T_{b0} of the radiation leaving the surface is then modelled as a linear combination of the measured brightness temperatures:

$$T_{b0} = a_0 + a_1 T_{b1} + a_2 T_{b2} \qquad (6.13)$$

The coefficients a_0, a_1 and a_2 are determined empirically. Because of the contribution from reflected sky radiation, the coefficients have different values during the day and during the night. The technique is reasonably accurate (to within 0.5 K) if the emissivity of the surface is constant, so it is reliable for sea-surface temperature measurements, but over land surfaces its usefulness is more limited.

Finally, we should mention the *two-look technique*. In this approach, the instrument views each rezel twice, at two different incidence angles. For example, the ATSR (Along-Track Scanning Radiometer) instrument carried by the ERS-1 and ERS-2 satellites uses a conical scanning technique such that rezels are viewed both at nadir (i.e. vertically) and at an angle of 52° to the nadir. Since $1/\cos(52°) \approx 1.6$, the oblique path looks through about 1.6 times as much atmosphere as the vertical path and the atmospheric corrections are correspondingly larger. By comparing the two brightness temperatures, the atmospheric correction can be estimated and eliminated.

Figure 6.17 illustrates schematically the principle of two-look correction. We can derive a simplified model as follows. We will assume that the brightness temperature of the radiation leaving the surface is T_{b0}, independent of direction, and that reflection of atmospheric radiation is insignificant. The brightness temperatures of the radiation reaching the sensor at positions 1 and 2, T_{b1} and T_{b2} respectively, can be found from the radiative transfer equation (section 3.4.1). To simplify this equation we will assume that scattering of radiation is unimportant (this is a very good assumption), that the physical temperature of the absorbing material has a constant value of T_a, and that T_{b0} and T_a are sufficiently similar that the dependence of the function B_f on the temperature T (equation (3.68)) can be assumed to be linear. With these simplifying assumptions, and despite the fact that the Rayleigh–Jeans approximation is not valid here, the solution of the radiative transfer equation is given by equation (3.72). Thus, we may put, approximately,

$$T_{b1} = T_{b0} \exp(-\tau) + T_a(1 - \exp(-\tau)) \tag{6.14}$$

where τ is the optical thickness of the atmosphere for a ray propagating vertically through it. The corresponding equation for the oblique path is obtained by replacing τ by the optical thickness for this path. Provided that the angle θ is not too close to 90°, this is given by $\tau \sec(\theta)$, and so

$$T_{b2} = T_{b0} \exp(-\tau \sec \theta) + T_a(1 - \exp(-\tau \sec \theta)) \tag{6.15}$$

Treating T_{b1}, T_{b2} and T_a as known variables, these two equations can be solved to find τ and T_{b0}.

6.4   Some imaging instruments

As we have done in section 5.7 for photographic systems, we will attempt in this section to describe some of the main types of electro-optical (visible, near-infrared and thermal infrared) imaging systems. Here the task is a much larger

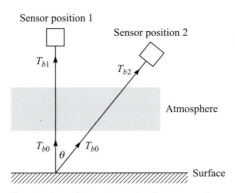

Figure 6.17. Two-look observation for atmospheric correction of a thermal infrared measurement.

one, as there is a far greater diversity of instruments. For this reason, we will concentrate on spaceborne systems. With the obvious exception of the very wide swath instruments, these will generally have airborne counterparts with spatial resolutions that are finer in proportion to the observing height.

We can approximately categorise spaceborne electro-optical imaging systems according to their spatial resolution. The highest spatial resolutions generally available from spaceborne imagers lie in the range 5 to 50 m, with swath widths typically between 30 and 200 km.[8] These imagers are intended mainly for land-surface studies, especially mapping, and sometimes also measurement of aerosol distributions over oceans. Thus, they generally concentrate on the optical and near-infrared parts of the spectrum, although some thermal IR imaging is often included. A good example of this type of instrument is the *Thematic Mapper* (TM) carried on board the Landsat-4 and Landsat-5 satellites. The TM is a mechanically-scanned imager, with seven wavebands as follows: 0.45–0.52 μm (blue); 0.52–0.60 μm (green); 0.60–0.69 μm (red); 0.76–0.90 μm (near-infrared); 1.55–1.75 μm (near-infrared, especially for discrimination between snow and cloud); 2.08–2.35 μm (mid-infrared); 10.4–12.5 μm (thermal infrared). From its operating altitude of 705 km it achieves a spatial coverage of 185 km × 185 km with a rezel size of 30 m × 30 m (120 m × 120 m for its thermal infrared band). Plate 4 shows a typical Landsat TM image, and figure 6.18 shows an enlarged extract of a single band of another image, to illustrate the spatial resolution.

Intermediate-resolution imagers (50 to 500 m) are generally similar to the high-resolution instruments, but with larger swath widths – typically several hundred kilometres. An example of this type of instrument is the *MSU-SK* carried on the Russian Resurs satellites. Like the TM, this is a mechanically scanned imager. It has five wavebands as follows: 0.50–0.60 μm (green); 0.60–0.70 μm (red); 0.70–0.80 μm (near-infrared); 0.80–1.10 μm (near-infrared); 10.4–12.6 μm (thermal infrared). From its operating altitude of 835 km it achieves a swath width of about 600 km and a rezel size of about 150 m.

Low-resolution (coarser than about 500 m) instruments show a greater diversity of application. General-purpose imagers have bands covering the optical, near-infrared and thermal infrared regions of the spectrum and swath widths typically between 500 and 3000 km. The disadvantage of comparatively coarse resolution is offset by the advantage of wide-swath imaging and the possibility of obtaining data on continental or ocean-wide scales. Such imagers have a very wide range of applications. For example, over land they can be used for measuring albedo (as an input to climate modelling), surface temperature, vegetation characteristics, and so on. Over oceans, applications include the monitoring of large-scale ocean circulation, sediment transport

---

[8] Instruments with even higher spatial resolutions are likely to appear in the non-military remote sensing domain in the early years of the twenty-first century. The successful launch of the Ikonos satellite in September 1999 is already providing black-and-white imagery with a resolution of 1 m and colour imagery with a resolution of 4 m.

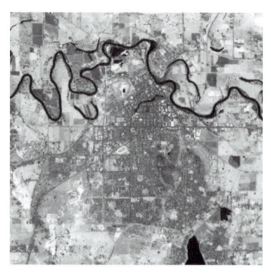

Figure 6.18. Extract (approximately 9.5 km × 9.5 km) of band 5 of a Landsat TM image, centred on the town of Wagga Wagga, Australia. (Reproduced by permission of the Australian Centre for Remote Sensing.)

from rivers, and sea-surface temperature. The inclusion of thermal infrared and optical bands means that this type of imager is also valuable for cloud monitoring. An example of such an instrument is the Advanced Very High Resolution Radiometer (*AVHRR*) carried by the NOAA satellites. This is a mechanically scanned imager with five wavebands: 0.58–0.68 μm (red); 0.73–1.10 μm (near-infrared); 3.55–3.93 μm (thermal infrared); 10.3–11.3 μm (thermal infrared); 11.5–12.5 μm (thermal infrared). From its operating altitude of 860 km it achieves a swath width of 2600 km and a rezel size of 1.1 km. To illustrate this spatial resolution, figure 6.19 shows an extract of a single band of an AVHRR image.

Similar to these general-purpose imagers, but with higher spectral resolution, are instruments intended principally for measuring ocean colour (see section 6.2.5). The bandwidths of these instruments are usually 20 nm [9] or less, and swath widths are generally in the range 1000 to 3000 km. An example of such an instrument is the Sea-Viewing wide field of view sensor, or *SeaWiFS*. This has eight wavebands, centred at 412, 443, 490, 510, 555, 670, 765 and 865 nm, each of width 20 nm except for the last two, which have widths of 40 nm. From its operating altitude of 705 km it achieves a spatial resolution of 1.1 km and a swath width of 1500 km.

Instruments whose primary purpose is to measure the temperature of the Earth's surface and of cloud-tops employ mainly thermal infrared bands,

---

[9] Note that, by contrast, multispectral imagers (e.g. Landsat TM) have bandwidths that are typically 100 nm. However instruments with significantly higher spectral resolution, for example 10 nm, have already been developed for airborne use and are likely to be deployed from space in the early years of the twenty-first century. Such instruments are generally known as *hyperspectral imagers* or *imaging spectrometers*.

Figure 6.19. Extract (approximately 560 km × 560 km) of band 2 of an AVHRR image, showing parts of the Northern Territory and the Gulf of Carpentaria, Australia. (Reproduced with permission of the Australian Centre for Remote Sensing.)

sometimes to the complete exclusion of visible-wavelength bands. An example of such an instrument is the Along-Track Scanning Radiometer (*ATSR*) carried on the ERS-1 satellite. This has wavebands at 1.6, 3.7, 11.0 and 12.0 μm, and from its operating altitude of 780 km it achieves a swath width of 500 km and a spatial resolution of 1.0 km.

Finally, we recall the spin-scan imaging technique described in section 6.2.2 and commonly employed for sensors carried on geostationary satellites. These sensors, which are used mainly for meteorological imaging, normally combine fairly broad-band coverage of the visible spectrum with thermal infrared channels. A typical example is the Visible and Infrared Spin-Scan Radiometer (*VISSR*) carried by the Japanese GMS (Geostationary Meteorological Satellite). This has wavebands at 0.5–0.75 and 10.5–12.5 μm. Since it is in geostationary orbit its view of the Earth's surface is fixed, and it can see approximately one quarter of the Earth's disc, which it scans in about 30 min. It achieves a spatial resolution of 1.25 km (visible) and 5 km (infrared) for rezels near nadir. Figure 6.20 shows an extract of the infrared band of a GMS VISSR image.

Figure 6.20. Extract (approximately 640 km × 640 km) of the visible band of a GMS VISSR image, showing a storm over the Yellow Sea. (Reproduced by permission of the Australian Board of Meteorology and the Japan Meteorological Agency.)

It is instructive to plot the spatial resolutions of typical spaceborne sensors against their swath widths, as in figure 6.21. The figure shows the strong correlation between spatial resolution and swath width, such that most systems image swaths having between 1000 and 10 000 pixels. For comparison, the figure also shows the typical performance of photographic systems, which have an equivalent number of pixels of between 3000 and 30 000. Assuming that the images generated by these systems are approximately square, the total data volume is calculated by squaring the number of pixels across the swath width, and then multiplying by the number of wavebands that are recorded. We can thus see that the typical data volume for a spaceborne electro-optical imaging system is a few megabytes to a few hundred megabytes. For a photograph, the equivalent data volume can be a few gigabytes.

6.5 Atmospheric sounding

Up to this point, this chapter has discussed systems for imaging the Earth's surface (or, where the surface is covered by cloud, for imaging the cloud itself).

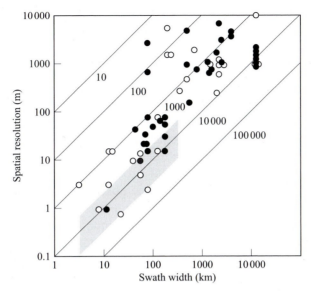

Figure 6.21. Spatial resolution plotted against swath width for the principal existing (filled circles) and planned (open circles) spaceborne electro-optical imaging systems. The diagonal lines show the number of pixels in a single row of data. The shaded area shows roughly the corresponding performance of airborne and spaceborne photographic systems.

Thus, the wavebands chosen for such imaging systems are within the atmospheric 'windows' discussed in chapter 4. However, by choosing to make observations at wavelengths at which the atmospheric attenuation coefficient is significant, information can be obtained about the composition of the atmosphere itself. The type of information that is required is the variation of temperature with altitude, and the variation of the density of atmospheric gases and aerosols with altitude. The most important of the gases are oxygen, because this is a good indicator of the total atmospheric pressure, and water vapour, because of its meteorological significance and also because of its importance for atmospheric correction of thermal infrared measurements (see section 6.2.6). However, many other gases are also routinely profiled by remote sensing methods, notably ozone and other radiatively active molecules.

Atmospheric sounding techniques exploit all three of the phenomena that were introduced in our discussion of the radiative transfer equation (section 3.4.1), namely absorption, scattering and thermal emission. Most observations are made in the thermal infrared or microwave bands (the latter will be discussed in chapter 7), although the optical and ultraviolet bands are used for some scattering measurements.

6.5.1 *Temperature profiling from observations at nadir*

In atmospheric temperature profiling observations, the significant terms in the radiative transfer equation are absorption and thermal emission – scattering can be neglected. The principle of nadir (vertical) atmospheric temperature profiling can be stated in words something like this. If a sensor views vertically downwards into the atmosphere at a wavelength at which the atmosphere is optically thick, the brightness temperature of the radiation that is received will be characteristic of the atmosphere at a depth below the sensor that is of the order of the absorption length. Thus, the greater the absorption coefficient, the smaller the absorption length, and hence the greater the altitude from which the temperature signal is received. Hence, by making observations at a number of wavelengths near a broad absorption line, different altitudes in the atmosphere can be investigated. Thermal infrared temperature profilers normally employ the broad and deep carbon dioxide absorption feature near $15\,\mu$m.

The foregoing explanation is of course oversimplified. The concept of absorption length is not strictly applicable since the absorption coefficient varies with altitude, and furthermore the brightness temperature received at the sensor is not characteristic of a single altitude but rather of a range of altitudes. Nevertheless, the essential principle is valid: the closer the observation is made to the centre wavelength of an absorption line, and hence the larger the optical thickness of the atmosphere, the greater will be the mean altitude from which the signal is obtained. We can dispose of the problem that the observed temperature will be derived from a range of altitudes by introducing a weighting function $w(h')$, where h' is the altitude of a layer of the atmosphere and the observed brightness temperature at altitude h is given by

$$T_{obs} = \int_0^\infty T(h')w(h')\,dh' \qquad (6.16)$$

Our task is to determine this weighting function $w(h')$.

Since only absorption and thermal emission are significant, the appropriate form of the radiative transfer equation is given by equation (3.71):

$$\frac{dL_f}{dz} = \gamma_a(B_f - L_f)$$

As discussed in section 3.4.1, the solution of this equation can be written as

$$L_f(\tau) = L_f(0)\exp(-\tau) + \int_0^\tau B_f(\tau')\exp(\tau' - \tau)\,dr' \qquad (3.73)$$

where τ and τ' are optical thicknesses measured from the Earth's surface to some point in the atmosphere. In fact, it will be much simpler to recast equation (6.16) in terms of optical thicknesses, so that

$$T_{obs} = \int_0^\tau T(\tau')w(\tau')\,d\tau' \qquad (6.17)$$

where τ is monotonically related to the altitude h' by

$$\tau' = \int_0^{h'} \gamma_a(h'') \, dh'' \tag{6.18}$$

With this substitution, equation (3.73) is almost in the form we need. We can simplify it greatly by making the same approximation that we used in section 6.3.6, namely that the physical temperatures in the atmosphere do not differ greatly from the brightness temperature $T_{b,0}$ of the radiation leaving the Earth's surface. In this case, we can make a Taylor expansion of equation (3.73) to obtain

$$T_{\text{obs}} \approx T_{b,0} \exp(-\tau) + \int_0^\tau T(\tau') \exp(\tau' - \tau) \, d\tau' \tag{6.19}$$

where τ is the optical thickness of the entire atmosphere. We should note, first, that this equation has exactly the same form as equation (3.74), and, second, that the approximation that we have used here in deriving equation (6.19) is not necessary – we have just used it to illustrate the way in which weighting functions are calculated. With this approximation, we see that the appropriate weighting function is just $\exp(\tau' - \tau)$, where the argument of the exponential is minus the optical depth below the sensor.

In order to estimate roughly the form of the altitude-based weighting function $w(h)$, we will make two more approximations. The first is that the absorption coefficient is proportional to the atmospheric pressure, and the second is that the pressure[10] varies with height according to the negative exponential form introduced in equation (4.4). With these assumptions, the relationship between τ' and h' becomes

$$\tau' = \tau\left(1 - \exp\left(\frac{-h'}{H}\right)\right) \tag{6.20}$$

where H is the atmospheric scale height, and the altitude-based weighting function is then given by

$$w(h) = \frac{\tau}{H}\exp\left(-\frac{h}{H} - \tau \exp\left(-\frac{h}{H}\right)\right) \tag{6.21}$$

Figure 6.22 illustrates weighting functions $w(h')$ calculated from equation (6.21) using a scale height H of 8 km. The principle of atmospheric temperature sounding is thus to make nadir observations of the brightness temperature at a number of wavelengths corresponding to different values of the total optical thickness τ, giving a family of weighting functions similar to those shown in figure 6.22. Inversion of the data allows the temperature profile to be retrieved. We can note from figure 6.22, first, the obvious point that if τ is small, there

---

[10] In fact, the absorption coefficient depends on the temperature as well, for the reasons discussed in section 4.2, so the procedure for solving the temperature profile is not quite as simple as outlined here. An iterative procedure must be used.

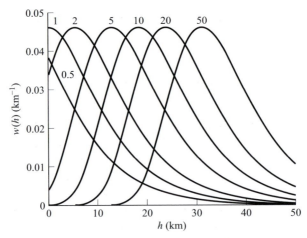

Figure 6.22. Weighting functions $w(h)$ calculated for profiling the atmospheric temperature distribution, using the simplified equation (6.21). The curves are labelled with values of τ, the total optical thickness of the atmosphere.

will be a significant contribution from the Earth's surface as well as from the atmosphere, and, second, that the vertical resolution of the technique is rather poor, being of the order of 10 km.

We have derived our simplified equation (6.21) on the assumption that the distribution of gas pressure with altitude is known. In fact, it is more usual to use the total atmospheric pressure as an altitude-like variable, rather than the altitude itself. The total pressure at altitude h is a measure of the number of molecules above the altitude h, as shown in equation (4.3), so provided that the absorbing gas (e.g. carbon dioxide) is well mixed in the atmosphere, the total pressure is a measure of the optical thickness of the atmosphere above h.

6.5.2   *Profiling of gas concentrations at nadir*

The essence of atmospheric temperature profiling is the assumption that the vertical distribution of the concentration of some absorbing gas, for example carbon dioxide, is known, and that therefore the vertical distribution of the absorption coefficient is also known. From this known distribution, the radiative transfer equation is solved to determine the temperature distribution. The essence of profiling gas concentrations is the converse of this procedure: if the temperature distribution is known, we can make a number of observations of the brightness temperature at wavelengths near the absorption line of a particular molecular species to determine the profile of the absorption coefficient, and hence concentration, of that species. As for temperature profiling, it is necessary to use an iterative procedure to solve for the concentration profile, since the absorption coefficient is dependent on temperature. Since most molecular spectral lines are much narrower than the $15\,\mu$m CO_2 feature commonly used for temperature profiling, high spectral resolutions are generally required for this technique. The methods by which they are achieved are discussed in section 6.5.5.

6.5.3 *Backscatter observations at nadir*

Nadir-looking observations can also be used to measure solar radiation that is scattered out of the atmosphere. From our discussion in section 4.2.2 (equation (4.12)), we expect the scattering coefficient to increase rapidly towards the short-wavelength (blue and ultraviolet) end of the spectrum, so it is in this region that backscatter measurements are made. Thermal emission can be neglected in this part of the electromagnetic spectrum, so it is only necessary to consider the scattering and absorption terms in the radiative transfer equation. Backscatter observations are used mainly for profiling of atmospheric ozone distributions.

6.5.4 *Limb-sounding observations*

We noted in section 6.5.1 that a nadir-sounding observation has a comparatively poor vertical resolution. Significantly higher resolutions can be achieved by limb-sounding, in which the sensor views in a direction that is almost tangential to the Earth's surface. Limb-sounding observations can be used to measure both absorption and emission. To illustrate the principle, we will derive a simple model of an absorption measurement.

Figure 6.23 shows the geometry of a limb-sounding absorption measurement. The sensor views a source of radiation (e.g. the Sun) in such a direction that the closest distance between the line of sight and the Earth's surface is h_0. We will measure distance x along the line of sight from this point, and write $h(x)$ for the altitude of the line of sight above the Earth's surface at x. Assuming that the sensor is located well outside the atmosphere, the optical thickness traversed by the ray is

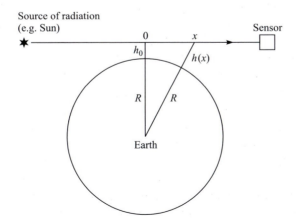

Figure 6.23. Geometry of a limb-sounding observation (schematic).

$$\tau = \int\limits_{-\infty}^{\infty} \gamma_a(h(x)) \; dx$$

where $\gamma_a(h(x))$ is the absorption coefficient at altitude $h(x)$. Provided that $h(x)$ remains small compared with the Earth's radius R, we can put

$$h(x) \approx h_0 + \frac{x^2}{2R} \tag{6.22}$$

and hence rewrite the equation for the optical thickness in terms of h:

$$\tau \approx \sqrt{2R} \int\limits_{h_0}^{\infty} \frac{\gamma_a(h)}{\sqrt{h - h_0}} \; dh \tag{6.23}$$

Thus, the appropriate weighting function in this case (the contribution to the total optical thickness from the altitude h) is proportional to $(h - h_0)^{-1/2}$ for $h > h_0$, and zero for $h > h_0$. Figure 6.24 illustrates this function.

In practice, the weighting function for a limb-sounding observation will not be quite as sharply peaked as implied by figure 6.24. This is because the finite angular resolution of the sensor will blur and broaden the response somewhat. As an example, we can consider a sensor at an altitude of 800 km arranged to observe the altitude $h_0 = 0$. In this case, the distance from the sensor to the tangent point is approximately 3300 km, so an instrument having an angular resolution of 0.1 milliradian[11] would introduce a vertical blurring of 0.33 km in the vertical response.

The improved vertical resolution of a limb-sounding observation, when compared with a nadir observation, is bought at the price of degraded horizontal resolution. This can be seen by considering an observation with a

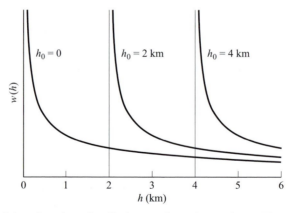

Figure 6.24. Weighting functions for limb-sounding observations. The curves are labelled with values of h_0, the minimum altitude of the ray.

---

[11] For comparison, an angular resolution of 0.1 milliradian is equivalent to a spatial resolution of 80 m in a nadir observation from an altitude of 800 km.

vertical resolution of Δh, so that the weighting function is significantly greater than zero for h between h_0 and $h_0 + \Delta h$. Equation (6.22) thus shows that the range of values of x corresponding to this range of altitudes is from $-(2R\,\Delta h)^{1/2}$ to $+(2R\,\Delta h)^{1/2}$. Taking a typical value of 1 km for Δh and setting $R \approx 6400$ km, we see that the sample volume is approximately 230 km long in the direction of the line-of-sight.

In order to illustrate the principle of limb-sounding observations, we have considered the case of an absorption measurement. As shown in figure 6.23, this requires that there is a source of radiation external to the atmosphere. Atmospheric limb-sounders commonly use the Sun as this source of radiation, in which case the observation is often referred to as a *solar occultation* observation. Since a satellite in low Earth orbit makes approximately 14 orbits of the Earth per day (see chapter 10), there are approximately 28 measurement opportunities per day (14 sunrises and 14 sunsets). The same is true for lunar occultation observations. However, stellar occultations (observations of stars near the Earth's limb) can be made from virtually anywhere on the Earth's night side. Thermal emission measurements can be made from any location.

6.5.5   *Spectral resolution for atmospheric sounding observations*

We noted in section 6.5.2 that high spectral resolutions are generally needed for atmospheric sounding measurements (we are attempting to resolve the fine spectral structure that could not adequately be shown in figure 4.5a). At thermal infrared wavelengths, the use of filters can achieve a spectral resolution of the order of 0.1 μm, and grating spectrometers can improve this by a factor of 10. However, resolutions of 0.1 nm or better may be required. One common way of achieving this degree of spectral resolution is through *Fourier-transform spectrometry*, in which a Michelson interferometer is used to resolve the fine spectral components present in prefiltered radiation. Figure 6.25 shows the principle of a Michelson interferometer. A parallel beam of radiation is incident on a beam-splitter at 45° to the beam. This sends half the amplitude of the

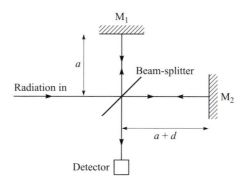

Figure 6.25.   Principle of a Michelson interferometer.

radiation towards mirror M_1 and half towards mirror M_2. The distance from the beam-splitter to M_1 is fixed and equal to a, but the distance to M_2 is equal to $a + d$, where d can be varied. After reflection from the mirrors, the rays are recombined and detected. It is clear that the radiation that has been reflected from M_2 has travelled a distance that is greater by $2d$ than the radiation that has been reflected from M_1. If the incident radiation consists of a single spectral component with intensity I_0 and wavenumber k, the intensity at the detector will be given by

$$I = \frac{I_0}{2}(1 + \cos 2kd)$$

Thus, as d is varied, the output of the device will exhibit interference fringes as shown in figure 6.26.

If we now add a second component to the incident radiation, with a different intensity and wavenumber, a second fringe system will be added to the output and the result will look something like figure 6.27. We can see that the fringes have now been modulated so that, in general, the minimum intensity of a fringe is no longer zero.

We define the *visibility* V of a fringe by

$$V = \frac{I_{\max} - I_{\min}}{I_{\max} + I_{\min}} \tag{6.24}$$

where $I_{\max}$ and $I_{\min}$ are, respectively, the maximum and minimum detected intensities. For the example shown in figure 6.27, the graph of V against d is as shown in figure 6.28. The visibility function $V(d)$ contains all the information about the spectral structure of the input radiation. If the intensity contained in an infinitesimal interval dk of wavenumber is $I(k)\,dk$, and we set $k' = k - k_0$, where k_0 is the mean wavenumber of the input radiation, it can be shown[12] that

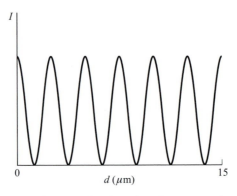

$d\,(\mu m)$

Figure 6.26. Output intensity from a Michelson interferometer when the input radiation consists of a single spectral component (in this example, the wavelength is $5\,\mu m$).

---

[12] See for example Lipson et al. (1995).

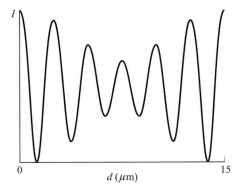

Figure 6.27. Output intensity from a Michelson interferometer when the input consists of two spectral components with different intensities.

$$V(d) = \frac{\int\limits_{-\infty}^{\infty} I(k') \exp(-2ik'd) \, dk'}{\int\limits_{-\infty}^{\infty} I(k') \, dk'} \tag{6.25}$$

This is a Fourier transform, so the function $I(k')$ can be retrieved from $V(d)$ by performing the inverse Fourier transform (see section 2.3).[13]

Thus, the procedure for determining the spectral structure of the input radiation is to scan the mirror M_2 in d and monitor the variations in output intensity. From these variations the visibility function can be obtained and then inverted to deduce $I(k)$. The maximum spectral resolution that can be obtained by this method depends on the largest value of d that can be achieved. If this value is D, the spectral resolution in terms of wavenumber is of the order

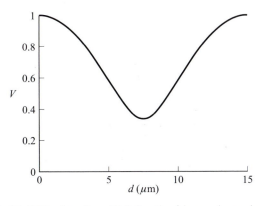

Figure 6.28. Visibility function $V(d)$ for the fringes shown in figure 6.27.

[13] Equation (6.25) implies that V can be complex. The definition of equation (6.24) therefore corresponds to the amplitude of the complex visibility. The phase is determined from the phase shift of the fringes.

of $2\pi/D$. For example, if $D = 0.1\,\text{m}$, the wavenumber resolution Δk will be about $60\,\text{m}^{-1}$. This corresponds to a wavelength resolution of about $2\,\text{nm}$ at a wavelength of $15\,\mu\text{m}$, and $0.01\,\text{nm}$ at $1\,\mu\text{m}$.

Before leaving the topic of spectral resolution, we will briefly describe an alternative technique. This depends on having an absorbing filter whose spectral absorption characteristics exactly match the spectral structure of the feature to be detected. It is clear that, by comparing the output obtained when this filter is present with the output when it is absent, a difference signal will be obtained that corresponds precisely to the intensity of the desired spectral feature. How can such a filter be constructed? The answer is beautifully simple. If we wish to detect radiation corresponding to a particular gas at a particular pressure, the filter consists of a transparent-walled cell containing the same gas at the desired pressure. By varying the pressure in the cell, different gas pressures can be detected. An instrument employing this technique is usually called a *pressure-modulated radiometer*, and it can achieve, in effect, remarkably high spectral resolutions (Δk down to $10^{-4}\,\text{m}^{-1}$). Chen (1985) and Elachi (1987) provide more information on this technique.

6.5.6   Some instruments

To conclude this section, we will describe a few spaceborne atmospheric profiling instruments that are representative of the main observing techniques.

As an example of a nadir temperature profiler, we can consider the AIRS (Atmospheric Infrared Sounder) instrument that will be included as part of the payload on the EOS-PM satellites. The principal waveband of this instrument is in the thermal infrared,[14] with coverage from $3.74\,\mu\text{m}$ to $15.4\,\mu\text{m}$. It uses a grating spectrometer to achieve a spectral resolution that varies from about 3 nm at the short-wavelength end to about 13 nm at the long-wavelength end, giving a total of 2300 spectral channels. In fact, the AIRS instrument can be scanned up to $49°$ either side of nadir, which will improve its ability to resolve vertical temperature structure in a manner similar to the discussion of two-look thermal infrared observations presented in section 6.3.6.

The IMG (Interferometric Monitor for Greenhouse Gases), carried on the ADEOS satellite, is an example of a nadir-looking instrument for profiling concentrations of atmospheric gases, particularly carbon dioxide, methane and nitrous oxide. As its name suggests, this instrument uses a Michelson interferometer. The waveband is from 3.3 to $14.0\,\mu\text{m}$, and the spectral resolution in wavenumber is $\Delta k = 60\,\text{m}^{-1}$, giving a wavelength resolution of 1 nm at a wavelength of $10\,\mu\text{m}$. The horizontal resolution of the instrument is about 8 km, but the 10-s scan time of the interferometer effectively degrades this to about 70 km. The vertical resolution is typically 2 km to 6 km, depending on the molecular species.

---

[14] It also has six broad-band channels spanning the wavelength range from $0.4\,\mu\text{m}$ to $1.7\,\mu\text{m}$.

The SBUV (Solar Backscatter Ultraviolet Radiometer) carried on the NOAA satellites is an example of a nadir-viewing instrument for profiling ozone concentrations by means of backscattered ultraviolet radiation. It has 12 comparatively broad wavebands, defined by filters, spanning the wavelength range from $0.16\,\mu m$ to $0.40\,\mu m$, and also views the Sun directly in order to obtain calibration data. The horizontal resolution of the instrument is 170 km. Ozone concentrations can be determined with an accuracy of about 5%, and the total (column-integrated) ozone concentration with an accuracy of about 1%.

Finally, we consider an example of a limb-sounding instrument. This is GOMOS (Global Ozone Monitoring by Occultation of Stars), which will be included on the Envisat satellite. Despite its name, GOMOS is intended to be used for profiling other things as well as ozone, notably atmospheric temperature, aerosols and water vapour. It has three wavebands: $0.25\,\mu m$ to $0.68\,\mu m$, $0.756\,\mu m$ to $0.773\,\mu m$ and $0.926\,\mu m$ to $0.952\,\mu m$. Spectral resolution is achieved using a diffraction grating, the dispersed spectra being projected onto a CCD. The spectral resolution is 1.2 nm for the ultraviolet-to-optical band, and 0.18 nm for the two near-infrared bands. The vertical resolution is 1.7 km.

PROBLEMS

1. The cross-section of a glass prism is an equilateral triangle. The refractive index of the glass is 1.601 at a free-space wavelength of $0.4\,\mu m$ and 1.569 at $0.7\,\mu m$. Show that if white light is incident on the prism at an angle of 40° to the surface normal, the spectrum from $0.4\,\mu m$ to $0.7\,\mu m$ will be dispersed over an angular width of 4.9°.

2. Sunlight is incident on a rough surface at an angle of 45° to the normal. Calculate the brightness temperature of the surface at wavelengths of $4\,\mu m$ and $10\,\mu m$, assuming that the surface is a perfect Lambertian scatterer (i.e. one for which the diffuse albedo is 1) at both wavelengths. Ignore atmospheric propagation effects.

3. Beginning with the equations (6.5) and (6.6) that describe one-dimensional heat flow in a solid, and the assumption that all time variations are sinusoidal with angular frequency ω, prove equations (6.7) and (6.8).

4. Thermal infrared measurements of the sea indicate a brightness temperature of 294.6 K at nadir and 293.3 K at an angle of 52° from nadir. Use the simple model developed as equations (6.14) and (6.15) to estimate the brightness temperature of the sea surface and the optical thickness of the atmosphere at nadir, assuming that the atmosphere has a physical temperature of 280 K.

5. Assume that the atmospheric absorption coefficient γ varies with height h as

$$\gamma = \gamma_0 \exp(-\beta h)$$

and use the simple model of limb-sounding geometry derived in section 6.5.4
to show that the reduction in intensity I is given by

$$\Delta \ln I = -\gamma_0 \, \exp(-\beta h_0) \sqrt{\frac{2\pi R}{\beta}}$$

where h_0 is the height at which the radiation path passes closest to the Earth's
surface. For atmospheric aerosols under clear-sky conditions, $1/\beta \approx 3 \, \text{km}$ and
$1/\gamma_0 \approx 14 \, \text{km}$ for visible radiation. Estimate the value of h_0 that will reduce the
intensity of a transmitted ray by a factor of 10.

Passive microwave systems

7.1 Introduction

In chapters 5 and 6 we considered passive remote sensing systems in which the diffraction resolution limit λ/D, while important, was not usually a critical parameter of the operation. In this chapter, we consider our last major class of passive remote sensing system, the passive microwave radiometer. This is a device that measures thermally generated radiation in the microwave (usually 5–100 GHz [1]) region. As we discussed in section 2.6, the long 'tail' to the Planck distribution at relatively low frequencies means that measurable amounts of radiation are emitted even in this range of frequencies.

Because microwave wavelengths are so much greater than those of visible or even of thermal infrared radiation, the resolution limit plays a much more important role, and we shall need to give careful attention to the factors that determine it. The treatment that follows in this chapter is similar to that of Robinson (1994), and is expanded upon by Ulaby et al. (1981, 1982, 1986). Much of the technology and nomenclature of passive microwave radiometry was originally developed in the field of radio astronomy, and further details can also be found in works on that subject.

7.2 Antenna theory

7.2.1 Angular response and spatial resolution

As we have remarked before, electromagnetic radiation is detected through its influence on electrons, which are excited to higher energy states by the incident photons. The energy of a microwave photon is typically only a few microelectron-volts, which is too small to excite an electron across an atomic or molecular band-gap. For this reason, electrical conductors (metals) are used. The incident electromagnetic wave induces a fluctuating current in the conductor, and this current can subsequently be amplified and detected. The *antenna* is a

---

[1] Frequencies much below 1 GHz are unsuitable because of the large signal contributed by the Galaxy, as well as the difficulty of achieving adequate spatial resolution.

structure that serves as a transition between the wave propagating in free space, and the fluctuating voltages in the circuit to which it is connected.

The usual form of a microwave antenna is a paraboloidal dish, although many other designs are possible. Figure 7.1 shows schematically the design of a simple microwave radiometer using such an antenna. The design of an antenna is dominated by two considerations: (1) the need to achieve a high sensitivity in the desired direction; (2) the need to achieve a high angular resolution (narrow beamwidth). Both of these requirements are met by making the antenna as large as possible. Let us consider a perfect antenna, in which there are no power losses. It has a characteristic *radiation resistance* R_r, a fictitious quantity that can most easily be understood in the case of a transmitting antenna. In this case, if an alternating current I is fed into the antenna at the appropriate frequency, the antenna will radiate a mean power $\langle I^2 \rangle R_r$ where $\langle I^2 \rangle$ is the mean square current. As far as the part of the circuit that feeds the antenna is concerned, the antenna 'looks' like a resistance R_r, in the sense that power is dissipated in it.

If the antenna (now in receiving mode again) is directed at a large, distant region that is emitting microwave energy, a voltage signal will appear at the output of the antenna. If the emission mechanism is thermal, this signal will be noise-like and will have the same characteristics as the *thermal noise* generated in a resistance R_r held at a temperature T_A. Thermal noise is in essence caused by the Brownian motion of electrons in the resistance, and is often referred to as Johnson noise or Nyquist noise. It can be shown that the power contained in a frequency interval Δf of Johnson noise is given by

$$P_N = kT_A \Delta f \tag{7.1}$$

where k is Boltzmann's constant. (Note that equation (7.1) implies that the power per unit frequency interval is constant.) The equivalent temperature T_A of the radiation resistance is then called the *antenna temperature*. If the distant emitting region is large enough, and has the radiation characteristics of a black body at a physical (absolute) temperature T, the antenna temperature is equal to T.

From the preceding discussion, we can provisionally describe the operation of a passive microwave radiometer something like this: the antenna is directed

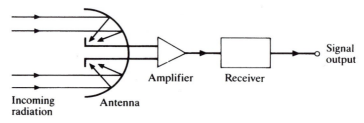

Figure 7.1. Schematic construction of a passive microwave radiometer. The antenna collects the incident radiation and generates a fluctuating voltage, which can then be amplified, detected and processed.

at some 'target', as a consequence of which noise-like voltage signal is generated in it. From equation (7.1), we can calculate the antenna temperature T_A, which will be some kind of weighted average of the brightness temperature viewed by the antenna. Clearly, our next task is to determine the nature of this weighted average.

The ideal antenna would receive radiation uniformly over a very small range of solid angle. To describe a real antenna, we introduce the concept of the *power pattern* $P(\theta, \phi)$. This is the power that would be detected by the antenna from a point transmitter of fixed strength, located in the direction (θ, ϕ) with respect to the antenna axis, at a fixed distance much greater than the Fresnel distance (see section 2.7). The power pattern is normalised so that neither the strength of the transmitter nor its distance from the antenna appear in its definition:

$$P_n(\theta, \phi) = \frac{P(\theta, \phi)}{P_{\max}(\theta, \phi)} \tag{7.2}$$

Figure 7.2 illustrates a typical power pattern (or, at least, its θ variation). It has a *main lobe* of maximum sensitivity, usually centred on the direction $(0, 0)$, and a number of undesirable *sidelobes*. The usual measure of the width of the main lobe is its half-power beam width (HPBW), defined as the angular width of the region in which $P_n \geq 1/2$. For antennas with large apertures (in comparison with the wavelength λ of the detected radiation), the power pattern is calculated by Fourier transform methods, since it is effectively the square of the diffraction pattern of the aperture distribution. For antennas that are smaller than a few wavelengths in any direction perpendicular to that of the incident radiation, the calculation is more complicated and requires the application of electrodynamics. Table 7.1 lists some of the properties of the power patterns of some common designs of antenna. The normal measure of the strength of the

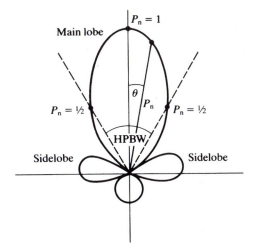

Figure 7.2. Power pattern of a typical antenna, plotted in polar coordinates.

sidelobes is the peak value of P_n, expressed in decibels. Thus, it may be said that a well-designed antenna will have sidelobe levels of -20 dB or lower.

From our definition of the power pattern, it follows that the antenna temperature will be given by

$$T_A = \frac{\int\limits_{4\pi} T_b(\theta, \phi) P_n(\theta, \phi)\, d\Omega}{\int\limits_{4\pi} P_n(\theta, \phi)\, d\Omega}$$

where $T_b(\theta, \phi)$ is the brightness temperature of the 'target' in the direction (θ, ϕ), and the integrals are carried out over all directions (i.e. over a range of Ω of 4π steradians). In fact, it is convenient to use the *beam solid angle*, defined as

$$\Omega_A = \int\limits_{4\pi} P_n(\theta, \phi)\, d\Omega \qquad (7.3)$$

so the formula for the antenna temperature becomes

$$T_A = \frac{\int\limits_{4\pi} T_b(\theta, \phi) P_n(\theta, \phi)\, d\Omega}{\Omega_A} \qquad (7.4)$$

Related to the concept of the beam solid angle is the *directivity D*, defined as

$$D = \frac{4\pi}{\Omega_A} \qquad (7.5)$$

This is a measure of the sensitivity of the antenna in its most sensitive direction. For a monopole antenna, which is completely isotropic, the beam solid angle is

Table 7.1. Properties of some common types of microwave antenna

The table shows the HPBW measured in two orthogonal directions ('iso' means isotropic; i.e. no variation of P_n in this plane), the directivity D and the maximum sidelobe level for some of the most important designs of antenna. For the last two types, it is assumed that the physical dimensions a, b and d are much greater than the wavelength λ of the radiation

Antenna type	HPBW (degrees)		D (dB)	Sidelobes (dB)	Notes
Monopole	(iso)	(iso)	0	(none)	
Short dipole	90	(iso)	1.76	(none)	
Half-wave Dipole	90	(iso)	2.15	(none)	
Six-element Yagi	42	(iso)	10.5	-10	TV-antenna type
Rectangular	$51(\lambda/a)$	$51(\lambda/b)$	$11 + 10\log_{10}(ab/\lambda^2)$	-13	a, b are sides of rectangle
Circular paraboloid	$72(\lambda/d)$	$72(\lambda/d)$	$\approx 10 + 20\log_{10}(d/\lambda)$	-25	d is diameter

4π steradians and the directivity is therefore 1. The directivity, which is often specified in decibels, is therefore a measure of maximum sensitivity relative to an isotropic antenna. Typical values of the directivity are shown in table 7.1.

Up to this point, we have been assuming that the antenna is loss-free. In a real antenna, resistive (ohmic) losses will reduce the detected power. These losses can be represented by specifying an efficiency η, which is the ratio of the detected power to the received power, or equivalently by specifying the *forward gain* G of the antenna, where

$$G = \eta D \tag{7.6}$$

It was mentioned earlier that the power pattern $P_n(\theta, \phi)$, and hence the beam solid angle Ω_A, can be determined from the dimensions of the antenna, either by Fourier transform methods or, for small antennas, by electrodynamic calculations. However, there is an alternative method of deriving Ω_A, and this introduces an important new concept – the *effective area* of the antenna.

From our discussion in section 2.6 we know that, provided the Rayleigh–Jeans approximation is valid, the spectral radiance corresponding to a brightness temperature T_b is

$$L_f = \frac{2kT_b}{\lambda^2} \tag{7.7}$$

where k is Boltzmann's constant and λ is the wavelength of the radiation. Thus, if we have a uniform source of brightness temperature T_b that subtends a small solid angle $\Delta\Omega$, the spectral flux density reaching the antenna is

$$F_f = \frac{2kT_b\,\Delta\Omega}{\lambda^2} \tag{7.8}$$

Next, we define the effective area A_e of the antenna such that, for radiation of spectral flux density F_f incident along its direction of maximum sensitivity, the power collected per unit frequency interval is $F_f A_e/2$. (The factor of $1/2$ arises because the radiation is assumed to be randomly polarised, and the antenna is assumed to receive only one polarisation state). If we now allow for the variation in the antenna's sensitivity that is expressed by the power pattern, and the possibility that the brightness temperature of the incident radiation can vary with direction, the power collected per unit frequency interval must be given by

$$\frac{kA_e}{\lambda^2} \int_{4\pi} T_b(\theta, \phi) P_n(\theta, \phi)\, d\Omega$$

From equation (7.1), it therefore follows that the antenna temperature must be given by

$$T_A = \frac{A_e}{\lambda^2} \int_{4\pi} T_b(\theta, \phi) P_n(\theta, \phi)\, d\Omega$$

and we can equate this to equation (7.4) to derive the following remarkably simple result:

$$\Omega_A A_e = \lambda^2 \tag{7.9}$$

This shows that an antenna with a large effective area will have a small beam solid angle, and conversely. For an antenna that is large compared with the wavelength λ, the effective area A_e will be approximately equal to the geometrical area of the antenna, and in this case equation (7.9) is obviously plausible. For example, if the antenna is a circular dish of diameter d, we know from the discussion of diffraction in section 2.7 that the main lobe of the power pattern has an angular radius of $1.22\lambda/d$. The beam solid angle is thus of the order of $(\lambda/d)^2$, and since the effective area is clearly of the order of d^2, we can see that equation (7.9) is at least plausible in this case.

We can discuss the implications of the foregoing theoretical development by considering a simple example. We will suppose that the antenna views a circular 'target' of brightness temperature T_1 against a background of brightness temperature T_0. To simplify equation (7.4), we will assume that the power pattern of the antenna is 1 for all values of θ up to some maximum value, and zero above this value, so that the antenna responds uniformly to a cone of directions subtending a solid angle of Ω_A, and does not respond at all to radiation arriving outside this cone. Figure 7.3 shows two situations: in figure 7.3a the target, which we assume subtends a solid angle of Ω_t, is not resolved, whereas in figure 7.3b it is fully resolved as a result of the antenna's smaller beam solid angle. From equation (7.4), we see that in the first (unresolved) case, the antenna temperature will be given by

$$T_A = \frac{(\Omega_A - \Omega_t)T_0 + \Omega_t T_1}{\Omega_A} = \left(1 - \frac{\Omega_t}{\Omega_A}\right)T_0 + \frac{\Omega_t}{\Omega_A}T_1$$

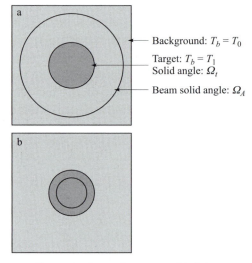

Figure 7.3. A microwave radiometer views of a target of brightness temperature T_1 against a background of brightness temperature T_0. In (a) the target is not resolved, and the antenna's field of view includes some of the background, whereas in (b) the target is resolved.

This is just a linearly weighted average of the brightness temperatures of the components in the antenna's field of view. If the antenna is made physically larger, thus reducing the size of its beam solid angle, the weighting will shift in favour of T_1. However, once the target becomes fully resolved (i.e. when $\Omega_A \leq \Omega_t$), the antenna temperature is just

$$T_A = T_1$$

It perhaps seems surprising that a further increase in the effective area of the antenna does not result in its detecting more power. However, this can be understood by realising that although a larger antenna can indeed collect more radiation, it will do so from a smaller range of directions.

7.2.2 Sensitivity

We have noted that it is usual to express the power per unit frequency interval that is detected by an antenna as a temperature (the antenna temperature). This facilitates calculation of the sensitivity of the system, since the noise power generated by the system itself is also normally expressed as a temperature. The system noise temperature depends on the detailed design of the receiver and other parts of the system, but it cannot be lower than the physical temperature of the receiver and will usually (at the frequencies typical of passive microwave radiometry, and for a well-designed receiver) be a factor of 1.2 to 2 times this value.

In order to improve its signal-to-noise ratio, the output from a radiometer is integrated (averaged) for some time Δt. If the bandwidth (frequency interval) over which the radiation is received is Δf, this is effectively an average over $N = \Delta t \, \Delta f$ independent samples. We thus expect the signal-to-noise ratio to be improved by a factor of $\sqrt{N}$, and hence that the sensitivity of the system should be given by

$$\Delta T = C \frac{T_{\text{sys}}}{\sqrt{\Delta t \, \Delta f}} \qquad (7.10)$$

In this expression, ΔT is the smallest change in antenna temperature that can be detected, T_{sys} is the system noise temperature, and C is a factor of the order of 1 that depends on both the design of the radiometer and the criterion used to define ΔT. Usually C has a value of 5 to 10, and the values of Δt and Δf are normally chosen to give a value of ΔT of the order of 1 K.

7.2.3 Scanning radiometers

Let us consider a passive microwave radiometer operating from a satellite at an altitude of 800 km, at a frequency of 10 GHz. The wavelength is therefore 3 cm, so even if the width of the antenna is as large as 1 m the spatial resolution at the Earth's surface will be of the order of 20 km. The region of sensitivity at the

Earth's surface is usually termed the *footprint*, although the concept is equivalent to the instantaneous field of view (IFOV) introduced in section 6.2.2.

Now suppose that we wish to image a strip of the Earth's surface, perhaps of the order of 1000 km wide, with a spatial resolution of 20 km. It is obvious that constructing an array of antennas, pointing in different directions, is an entirely impractical approach since 50 antennas, each occupying an area of the order of $1\,m^2$ and each adding significantly to the mass of the instrument, would be needed. However, scanning of the radiometer footprint can be achieved, either by mechanical or electrical means.

The obvious method of scanning a radiometer footprint is by mechanical steering of the antenna. The antenna itself (or part of it – e.g. a reflector) may rotate or oscillate with respect to the rest of the instrument, or the whole platform can be made to rotate. The latter approach is clearly more appropriate for spaceborne than for airborne platforms. One particularly common form of mechanical scanning is the *conical scan*, in which the antenna beam is rotated around the nadir direction at a fixed angle to the nadir direction. This is illustrated in figure 7.4. It differs from the type of conical scanning used by the Along-Track Scanning Radiometer (section 6.3.6) in that the received radiation makes a constant angle with the nadir direction. This simplifies the task of atmospheric correction.

One possible disadvantage of mechanical scanning is that it may cause some undesirable oscillation or vibration of the instrument, leading to pointing errors and loss of spatial resolution. An alternative approach to scanning the beam that avoids this problem is to use an electrically scanned antenna. This has no moving parts. It consists of a closely spaced array of smaller antennas (e.g. waveguides, horns or dipoles). The signals detected at each of these elements can be advanced or retarded in phase under electronic control (hence the alternative name of *phased array* for this type of system), and in this way beam-steering is achieved.

Figure 7.5 illustrates the principle of electrical steering in one dimension. It shows a simplified array, having only eight detectors at a regular spacing *d*. The

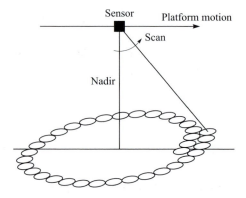

Figure 7.4. Conical scanning of a passive microwave radiometer.

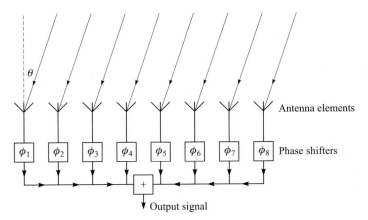

Figure 7.5. An electrically steered (phased) array. The signals from the antenna elements are given phase shifts ϕ_1 to ϕ_8 so that radiation from the direction θ is combined in phase.

signals from these detectors are shifted in phase and then added together. If no phase shifts are applied, and the elements abut one another so that there are no holes in the array, the device clearly functions as an ordinary antenna of width $8d$, having a power pattern with a main lobe pointing in the direction $\theta = 0$ and a width of approximately $\lambda/8d$ radians.

If, however, we consider radiation arriving at angle θ, it is clear that the phase of the signal received at detector 2 is $kd \sin \theta$ in advance of that at detector 1 (k is the wavenumber $2\pi/\lambda$), the phase at 3 is in advance of that at 2 by the same amount, and so on. In order to add these signals constructively, the phase shifters are set to introduce phase delays ϕ_1 to ϕ_8 that will exactly compensate for this. Thus, by introducing a phase gradient across the ray, the direction of the main lobe can be shifted.

Electronic steering of an array has some disadvantages. One is the increased complexity of the system, since in practice it will have many more than eight elements, probably steerable in two dimensions instead of just one. A second disadvantage is the decrease in performance caused by errors in setting the phases and gains of the phase shifters. However, a problem that is fundamental rather than technological is caused by the fact that the *projected* length of the antenna, measured perpendicularly to the direction of the incoming radiation, is proportional to $\cos \theta$, so that as the scan angle is increased the HPBW will widen accordingly. This can be an important consideration if the instrument is required to have a very wide imaging swath.

7.3   Major applications of passive microwave radiometry

Passive microwave radiometry is applied to both observation of the Earth's surface and to atmospheric sounding. In this section we discuss surface observations: atmospheric applications are described in section 7.6.

7.3.1   Oceanographic applications

The main applications of surface imaging by passive microwave radiometers
are oceanographic, particularly the determination of sea-surface temperature
(SST). This can be determined with a relative accuracy of about 0.2 K, and an
absolute accuracy of about 1 K if careful atmospheric correction is performed.
Since the brightness temperature of the ocean surface is influenced not only by
the physical temperature (the SST) but also by the observation frequency and
polarisation, the salinity, surface roughness, and the presence of any surface
materials such as slicks or foam, an accurate measurement of the SST requires
a multifrequency observation, usually in two polarisations. The absorption
length for microwaves in sea water is of the order of 1 cm – considerably
greater than for thermal infrared radiation. A comparison of the SST measured
using the two techniques thus has potential for investigating the surface temperature anomaly discussed in section 6.3.5.2, although at present neither
technique has sufficient accuracy for this to be a practical proposition. Plate
5 shows an example of SST data derived from passive microwave radiometry.

   At low frequencies, passive microwave radiometry can be used to determine
ocean salinity. Below about 5 GHz, the ionic conductivity of sea water
increases the imaginary part of the dielectric constant significantly with respect
to the value for pure water, thus increasing the Fresnel reflection coefficients
and decreasing the emissivity (figure 7.6). This is not really a practicable technique for spaceborne use; a 1-m diameter antenna operated from an altitude of
800 km at a frequency of 1 GHz, low enough for the effect of salinity to be
appreciable, would have a footprint of over 200 km.

   Passive microwave observations of ocean surfaces can also be used to
deduce the surface roughness of the ocean. In this context, of course, 'surface
roughness' means surface waves, and these are generated by the action of wind
on the ocean surface. Thus, indirectly, wind speed can be inferred from passive
microwave measurements. In fact, the effect of wind speed appears to be
mediated by two mechanisms: an increase in surface roughness, and an increase

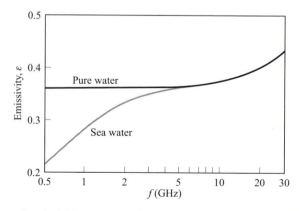

Figure 7.6. Normal emissivities at 20°C of pure water and sea water (35‰ salinity) as functions of frequency.

in the proportion of the surface covered by foam. Modelling the surface roughness effect is rather complicated because the ocean surface contains a whole
spectrum of waves rather than a single dominant component.[2] It is observed
that while the horizontally polarised component of the emitted radiation is
sensitive to the wind speed, the vertically polarised component is independent
of surface roughness for viewing angles near 50° from nadir.[3] Thus, a viewing
angle of about 50°, and a dual-polarisation observation, offers scope for discriminating the effect of surface roughness from other influences on the
observed brightness temperature. In practice, estimation of wind speed from
passive microwave radiometry is accurate to about $\pm 2\,\mathrm{m\,s}^{-1}$.

The final oceanographic application of passive microwave radiometry that
we shall mention is the identification of *sea ice*. At frequencies below about
30 GHz the emissivity of sea ice is significantly greater than that of sea water
(see figure 7.7), so the proportion of the ocean surface that is covered by ice can
be determined rather easily. In fact, as shown in figure 7.7, different types of
sea ice generally have different emissivities, so a multifrequency observation
can be used to estimate the proportions of different ice types and of open water
present in the antenna's footprint. Plate 6 shows an example of this
application.

7.3.2   *Land-surface applications*

The poor angular resolution of passive microwave radiometry implies that,
except from low-altitude airborne platforms, the technique is not always par-

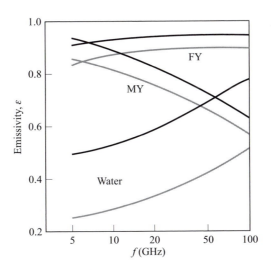

Figure 7.7. Emissivities of water, first-year sea ice (FY) and multi-year sea ice (MY) at 50° to
nadir. The black curves show vertical polarisation, the grey curves horizontal polarisation.
The data for this figure have been adapted from Eppler et al. (1992).

---

[2] See, for example, Stewart (1985).
[3] See, for example, Robinson (1994).

ticularly useful over land surfaces. As we have already remarked, a spaceborne system will have a spatial resolution of the order of 20 km or so, and while this is adequate for characterising many ocean features, the scale of spatial variability on land surfaces is generally finer than this, so that the technique will often fail to resolve interesting features. There are, however, a few promising lines of research. Most obviously, passive microwave observations, even from space, can be used to measure the surface temperature of large, physically homogeneous regions. Soil moisture content can also be estimated from low-frequency measurements, since the presence of liquid water raises the dielectric constant, and hence lowers the emissivity, of soil. The superficial extent and depth (up to about 0.2 m) of snowfields can also be estimated. However, applications that depend principally on the ability to estimate the dielectric constant of the material will also be influenced by the surface roughness. In such cases, more information can usually be obtained by the use of an *imaging radar* technique, discussed in chapter 9.

7.4   Atmospheric correction

As we have seen in chapter 4, the Earth's atmosphere is not entirely transparent in the microwave region of the electromagnetic spectrum, which means that the brightness temperature recorded at a spaceborne sensor will not correspond exactly to the quantity εT, where ε is the emissivity of the surface and T its physical temperature. The measured brightness temperature can be considered to have three components: (1) the emitted component εT, which will be reduced by atmospheric attenuation as it propagates upwards to the detector; (2) downwelling atmospheric radiation that has been reflected from the surface (and is of course also attenuated on its upward journey through the atmosphere); (3) upwelling atmospheric emission. It is necessary to correct for components (2) and (3) if the first component is to be retrieved from the measured brightness temperature.

These effects can be modelled using the radiative transfer equation (section 3.4.1). Since we are considering microwave radiation and ordinary terrestrial temperatures, the Rayleigh–Jeans approximation provides a considerable simplification. Let us first consider component (3), the contribution from upwelling atmospheric radiation, in the case of a clear atmosphere (one without any cloud, rain, snow, etc., in it). From equation (3.74) we can write this upwelling component as

$$T_b = \int_0^\tau T(\tau') \exp(\tau' - \tau) \, d\tau'$$

where τ is the total optical thickness of the path and τ' is the optical thickness between the surface and some point on the path where the physical temperature is $T(\tau')$. To obtain a rough idea of the magnitude of this contribution, we will assume that $T(\tau')$ is constant, so that the contribution is just

$$T_b = T(1 - \exp(-\tau)) \tag{7.11}$$

Figure 7.8 shows the result of applying equation (7.11) to the atmospheric absorption data of figure 4.5, assuming an atmospheric temperature T of 250 K. Two curves are shown in the figure, one for radiation propagating vertically upwards (towards the zenith), and one for radiation propagating at an angle of 50° to the zenith.[4] We see that for frequencies below about 15 GHz the contribution is only a few kelvin, although it is significantly increased near the water vapour absorption line at 22 GHz and the oxygen line at 60 GHz. Thus, for frequencies below about 15 GHz, a fairly crude knowledge of the atmospheric temperature profile is sufficient to correct for this term.

Next, we consider the downwelling atmospheric radiation. Provided that no radiation enters the atmosphere from above, it is clear that this is given by the same calculation as for the upwelling radiation, so the data in figure 7.8 also show the downwelling component. Again, for frequencies below about 15 GHz, this is just a few kelvin. If, in addition, the surface reflectance is small (i.e. the emissivity is large), only a small proportion of this already small component will be reflected. We should, however, make two remarks about the reflection of downwelling radiation. The first is that downwelling radiation will reach the surface from all directions, so it is necessary to integrate the contribution over all incidence directions. The radiation will in fact be most intense at large angles to the vertical, where the optical thickness of the path through the atmosphere is greatest. The second remark is that our assumption that no radiation enters the atmosphere from above is only valid

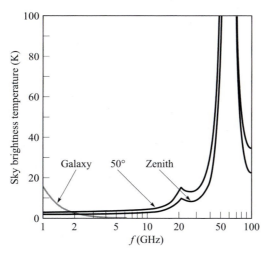

Figure 7.8. Typical upwelling brightness temperature (black) of the atmosphere for radiation propagating vertically and at 50° to the zenith. The grey curve shows typical values of the brightness temperature of radiation from our galaxy.

---

[4] As we saw in section 4.2.2, the total optical thickness for a path that makes an angle of θ to the vertical is $\tau/\cos\theta$, where τ is the optical thickness of the vertical path, provided that θ is less than about 75°.

for frequencies above about 3 GHz. Below this point, radiation from our galaxy is significant, and this is shown in figure 7.8.

We have discussed corrections due to the clear atmosphere. We ought now also to consider the effects of particulate material. At microwave frequencies the effects of aerosols are negligible. The larger meteorological particles were discussed in section 4.4, where it was shown that fog and cloud also have negligible effects for frequencies below about 15–20 GHz. Heavy rain, on the other hand, can introduce significant absorption and scattering effects at frequencies as low as 10 GHz. Since these are very strongly dependent on frequency (figure 4.6), they can be estimated and corrected for by a suitable multifrequency observation.

Returning to figure 7.8, or equivalently to figure 4.5, we can say that at frequencies between about 3 and 15 GHz the signal detected by a passive microwave radiometer will normally be dominated by surface emission, with a correction of a few kelvin due to atmospheric water vapour. Between about 15 and 35 GHz, surface emissions are still generally dominant although the water vapour signal is significantly larger. Above 35 GHz, the effects of molecular absorption become dominant, and observations at these frequencies will be more useful for atmospheric sounding than for surface sensing.

7.5   Example: the SSM/I

In this section, we describe a typical spaceborne passive microwave imaging radiometer. This is the Special Sensor Microwave Imager (SSM/I) carried on the DMSP (Defense Meteorological Satellite Program) satellites of the United States Air Force. It operates at four frequency bands, at 19.4, 22.2, 37.0 and 85.5 GHz, recording both horizontally and vertically polarised radiation in each band. It achieves a sensitivity to brightness temperature of ±0.8 K.

The SSM/I is a conically scanned radiometer. It operates from an altitude of 833 km and the antenna axis makes an angle of 45° to the nadir. This gives an incidence angle at the Earth's surface of 53°, as shown in figure 7.9. The antenna diameter is 1 m, giving a beamwidth of approximately 0.035 radians (2°) at 19.4 GHz, so at a distance of 1270 km and at an incidence angle of 53° this results in an elliptical footprint with dimensions of approximately 70 × 45 km. The corresponding footprint dimensions at the higher frequencies are about 60 × 40 km, 35 × 25 km and 15 × 10 km, respectively. The diameter of the circular path, shown schematically in figure 7.4, is about 1800 km for the scanning geometry shown in figure 7.9. However, the SSM/I does not make full use of this width, and it records data from a swath of 1400 km. One revolution of the conical scan takes 1.9 sec, during which time the footprint moves forward by 12.5 km. The sampling interval is 8.4 ms, giving a spatial sampling interval of 25 km and hence 64 samples per swath. (For the 85.5 GHz band, the sampling interval is halved and the number of samples per swath thereby doubled.)

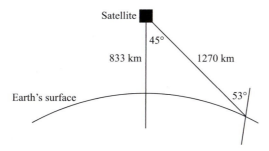

Figure 7.9. Scanning geometry of the SSM/I.

The two lower-frequency bands of the SSM/I are used mainly for ocean-surface and soil moisture measurements. The 22 GHz band is also used for measuring the total (integrated) water-vapour content of the atmosphere. The 37 GHz band is mainly used for estimating rainfall and the liquid water content of clouds, but is also used to determine sea-ice cover; and the 85 GHz band is used primarily for atmospheric profiling.

The SSM/I operates entirely above 15 GHz, where the effects of atmospheric propagation are at least appreciable. Other imaging microwave radiometers also include lower-frequency bands, for example the SMMR (Scanning Multichannel Microwave Radiometer), which has bands at 6.6, 10.7, 18, 21 and 37 GHz, and the MIMR (Multifrequency Imaging Microwave Radiometer), planned to be included on the EOS-PM and Metop satellites, which will have bands at 6.8, 10.7, 18.7, 23.8, 36.5 and 89 GHz.

7.6   Atmospheric sounding using passive microwave observations

Passive microwave radiometry has important applications in atmospheric sounding as well as in surface imaging, for the obvious reason that the microwave region contains a number of important absorption lines. The principles involved are not essentially different from those discussed in section 6.5 for the optical and infrared regions of the electromagnetic spectrum.

Temperature profiling from nadir-viewing observations is normally performed by using one of the deep absorption lines of oxygen, for example at 60 or 118 GHz. The weighting functions have typical widths of 10 km in altitude at 60 GHz, and somewhat less than this at 118 GHz. As we saw in section 6.5.1, the requirement for this type of profiling is an instrument having a number of frequency bands near the absorption line. An example of such an instrument is the SSM/T (Special Sensor Microwave Temperature Sounder), carried, like the SSM/I that we discussed earlier, on the DMSP satellites. This has seven bands at frequencies between 50.5 and 59.4 GHz, and achieves a temperature precision of approximately 0.5 K.

At frequencies below about 200 GHz, profiling of atmospheric molecular species is obviously limited to oxygen and water vapour, since no other absorption lines occur in this region. Since oxygen is well mixed in the atmosphere, profiling of this molecule is equivalent to establishing the atmospheric density profile, and determination of the total (column-integrated) oxygen concentration is equivalent to establishing the surface atmospheric pressure. As an example of an instrument for profiling water vapour, we can consider the MHS (Microwave Humidity Sounder) that will be carried on the Metop satellite, amongst others. This has five frequency bands: 89 GHz, 157 GHz and three at 183.3 GHz. The three 183-GHz bands (this is a major absorption line of water vapour) have bandwidths of 0.5, 1.0 and 2.2 GHz, and their centre frequencies can be scanned through a range of 1, 3 and 7 GHz, respectively.

Limb-sounding observations can also be made using passive microwave radiometry. In general, the principles are the same as those discussed in section 6.5.4. The vertical resolution of the technique is substantially better than for nadir-viewing observations, but this is at the expense of the horizontal resolution[5] and is also dependent upon the sensor having sufficient angular resolution. Since the distance from a spaceborne sensor to the point in the atmosphere from which most of the signal is derived is about 3300 km, a beamwidth of 0.3 milliradians is needed to achieve a vertical resolution of 1 km, and this of course is much more difficult to arrange for a microwave system than for one operating in the infrared. As an example, we can consider the MLS (Microwave Limb Sounder) instrument carried on UARS (Upper Atmosphere Research Satellite). This has channels at 63, 183 and 205 GHz, and is used to profile atmospheric temperature, oxygen, water vapour and other constituents. It has a vertical resolution of about 4 km.

In figure 4.5, we saw that the part of the microwave spectrum below about 200 GHz is rather sparsely populated by molecular absorption lines. Above this frequency, and especially above 300 GHz (submillimetre wavelengths), the situation changes dramatically and the region is very densely populated with molecular rotational transitions. These include H_2O, O_2, CO, SO_2, N_2O and NO_2, as well as a large number of less abundant but nevertheless significant molecules and ions such as ClO and HCO^+. Limb sounders have been designed to exploit this abundance of absorption lines. A spaceborne example is the AMAS (Advanced Millimetre-wave Atmospheric Sounder) instrument that is planned to be included on the payload of the Russian Meteor-3M satellite. This has eight frequency bands between 298 and 626 GHz. The design of instruments that can operate in the millimetre and submillimetre bands presents considerable technical challenges. This is especially true in the submillimetre range, where the instruments will typically use a hybrid of radio and optical techniques. A discussion of these is beyond the scope of this book, although Elachi (1987) provides some technical details.

---

[5] The long path through the atmosphere does, however, mean that constituents with very low concentrations can be detected.

PROBLEMS

1. Use the data of table 7.1 to show that the effective area of a rectangular or circular paraboloidal antenna is roughly equal to its geometrical area.

2. A phased array is designed to operate at a wavelength of 6 cm. The antenna elements are spaced at intervals of 4 cm. (a) Show that if the beam is steered more than $30°$ away from the normal ($\theta = 0$) direction, it will respond to radiation from *two* directions instead of just one. (b) Show that this phenomenon of multiple responses can eliminated if the antenna elements are less than half a wavelength apart.

3. At 10 GHz and a certain incidence angle, the apparent brightness temperatures of sea water, first-year ice and multi-year ice are 80 K, 252 K and 200 K, respectively. At 37 GHz these figures become 119 K, 253 K and 168 K. If a microwave radiometer measures a brightness temperature of 180 K at both frequencies, what are the fractions of open water and multi-year ice present in the IFOV?

4. A passive microwave radiometer operating at 37 GHz has an effective area of 0.3 m^2, and can detect a change of 0.9 K in its antenna temperature. It is operated from an altitude of 800 km to observe a single ice floe (whose brightness temperature is 253 K) surrounded by water (brightness temperature 119 K). Calculate the area of the smallest *detectable* floe.

5. If the atmospheric microwave absorption coefficient γ varies with height z as

$$\gamma = \tau\beta\exp(-\beta z)$$

where τ is the optical thickness of the entire atmosphere, show that the brightness temperature measured by a passive microwave radiometer looking vertically down through the atmosphere is given by

$$T(0)\exp(-\tau) + \int_0^\infty a(z)T(z)\,dz$$

where the weighting function $a(z)$ is given by

$$a(z) = \tau\beta\,\exp(-\beta z)\exp(-\tau\,\exp(-\beta z))$$

Show that $a(z)$ takes its maximum value at $(\ln\tau)/\beta$, and that, for a fixed value of β, the only effect of changing τ is to shift $a(z)$ along the z-axis without change of scale. Interpret this result.

8

Ranging systems

8.1 Introduction

Chapters 5 to 7 considered *passive* sensors, detecting naturally occurring radiation. In this chapter and the next we shall discuss *active* sensors, which emit radiation and analyse the signal that is returned by the Earth's surface or atmosphere. We have already identified three possible classifications of remote sensing systems, distinguishing between passive and active and between imaging and non-imaging, as well as classifying them according to the wavelength of radiation employed. We can also classify active systems according to the use that is made of the returned signal. If we are principally concerned with the time delay between transmission and reception of the signal we shall call the method a *ranging technique*, whereas if we are also (or mainly) interested in the strength of the returned signal we shall call it a *scattering technique*. The distinction between the two cannot be made entirely rigorous, but it provides a useful way of thinking about active remote sensing systems. It is clear that ranging systems are simpler both to visualise and, because of their less stringent technical demands, to construct, and we shall therefore consider them first. In chapter 9 we discuss the scattering techniques.

8.2 Laser profiling

Laser profiling (or *laser altimetry*) is the simplest application of the lidar (LIght Detection And Ranging) technique.[1] Conceptually, it is extremely straightforward. A short pulse of 'light' (visible or near-infrared radiation) is emitted towards the Earth's surface by the instrument, and its 'echo' is detected some time later. By measuring the time delay and knowing the speed of propagation of the pulse, the range (distance) from the instrument to the surface can be determined. By transmitting a continuous stream of pulses, a profile of the range can be built up, and if the position of the platform as a function of time is accurately known the surface profile may then be deduced.

---

[1] Other applications of lidar are discussed in chapter 9.

The operation and construction of a typical laser profiler are shown schematically in figures 8.1 and 8.2. The transmitter is a semiconductor laser, usually Nd:YAG (neodymium:yttrium–aluminium–garnet) operating at $0.53\,\mu m$ or $1.06\,\mu m$, or GaAs (gallium arsenide) operating at $0.9\,\mu m$. This is capable of producing a short (of the order of 1 ns), intense pulse with a small angular width. The receiver is a photodiode (see chapter 6). An interval timer with a resolution of the order of 1 ns is started by the signal that generates the transmitted pulse, and stopped by the received pulse. The travel time of the pulse, T_t, is given by

$$T_t = \frac{2H}{v_g} \tag{8.1}$$

where H is the range and v_g is the group velocity (section 3.1.3) of the pulse. As we saw in figure 3.4, the group velocity for optical and near-infrared radiation propagating in dry air differs from c, the speed of light *in vacuo*, by at most 0.03%, so the error caused by setting $v_g \approx c$ in equation (8.1) is small. Atmospheric correction of laser profile data is discussed in section 8.2.1.

The desirable features of such a system are that it should achieve a high spatial resolution at the surface (i.e. the sampled points should be close together) and a high-range resolution, and that the sensitivity should be great enough to detect signals returned from weakly reflecting surfaces.

The accuracy ΔT_t with which the travel time can be determined is normally governed by the *rise time t_r* of the received pulse, and its signal-to-noise ratio S. This can be understood from figure 8.3. If V_s is the voltage amplitude of the received pulse and V_n is the amplitude of its variation due to noise, the (voltage) signal-to-noise ratio is defined as $S = V_s/V_n$. It is evident from the figure that the greatest accuracy with which the timing of the received pulse can be determined is given by

$$\Delta T_t = \frac{t_r}{S} \tag{8.2}$$

although the precision of the system may in fact be limited by that of the interval timer.

A typical pulse transmitted by a laser profiler will have a rise time t_r of a few nanoseconds, although the received pulse may be somewhat longer if the sur-

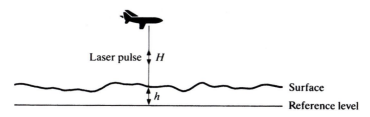

Figure 8.1. Principle of operation of a laser profiler.

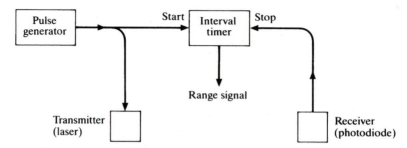

Figure 8.2. Construction of a laser profiler (schematic).

face that is being profiled is particularly rough. The signal-to-noise ratio S of the received pulse will depend on the reflectivity of the surface and the range H, as well as on system parameters such as the transmitted power, and less easily calculated influences such as the amount of incident sunlight, the weather, and atmospheric attenuation.

If pulses are transmitted at a frequency p (called the *pulse repetition frequency* or PRF) and the platform velocity is v, the linear sampling interval is v/p. If the angular beamwidth of the system is $\Delta\theta$, the linear dimension of the footprint is $H\Delta\theta$. It is obviously desirable that $H\Delta\theta$ should be no larger than some maximum value set by the nature and type of the surface under investigation. It might also be imagined that there would be no point in reducing the value of v/p below $H\Delta\theta$, but this is not so. Any decrease below this value means that, in effect, a number of independent measurements of the range is being made over a single footprint, and averaging over these measurements will therefore improve the range accuracy. Since the number of independent measurements is

$$N = \frac{H\Delta\theta}{v/p}$$

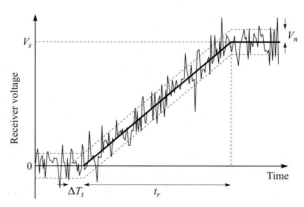

Figure 8.3. Determination of the arrival time of a noisy pulse. The accuracy ΔT_t with which the pulse can be located is $t_r V_n / V_s$.

and the improvement in range accuracy is proportional to $\sqrt{N}$, we may (provided that $N > 1$) write the range accuracy as

$$\Delta H = \frac{v_g t_r}{2S} \left(\frac{v}{pH\Delta\theta} \right)^{1/2} \tag{8.3}$$

For example, a typical airborne system might have $t_r = 5$ ns, $S = 1$, $v = 50\,\mathrm{m\,s^{-1}}$, $H = 200\,\mathrm{m}$ and $\Delta\theta = 0.001$ radian. At a PRF $p = 1000\,\mathrm{s^{-1}}$, this would give $N = 4$ and $\Delta H = 0.38$ m. Of course, a further increase in range resolution can be obtained by averaging over more than four pulses (i.e. over more than one footprint, and hence at the expense of the horizontal resolution). The system we have just described has a horizontal resolution $H\Delta\theta$ of 0.2 m. By averaging over a horizontal distance of, say, 1 m, the range resolution can be improved to 0.17 m.

Equation (8.3) shows that the measurement accuracy is increased by increasing the PRF. However, if p is increased beyond a certain point, the measured range will become ambiguous, for if the travel time T_t exceeds the interpulse period $1/p$ it will not be certain which echo belongs to which transmitted pulse. In this case, the calculated range will suffer from a *range ambiguity* of

$$H_{\mathrm{amb}} = \frac{v_g}{2p} \tag{8.4}$$

in the sense that the calculated range H may be increased or decreased by integer multiples of H_{amb} without changing the apparent travel time.[2] For this reason, it is desirable to operate the system in such a way that $H_{\mathrm{amb}} > H$, namely that

$$p < \frac{v_g}{2H} \tag{8.5}$$

For an airborne system this imposes an upper limit of tens or hundreds of thousands of pulses per second, whereas for a spaceborne system it is a few hundred pulses per second. In practice, laser profilers are normally operated well within these limits and no range ambiguities arise.

Airborne laser profiling has found applications in topographic mapping, and its high range accuracy (of the order of 0.1 m) has made it suitable for civil engineering applications. It has also proved particularly useful in reconnaissance of sea ice, where a knowledge of the freeboard (the height of the ice surface above the water level) allows the extent of the submerged portion to be estimated. Figure 8.4 illustrates typical output from a laser profiler.

The first laser profiler to be operated from space is the *Balkan-1* instrument, carried on the *Mir* space station ($H \approx 350\,\mathrm{km}$) since 1995. This operates at a PRF of $0.18\,\mathrm{s^{-1}}$, giving it an along-track sampling interval of about 45 km. The footprint width is 150 m and the range resolution 3 m. In the near future, the *GLAS* (Geoscience Laser Altimeter System) should be operated from the *EOS-*

---

[2] This is an example of the phenomenon of *aliasing*, discussed in greater detail in section 10.3.4.5

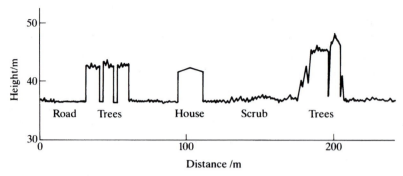

Figure 8.4. Typical output from an airborne laser profiler (modified from Jepsky, 1985).

Laser Alt satellite ($H \approx 700$ km). This instrument, which will have a PRF of $40\,s^{-1}$, will have a footprint width of 200 m and a range resolution of 0.1 m.

8.2.1 *Atmospheric correction of laser profiler data*

As we remarked in the previous section, the propagation speed of the pulses is given by the group velocity, and this is different from (slightly less than) the speed of light *c in vacuo*. Under normal atmospheric conditions the group velocity differs from *c* by at most 0.03%, so for an observation made over a range *H* of 1000 m the error incurred by assuming $v_g = c$ will be at most 0.3 m. For very precise measurements, the correct value of v_g should be used in equation (8.1). This depends on the wavelength, the atmospheric pressure, temperature and water vapour content.

For spaceborne or high-altitude airborne measurements, the atmospheric properties (pressure, temperature and water vapour content) are not constant along the path of the laser and it is necessary to integrate to find the travel time. Specifically, the *one-way* travel time for a path of length *z* is given by

$$T_t = \int_0^z \frac{1}{v_g}\, dz'$$

where v_g is the group velocity as a function of the distance z' along the path. It is often convenient to use instead the quantity *P*, defined by

$$P = \int_0^z \left(\frac{c}{v_g} - 1 \right) dz' \tag{8.6}$$

so that

$$P = cT_t - z \tag{8.7}$$

The quantity *P* thus has the dimensions of a length, and is in fact equal to the one-way range error that would be incurred if we assumed that the pulse had travelled at the speed *c* rather than at v_g.

The quantity P is useful because, at a given wavelength, it is proportional to the integral of the number density of molecules along the path. For practical purposes, we can assume that the atmosphere consists of two components: the *dry atmosphere* (mainly nitrogen, oxygen and carbon dioxide) and the *water vapour* component. To determine the integrated number density of molecules in the dry atmosphere for a vertical path, all we need to know is the atmospheric pressure difference (specifically the difference in the partial pressures of dry air) between the top and bottom of the path. Figure 8.5 shows the value of P as a function of wavelength when this pressure difference is equal to the standard atmospheric pressure of 101 325 Pa, namely when the path traverses the entire atmosphere. For a path that does not traverse the whole atmosphere, or that makes an angle θ to the vertical, the values shown in figure 8.5 should be multiplied by

$$\frac{\Delta p}{\cos \theta}$$

where Δp is the pressure difference between the two ends of the path, expressed in atmospheres. This expression is only valid for values of θ up to about 75°.

For the water vapour component, the integrated number density of molecules for a vertical path is usually expressed as the thickness of the layer of water that would result if the water vapour were precipitated (condensed). Figure 8.6 shows the value of P as a function of wavelength for 1 m of precipitable water.[3] In table 4.1 we noted that the total mass of water vapour in the atmosphere varies typically between 6.5 and 180 kg m^{-2}. This corresponds to a range of 6.5 to 180 mm of precipitable water, so we see that a typical value of the correction due to water vapour will be of the order of 0.05 m.

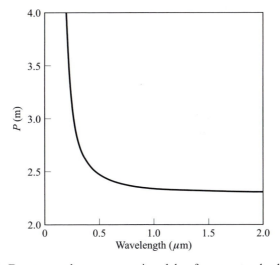

Figure 8.5. Dry-atmosphere propagation delay for one standard atmosphere.

---

[3] Again, for a path that is not vertical, the value of P is divided by cos θ.

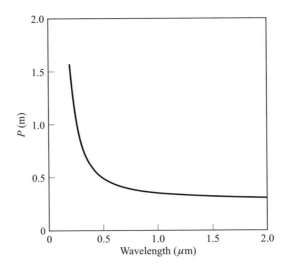

Figure 8.6. Propagation delay due to water vapour for 1 m of precipitable water.

8.3 Radar altimetry

The radar altimeter is similar in operation to the laser profiler. The basic principle, that of timing a short pulse over its round trip from the instrument to the surface and back again, is the same. Most of the differences between the kind of information obtainable from the two instruments can be ascribed to the larger beamwidth of the radar altimeter, which results from the fact that it operates at a much longer wavelength.

Figure 8.7 shows, extremely schematically, the construction of a radar altimeter. A pulse generator produces extremely short pulses at a frequency of, typically, around 10 GHz. These are fed to an antenna that radiates pulses of microwave electromagnetic radiation towards the Earth's surface. The same antenna collects the reflected pulse, and the signal is fed to a detector for subsequent analysis. The primary variable of interest is the time delay between the transmitted and received pulses, but, as we shall see, the shape of the received pulse also contains useful information.

We will first develop a very simple model of the operation of a radar altimeter, to illustrate the main features. In this model, we shall assume that the Earth's surface is flat, and that it consists of a uniform density of isotropic, incoherent, point-like scatterers. We shall also neglect the operation of the inverse square law, which we can justify if the ranges of all the scatterers

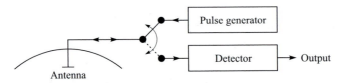

Figure 8.7. Operation of a radar altimeter (schematic).

that make a significant contribution to the received signal do not differ very much, and neglect the fall-off in sensitivity away from the axis of the antenna's main lobe. With these assumptions, the power that would be received if the antenna were to transmit *continuously* would just be proportional to the area of the Earth's surface illuminated by it. Of course, the antenna does *not* transmit continuously, and we therefore need to consider the time-structure of the received pulse.

It is clear that the return signal, if any, received at a time t after the emission of a pulse must arise from those scatterers situated at a distance $ct/2$ from the altimeter.[4] It will be convenient to describe this in terms of a 'scattering zone' that propagates away from the altimeter at a speed of $c/2$, such that any scatterer within the scattering zone contributes to the received signal at that time. This is shown schematically in figure 8.8.

If the distance from the altimeter to the surface is H, it is clear that no return signal will be received until

$$t = t_0 = \frac{2H}{c} \tag{8.8}$$

A short time Δt after this, the intersection of the scattering zone with the surface will be a circular disc of radius r, as shown in figure 8.9. Provided that $r \ll H$, this radius is given by

$$r \approx \sqrt{cH\Delta t}$$

so the area of the disc is $\pi c H \Delta t$. According to our simplifying assumptions, the received power is proportional to this area and hence to Δt. Thus, the received power will at first increase linearly with time. However, at time $t = t_0 + t_p$, where t_p is the duration of the pulse, the trailing edge of the scattering zone will just reach the surface. At times later than this the scattering zone will

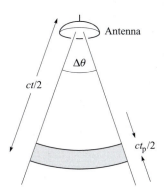

Figure 8.8. A radar altimeter emits a pulse of duration t_p beginning at time $t = 0$. $\Delta\theta$ is the beamwidth of the antenna. Any scatterer within the scattering zone (shaded) will contribute to the signal received at time t.

---

[4] For the time being, we will ignore the fact that the pulses propagate at the relevant group velocity and simply assume that they propagate at the speed of light c.

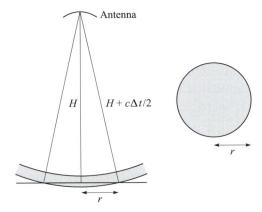

Figure 8.9. At a time $\Delta t < t_p$ after the first return signal is received, the scattering zone (shaded) intersects the surface in a disc of radius r. Left: elevation; right: plan view.

intersect the surface in an annulus, as shown in figure 8.10. The inner radius of this annulus is

$$r_1 \approx \sqrt{cH(\Delta t - t_p)}$$

and the outer radius is

$$r_2 \approx \sqrt{cH\Delta t}$$

so the area of the annulus is $\pi c H t_p$. This is independent of the time, so we can now see that the prediction of our simplified model is that the received power will increase linearly from zero at $t = t_0$ until $t = t_0 + t_p$, whereafter it will remain constant. This is shown in figure 8.11.

Inspection of figure 8.11 shows that the received power contains useful information only during the period $t_0 \le t \le t_0 + t_p$. It is clear that no further

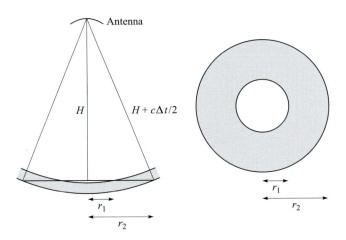

Figure 8.10. At a time $\Delta t > t_p$ after the first return signal is received, the scattering zone intersects the surface in an annulus. Left: elevation; right: plan view.

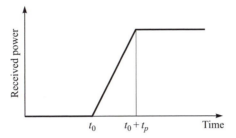

Figure 8.11. Time-dependence of the power received by a radar altimeter above a flat surface, according to the simple model derived in the text.

information is obtained at later times. This means that, in effect, the footprint of the instrument is a disc of radius r_p, where

$$r_p = \sqrt{cHt_p} \qquad (8.9)$$

is the radius of the scattering disc at time $t_0 + t_p$.

   In deriving this model, we have assumed that the beamwidth of the antenna is sufficiently large that variations in its response at increasingly large angles from the beam axis can be neglected. In general this will not be true, and the effect of the declining power pattern of the antenna (and also of the increasing distance of the scatterers from the antenna) as the time increases beyond t_0 will be to cause the received power to be less than predicted by our simple model, by a factor that increases with time. Thus, figure 8.11 should in fact be modified to look something like figure 8.12. We can, however, distinguish two limiting cases. The first of these is where the reduction in power as a result of the beam power pattern is negligible. The condition for this to be true is that

$$H\Delta\theta \gg 2r_p$$

in which case the altimeter is said to be *pulse-limited* and equation (8.9) correctly describes its spatial resolution. However, if

$$H\Delta\theta \ll 2r_p$$

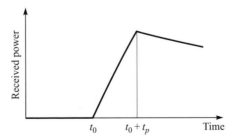

Figure 8.12. Modification of figure 8.11 to take account of the antenna power pattern and the increasing distance of scatterers from the antenna.

the altimeter is *beam-limited* and the spatial resolution is just $H\Delta\theta$. Most radar altimeters are pulse-limited, while laser profilers are beam-limited. As an example, we can consider a radar altimeter with a pulse length of 3 ns operating from a height of 800 km. Equation (8.9) shows that the radius of the beam-limited footprint will be approximately 850 m. From section 2.7 we know that the angular beamwidth (in radians) of an antenna of diameter D at wavelength λ is approximately 1.22 λ/D, so if $\lambda = 3$ cm and $D = 1$ m the radius of the beam-limited footprint will be about 30 km. In this case, clearly, the effect of the beam power pattern can be ignored and the altimeter is pulse-limited.

8.3.1 *Effect of the Earth's curvature*

Our model of the operation of a pulse-limited altimeter involved a number of simplifications, which we should now examine. One obviously incorrect assumption is that the Earth's surface is flat, and while this is entirely adequate for airborne radar altimeters it is not self-evident that it is valid for a space-borne instrument. Fortunately, the effect of the Earth's curvature can be dealt with rather simply. Figure 8.13 shows how figure 8.9 can be modified to take the Earth's curvature into account. The Earth is assumed to be a sphere of radius R, in which case simple trigonometry shows that

$$(H + c\,\Delta t/2)^2 = (H + R)^2 + R^2 - 2R(H + R)\cos\alpha$$

Assuming that $\alpha \ll 1$, this can be simplified to give

$$\alpha^2 \approx \frac{c\,\Delta t}{\dfrac{R^2}{H} + R}$$

and since $r = R\alpha$, we must have

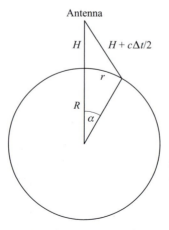

Figure 8.13. Geometry of a radar altimeter measurement when the Earth's curvature is taken into account.

$$r^2 \approx \frac{c \, \Delta t}{\dfrac{1}{H} + \dfrac{1}{R}}$$

Thus, the radius of the pulse-limited footprint can still be calculated using equation (8.9), provided that the range H is replaced by the effective height

$$H_{\text{eff}} = \left(\frac{1}{H} + \frac{1}{R}\right)^{-1} \tag{8.10}$$

Using the same example as before ($H = 800 \, \text{km}$, $t_p = 3 \, \text{ns}$), we see that the effective height is about 711 km, so the value of r_p is reduced from 850 m to about 800 m.

8.3.2 *Effect of coherence: range accuracy*

Our model of the response of a radar altimeter is still rather crude. We can improve on the assumption that the surface consists of a uniform distribution of isotropic scatterers by incorporating the form of the BRDF, if it is known. However, there is a more important respect in which the model fails. We have assumed that the power received by the altimeter is proportional to the number of scatterers (and hence the area) visible to it, and have thus added together the powers scattered by the various scatterers. If the radiation reaching the antenna from two scatterers is *coherent*, (i.e. if there is a definite phase relationship between the two waves), the signals are capable of *interfering* with one another. They should thus be added as vector (or more precisely phasor) quantities, with due regard to amplitude and phase.

Two points will be coherently illuminated with respect to each other if the difference between their distances from the source of illumination (the radar altimeter) is less than the *coherence length* l_c of the radiation, and their separation measured in a direction perpendicular to the propagation direction of the radiation is less than the *coherence width* w_c of the radiation. These quantities are given by

$$l_c \approx \frac{c}{\Delta f} \tag{8.11}$$

and

$$w_c \approx \frac{cH}{Df} \tag{8.12}$$

where c is the speed of light, f is the frequency of the radiation, Δf its bandwidth (which must be at least $1/t_p$, and is usually equal to it), and D is the diameter of the antenna. As before, H is the distance from the antenna to the surface. A typical spaceborne radar altimeter will have $l_c \approx 1 \, \text{m}$ and $w_c \approx 10 \, \text{km}$, so in practice most of the scattering zone will be coherently illuminated. The consequence of this is that the power received from a flat surface will not have the simple form shown in figure 8.11, but will instead be very noisy, in the manner of figure 8.3 but more so, with a signal-to-noise ratio of the order of 1. However, by averaging together many pulses, something resembling figure 8.11 can be obtained. An important consequence of the fact that the signal-to-noise ratio for a single pulse is ≈ 1 is that the range accuracy of a single pulse is approximately

$$\Delta H \approx \frac{ct_p}{2} \tag{8.13}$$

consistently with equations (8.1) and (8.2).

8.3.3   Response from a rough surface

The last of the simplifying assumptions that we shall examine is that the surface is flat. Let us suppose instead that the surface is rough, for example an ocean surface with waves. Considering figure 8.9 again, we can see that the first return signal will now be received earlier than before, when the scattering zone just touches the tops of the waves. Similarly, the time taken for the received pulse to rise to its maximum value will be increased, because this will now correspond to the time taken for the trailing edge of the scattering zone to reach the lowest scatterers (the lowest troughs of the waves). This is illustrated in figure 8.14.

The time taken for the received signal to rise from zero to its maximum value will thus be increased beyond the value of t_p (the duration of the transmitted pulse) that we derived in section 8.3. The shape of the time-variation (the *waveform*) of the received signal will also be altered, depending on the distribution of the surface scatterers in height. In general, it can be shown that (at the same degree of approximation that we used to derive figure 8.11)

$$\frac{dP_r}{dt} \propto \int_{-\infty}^{\infty} P_t\left(t - \frac{2H}{c} + \frac{2h}{c}\right) f(h)\, dh \tag{8.14}$$

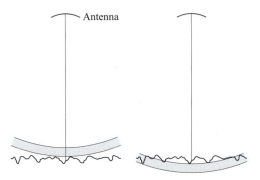

Figure 8.14. The scattering zone of a radar altimeter encounters a rough surface. Left: the instant when the first return signal is received; right: the instant when the received power reaches its maximum value.

where $P_r(t)$ and $P_t(t)$ are, respectively, the received and transmitted powers at time t, and $f(h)\,dh$ is the proportion of scatterers between heights h and $h + dh$ above the mean height of the surface.

The right-hand side of equation (8.14) is in fact a *convolution* of the surface height distribution and the shape of the transmitted pulse. Approximately, we may state that if the surface height is distributed over a range Δh, the time t'_p taken for the received power to increase from near zero to near maximum will be given by

$$t'^2_p \approx t^2_p + k\frac{\Delta h^2}{c^2} \tag{8.15}$$

where k is a dimensionless constant that depends on how Δh, 'near zero' and 'near maximum' are defined. Of course, equation (8.14) allows a more quantitative statement to be made if the shapes of the transmitted pulse and of $f(h)$ are known. As an example, figure 8.15 shows the waveform for a rectangular transmitted pulse (i.e. one in which the pulse is switched on abruptly, remains constant for time t_p, then is switched off again) incident on a surface for which $f(h)$ is a Gaussian distribution.

An important consequence of the broadening of the pulse received from a rough surface, from t_p to t'_p, is that the radius r_p of the pulse-limited footprint is no longer given by equation (8.9). Instead, the effective resolution is coarsened to

$$r'_p = \sqrt{cHt'_p} \tag{8.16}$$

In the case of figure 8.15, for example, the rise time of the received pulse is of the order of 8 ns, so if $H_{\mathrm{eff}} = 711\,\mathrm{km}$ the radius of the pulse-limited footprint has been broadened from $800\,\mathrm{m}$ to about $1300\,\mathrm{m}$. Pulse-broadening also implies a coarsening of the range resolution, which can be seen by substituting t'_p for t_p in equation (8.13).

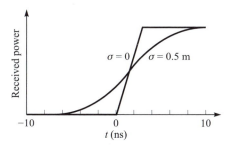

Figure 8.15. Waveform of the received power for a rectangular pulse of duration 3.0 ns incident on a flat surface and a surface with a Gaussian distribution of heights with a standard deviation of 0.50 m. The zero of the time axis is the time at which the leading edge of the pulse is received from the mean height of the surfaces.

8.3.4   Applications of radar altimetry

8.3.4.1   Sea-surface topography

Radar altimeter measurements are used extensively for characterising the topography of the ocean surface. The long-term average of the ocean surface (i.e. the average after the effects of tides, surface gravity waves, atmospheric pressure variations and wind-driven disturbances have been removed) is, with a proviso we shall discuss later, coincident with the *geoid*. This is a surface of constant gravitational potential (an equipotential surface) that is a close approximation to an *ellipsoid* of revolution with its shortest axis along the Earth's polar axis and circular symmetry about this axis. The geoid differs from this ellipsoid by distances of the order of 100 m, these variations being due to variations in the density of the Earth's mantle and lithosphere and to variations in the topography of the Earth's solid surface. Determination of the geoid is thus important for understanding the structure of the Earth and its gravity field, and also for making accurate predictions of satellite orbits (see chapter 10).

As we have already mentioned, the derivation of the mean sea surface requires the averaging of repeated measurements to reduce the magnitude of time-dependent effects. Where these phenomena vary more or less randomly in time, it is merely necessary to take the average over a time that is much longer than the correlation time of the phenomenon. However, for phenomena that are strongly periodic, such as tides, a potential problem is introduced by the fact that the sampling is also periodic, at least for a spaceborne system. It is clear that if the sampling frequency exactly matches the frequency of the tide, the tidal variation will be sampled at the same point in its cycle for each measurement, and consequently no information at all will be obtained about the temporal variability of the tide. This is another example of the phenomenon of *aliasing*, which will be discussed in greater detail in section 10.3.4.5.

Figure 8.16 is a visualisation of the global mean sea surface, using data from the radar altimeter carried on the ERS-1 satellite. It clearly shows topographic variations at a wide range of scales, but perhaps the most strikingly obvious features are those that correspond to the deep ocean-floor trenches, for example the Kermadec Trench at the lower left of the figure. The surface features corresponding to these trenches are typically 10 m deep and 200 km wide. They reflect the fact that the gravitational field strength is reduced above a trench, so that the equipotential surface moves closer to the Earth's centre in this region.

The one situation in which long-term averaging of the sea-surface topography does not result in a surface that corresponds to the geoid is where the surface has a steady motion as a result of an ocean current. In the northern hemisphere, the surface of a stream of water moving at a constant velocity v relative to the Earth's surface will be tilted such that the right-hand edge of the stream is higher than the left (in the southern hemisphere the left side is higher than the right), the angle of tilt being given by

$$\frac{2\Omega v \sin \phi}{g}$$

Figure 8.16. Visualisation of the global mean sea surface, using data from the ERS-1 radar altimeter. (© ESA 1994.)

where Ω is the Earth's angular velocity, ϕ is the latitude and g the gravitational field strength. This phenomenon, called *geostrophic balance*, is a consequence of the Coriolis force, and the slopes are very small. For example, the Gulf Stream has a typical velocity of $2\,\text{m\,s}^{-1}$, so at a latitude of 45°N the slope is 2×10^{-5} radians, but over a typical stream width of 100 km this accounts for a difference in height of 2 m across the stream. This surface tilt can be discerned in figure 8.16.

8.3.4.2 Sea-surface roughness

The sensitivity of the waveform (the shape of the received pulse) to surface roughness, discussed in section 8.3.3, means that radar altimeter measurements can be used to determine the sea state. The simplest and most widely reported measure of sea state is the *significant wave height* $H_{1/3}$, defined as the mean height (from trough to crest) of the highest third of the waves. It is approximately related to the variance σ^2 of the surface height distribution by

$$H_{1/3} \approx 4\sigma$$

The significant wave height is related to the wind speed, and can be used to determine it, although it is also dependent on the *fetch* (the distance from land, measured in the direction in which the wind is blowing) and the *duration* (the time for which the wind has been blowing). For sufficiently large fetch and duration, the sea is said to be *fully developed*, in which case the significant wave height depends only on the wind speed, and is given by

$$\frac{H_{1/3}}{\text{m}} \approx 0.02 \left(\frac{v_w}{\text{m\,s}^{-1}}\right)^2$$

where v_w is the wind speed 10 m above the surface. On this basis, wind speeds over oceans can be determined to an accuracy of typically $\pm 2\,\text{m\,s}^{-1}$.

In fact, determination of the significant wave height is also important for measuring the sea-surface topography. The reason for this is that the height distribution $f(h)$ of scatterers, introduced in section 8.3.3, is not symmetric about the mean surface height. The consequence of this is that the mean surface height is underestimated by an amount that is typically 2–3% of the significant wave height. This effect is known as *electromagnetic bias* or *sea-state bias*.

8.3.4.3   *Land and ice-sheet topography*

Topographic measurements over land surfaces are considerably more difficult to make using a spaceborne radar altimeter. There are essentially two technological reasons for this, both related to the fact that land surfaces exhibit considerably larger slopes than ocean surfaces. The first of these relates to a phenomenon usually referred to as 'loss of lock' or 'loss of tracking', and it can be explained roughly as follows. If we have a spaceborne radar altimeter at an altitude of 800 km, the time interval between the emission and reception of a pulse is of the order of five million nanoseconds. If the instrument is to be capable of resolving the waveform of the returned pulse, it will have to sample the pulse at intervals of the order of 1 ns or even less. In order to keep the volume of data collected by the instrument within manageable limits, the receiving and detecting part of the instrument is therefore only activated a short time before the expected return of the pulse. This clearly requires an accurate prediction of the arrival time of the next pulse, based on the last pulse or last few pulses received. Over very smoothly varying surfaces with small slopes this is not much of a problem. However, land-surface slopes can often be so steep that the next pulse will arrive well outside the time during which the instrument is ready to accept it.

The second problem that can arise in radar altimetry of comparatively steep slopes is that of *slope-induced error*. In effect, the return pulse is derived, not

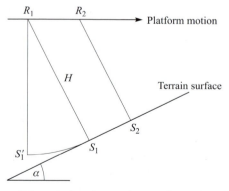

Figure 8.17. Illustration of slope-induced error in one dimension. When the radar altimeter is located at R_1, scattering actually occurs from the point S_1 but is assumed to occur at the point S_1'.

from the nadir point (the point directly beneath the instrument), but from the closest point to the altimeter. This is illustrated in figure 8.17.

When the radar altimeter is located at the position R_1, the closest point of the surface is S_1. If no correction is made for the slope-induced error, the scattering point will be assumed to be at the position S_1', where the distances R_1S_1 and R_1S_1' are both equal to the range H and S_1' is directly below R_1. For a small slope α (radians), the horizontal error is approximately $H\alpha$ and the vertical error is approximately $H\alpha^2/2$. Thus, for a spaceborne altimeter ($H \approx 800$ km) observing the ocean surface, where slopes are unlikely to exceed 10^{-4} radians, the errors are at most 80 m horizontally and 4 mm vertically, and can safely be ignored. However, over a land surface with a slope of, say, 0.03 radians, the errors are 24 km horizontally and 360 m vertically, and hence far from negligible.

Provided that the slope α does not change too rapidly, slope-induced error can be corrected. Figure 8.17 indicates how this can be achieved. If we have a second measurement from position R_2, such that the true location of the scattering point is S_2, the slope α can be deduced from the rate at which the range H changes with the along-track distance that the instrument has travelled. Once α is known, the scatterers can be assigned to their correct locations.

Two more remarks should be made about slope-induced error. The first is that the problem is in fact a two-dimensional one, so that correction requires two-dimensional coverage of the area rather than the one-dimensional transect illustrated in figure 8.17. The second is that if the angle α is large compared with the beamwidth of the antenna, little or no signal will be received from the surface unless the antenna is tilted so that its beam axis is normal to the surface.

Despite these difficulties, spaceborne radar altimetry has been applied with some success to mapping land-surface topography in comparatively flat areas, and with notable success to mapping the Antarctic and Greenland ice sheets (where the surface slopes are generally a few degrees at most).

8.3.5 *Atmospheric and ionospheric correction of radar altimeter data*

For simplicity, we have been assuming throughout our discussion of radar altimetry that the pulses propagate at the speed of light c. However, as for the laser profiler, we should really use the appropriate group velocity. The principles of correcting the measured time delay for atmospheric (principally tropospheric) delay are very similar to those for the laser profiler, discussed in section 8.2.1. There is, though, a major simplification in the case of the microwave frequencies employed for radar altimeters. The tropospheric propagation is practically non-dispersive, which means that the group velocity is equal to the phase velocity and independent of the frequency. The corresponding values of P (defined by equation (8.7)) are 2.33 metres per atmosphere for the dry tropospheric component, and 7.1 metres per metre of precipitable water. Thus, the tropospheric delay for a spaceborne observation is typically between 2.4 m and 3.6 m depending on the amount of water vapour present in the atmosphere. Correction for this delay therefore requires that both the atmospheric pressure and water vapour distributions be known.

For a spaceborne observation we must, however, also consider the effect of the ionosphere. This was discussed in section 4.5. Rearranging equation (4.23), we find that the quantity P is given by

$$P = \frac{e^2}{8\pi^2\varepsilon_0 m_e} \frac{N_t}{f^2} \tag{8.17}$$

where e and m_e are, respectively, the charge and mass of an electron, N_t is the total electron count, and f is the frequency. For a typical daytime total electron count of 3×10^{17} m^{-2} and a frequency of 10 GHz, we find that $P = 0.12$ m. If necessary (for high-precision measurements), this effect can be corrected if the total electron content is known, or, more conveniently, by using a dual-frequency radar altimeter.

8.3.6   *Example: the Envisat radar altimeter*

As an example of a typical spaceborne radar altimeter, we consider the RA-2 instrument that will be included as part of the payload of the Envisat satellite, orbiting at an altitude of 800 km. This is a dual-frequency altimeter, operating at both K_u-band (13.6 GHz) and S-band (3.2 GHz), so, with an antenna diameter of approximately 1 m, it has beam-limited footprints of 26 km (K_u-band) and 90 km (S-band). In its highest-resolution mode of operation, the K_u-band subsystem generates pulses of length 3.1 ns, so the diameter of the pulse-limited footprint over a smooth surface is 1.7 km. The S-band pulse length is 6.3 ns, giving a pulse-limited footprint of 2.4 km. The instrument is thus pulse-limited at both frequencies.

The range resolution for a single measurement over a flat surface is approximately $ct_p/2$, or about 0.5 m for K_u-band and 1 m for S-band. However, 100 K_u-band waveforms and 25 S-band waveforms are averaged in the instrument, improving these resolutions to approximately 5 cm and 20 cm respectively. The pulse repetition frequency for the K_u-band measurements is 1800 s^{-1} so each 100-waveform average is acquired in 56 milliseconds, during which time the satellite moves a distance of approximately 400 m relative to the Earth's surface. This, therefore, is the spatial sampling interval. For the S-band measurements the PRF is 450 s^{-1}, so the spatial sampling interval is again 400 m. The principal reason for having two operating frequencies is to allow for ionospheric corrections, which are clearly necessary if range accuracies as fine as 5 cm are to be achieved.

The K_u-band subsystem can also be operated in two lower-resolution modes, with pulse lengths of 12.5 and 50 ns, respectively. These give pulse-limited footprint diameters of 3.5 and 7 km (so they are still pulse-limited), and range resolutions, after averaging, of approximately 20 and 75 cm, respectively. The purpose of these modes to allow tracking of rougher terrain.

In passing, we can make two final remarks on the technical difficulties associated with achieving accuracies of the order of 10 cm in the range measured from a spaceborne radar altimeter. The first of these is a rather obvious one. If it is desired to measure the Earth's surface topography with an absolute accuracy of 10 cm, it will be necessary to know the position of the satellite to this accuracy.

The second point concerns the production of the extremely short (nanosecond) pulses needed to obtain high range precision. It would in fact be extremely difficult or impossible to put enough energy into a pulse only a few nanoseconds long, such that the pulse would be detectable after its 1600-km round trip to the Earth's surface. Instead, *pulse compression* is used. The real pulses are, in the case of the Envisat RA-2, 20 μs long, but they are modulated in frequency. We saw in section 2.3 that a pulse of duration T must contain a range of frequencies $\Delta f \approx 1/T$. Pulse compression effectively uses the converse of this principle, which states that if we wish to construct a pulse of length t_p, we must use a range of frequencies of $\Delta f \approx 1/t_p$. The frequency modulation necessary to achieve a synthetic pulse length of 3.1 ns thus requires a bandwidth of approximately 320 MHz. In practice, this is achieved by generating a '*chirp*', which is a pulse whose frequency rises from $f_0 - \Delta f/2$ to $f_0 + \Delta f/2$ over a period of 20 μs (f_0 is the central frequency, i.e. 13.6 GHz for the K_u-band subsystem).

8.4   Other ranging systems

We have now discussed the two main types of ranging system that conform to the definition of remote sensing enunciated in chapter 1. We should also, however, briefly mention *radio echo-sounding*. This is a technique for measuring the thickness of ice sheets and glaciers, and it relies on the large attenuation length of VHF (ca. 100 MHz) radio waves in ice. Since the attenuation length in ice at these frequencies is of the order of 100 m to 1000 m (see figure 3.1), it is feasible to transmit a signal through a large body of ice and to detect the echo from the bedrock beneath it, even at a range of several thousand metres, which is typical of the Antarctic and Greenland ice sheets. This technique has been extensively and successfully employed (see e.g. Cracknell, 1981; Drewry, 1983; Rees, 1988) for mapping ice sheets and glaciers, with a range resolution approaching 1 m. However, because of the long wavelength (≈ 3 m) in free space, narrow-beam antennas are not yet technologically feasible and such remote sensing has so far been confined to observations from platforms on or relatively close to the ice surface. Satellite observations will be precluded until narrow-beam instruments can be devised and placed in orbit.

Similar techniques are used for determining the thickness of saline ice, although in this case the higher electrical conductivity and inhomogeneous structure of the medium greatly reduce the attenuation length. For this reason, high-power systems must be employed, and the distance from the platform to the ice surface must be kept small. Satellite-based remote sensing of sea-ice thickness is an even more distant prospect than for ice sheets and glaciers, whose ice is comparatively pure and homogeneous.

Finally, we mention the use of *soil-sounding radars* in archaeological investigations. Again the technique is similar to radio echo-sounding, although the frequencies used are typical radar (microwave) frequencies rather than VHF. It has achieved a limited degree of success over dry soils in which buried artefacts produce a strong electromagnetic contrast.

PROBLEMS

1. A laser profiler is operated at a wavelength of $1 \, \mu$m from an altitude of 10 000 m, and views at 45° to the nadir. Estimate the range corrections needed to account for the dry atmosphere and the water vapour component if the atmosphere contains 50 mm of preciptable water. The atmospheric pressure at 10 000 m altitude can be taken as 0.26 atmospheres.

2. (For enthusiasts.) Prove equation (8.14).

3. A radar altimeter emits a pulse whose variation of power with time is Gaussian with a length of 3.00 ns between $1/e$ points. The pulse is reflected from a surface whose height distribution is Gaussian with a range of 1.00 m between the $1/e$ points. Calculate the time for the reflected pulse to rise from 8% to 92% of its final value, neglecting coherence effects. [Note that 84% of the area under a Gaussian curve is enclosed between the $1/e$ points.]

4. (For Fourier transform enthusiasts only!) A chirp signal, in which the angular frequency rises uniformly from $\omega_0 - \Delta\omega/2$ to $\omega_0 + \Delta\omega/2$ over a time T, can be written as $\exp(i\phi(t))$, where the phase $\phi(t)$ is given by

$$\phi(t) = \omega_0 t + \frac{\Delta\omega}{2T} t^2$$

for $|t| \leq T/2$. This signal is then passed through a delay-line, whose effect on a component of angular frequency ω can be represented as multiplication by the factor

$$\exp\left(-i\omega\left(\frac{\omega_0 T}{\Delta\omega} + \frac{\omega T}{2\,\Delta\omega}\right)\right)$$

Show that the signal that emerges from the delay-line is a carrier of angular frequency ω_0, modulated by an envelope of width $2\pi/\Delta\omega$.

9

Scattering systems

9.1 Introduction

In this chapter, we complete our survey of the principal types of remote sensing instrument by discussing those active systems that make direct use of the backscattered power. Optical (lidar) systems are used for sounding clouds, aerosols and other atmospheric constituents, for characterising surface albedo, and for measuring wind speeds. These are discussed briefly in section 9.2. However, the bulk of this chapter is concerned with microwave (radar[1]) systems.

In section 9.3 the ground-work established in chapter 3 is extended to a derivation of the radar equation, which shows how the power detected by a radar system is related to the usual measure of backscattering ability, the differential backscattering cross-section σ^0. The remainder of the chapter discusses the main types of system that employ this relationship. The first and simplest is the microwave scatterometer (section 9.4), which measures σ^0, usually only for a single region of the surface but often for a range of incidence angles. As described here, this is not an imaging system, although the distinction between microwave scatterometers and imaging radars is not a precise one.

The last two sections discuss the true imaging radars. Section 9.5 describes the side-looking airborne radar (SLAR), or real-aperture radar, which achieves a usefully high spatial resolution in one dimension by time-resolution of a very short pulse. Resolution in the perpendicular direction is achieved by using an antenna with a narrow beamwidth, namely a large antenna. This approach is not feasible for satellite-borne radars, since the antenna needed to produce a useful spatial resolution would be impractically large. Instead, a large aperture (antenna) is synthesised, and the technique is thus known as synthetic aperture radar. This technique, which is conceptually the most complicated in remote sensing, is discussed in section 9.6.

---

[1] 'Radar' is an acronym, originally standing for 'Radio Detection And Ranging'. The functions that can be performed by microwave scattering systems now extend far beyond detection and ranging, but the term continues to be in very general use.

9.2   Lidar

Lidar techniques were introduced in section 8.1, where the simplest form, the laser profiler, was discussed. In that case, the delay time of the returned pulse was the principal variable to be measured. However, it is also, of course, possible to analyse the temporal structure of the returned signal, and this is the principle of the *backscatter lidar*. Backscatter lidars are used to calculate vertical profiles or column-integrated values of the backscattering coefficients due to atmospheric constituents such as aerosols and cloud particles. The horizontal resolution of a backscatter lidar is similar to that of a laser profiler, being set by the angular width of the laser beam, but the vertical resolution is somewhat poorer, being typically 10 m to 200 m. This is because the backscattered signal must be integrated over a range of heights to give a detectable output.

Enhancements of the basic backscatter lidar are possible. The *differential absorption lidar* (DIAL) uses a tuneable laser to measure the spectral variation of the backscattered signal, so that atmospheric absorption lines can be distinguished. The *Doppler lidar* or *wind lidar* measures the Doppler shift (section 2.4) of the backscattered signal. This adds to the backscatter lidar the ability to determine the component of the scattering medium's velocity along the line of sight (i.e. the vertical component for a downward-looking lidar).

Lidars have been operated extensively from aircraft for meteorological sounding. The first spaceborne lidar was *Alissa*, a French system carried on the *Mir* space station. This uses a fixed-wavelength Nd:YAG laser (532 nm) for profiling cloud and aerosol structure. It has a pulse repetition frequency of $8s^{-1}$ and achieves a spatial resolution of 300 m horizontally and 150 m vertically. The *GLAS* instrument described in section 8.2 will also operate as a backscatter lidar, with a spatial resolution of 75 m to 200 m horizontally and 50 m to 150 m vertically.

9.3   The radar equation

The remainder of this chapter discusses radar systems, namely scattering systems that operate at microwave frequencies. We have already developed, in chapters 3 and 7, most of the theory necessary to calculate the response of a radar. In this section we will develop this theory a little further in order to derive the radar equation.

Figure 9.1 shows, schematically, an antenna transmitting microwave radiation towards a surface. Some of this incident radiation is scattered into a range of directions, and some of the scattered radiation is collected by a receiving antenna. The distance from the transmitting antenna to the surface is R_t and the distance from the surface to the receiving antenna is R_r. We begin by supposing that the transmitting antenna has gain G_t and is transmitting a power P_t. If this power were radiated isotropically, the flux density (power per unit area) at a distance R_t would be given by

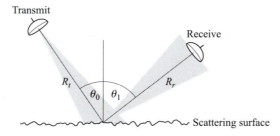

Figure 9.1. Geometry for deriving the bistatic radar equation

$$\frac{P_t}{4\pi R_t^2}$$

so from the definition of antenna gain (equations (7.5) and (7.6)) we can see that the flux density in the direction of the antenna's main beam axis must be given by

$$F = \frac{G_t P_t}{4\pi R_t^2}$$

If the angle that this radiation makes with the surface normal is θ_0, the irradiance at the surface is, following the discussion in section 3.3.1,

$$E = F \cos\theta_0 = \frac{G_t P_t}{4\pi R_t^2} \cos\theta_0$$

Thus, from the definition of the bistatic scattering coefficient γ (equation (3.36)), the radiance scattered into the direction[2] θ_1 is

$$L = \frac{\gamma E}{4\pi \cos\theta_1} = \frac{G_t P_t}{(4\pi)^2 R_t^2} \frac{\cos\theta_0}{\cos\theta_1}\gamma$$

Next, we suppose that the receiving antenna has an effective area (section 7.2.1) of A_r and is directed so that its main beam axis points directly at the region illuminated by the transmitting antenna. The solid angle subtended by this antenna at the distance R_r is given by A_r/R_r^2, so the power that it will collect from an area A of the scattering surface is given by

$$P_r = LA\frac{A_r}{R_r^2}\cos\theta_1 = \frac{A_r G_t P_t}{(4\pi)^2 R_t^2 R_r^2}\cos\theta_0\,\gamma A \tag{9.1}$$

This is one form of the *bistatic radar equation*, showing how the received power is related to the transmitted power, the radar geometry and the scattering properties of the surface. However, in all cases that will concern us in this chapter, we shall be interested only in the *monostatic radar equation*, where the same antenna is used for both transmitting and receiving radiation. In this case, we can put

---

[2] We should really include the azimuthal components ϕ_1 and ϕ_2 as well as the components θ_1 and θ_2 in specifying the incident and scattered directions. However, we omit them for clarity.

$$R_t = R_r = R$$
$$\theta_0 = \theta_1 = \theta$$
$$G_t = G$$

and

$$A_r = A_e$$

and the radar equation simplifies to

$$P_r = \frac{A_e G P_t}{(4\pi)^2 R^4} \cos\theta \; \gamma A \tag{9.2}$$

This can be rewritten in terms of the backscattering coefficient σ^0 (equation (3.37)) as

$$P_r = \frac{A_e G P_t}{(4\pi)^2 R^4} \sigma^0 A \tag{9.3}$$

We can simplify this equation even further by noting the relationship between the gain G of an antenna and its effective area A_e (from equations (7.5), (7.6) and (7.9)):

$$A_e = \frac{\lambda^2 G}{4\pi\eta}$$

(where λ is the wavelength and η the efficiency of the antenna), so that

$$P_r = \frac{\lambda^2 G^2 P_t}{(4\pi)^3 \eta R^4} \sigma^0 A \tag{9.4}$$

Thus, equation (9.4) (or equivalently (9.2) or (9.3)) shows the power received from an area A of scattering surface in the monostatic case. The backscattering coefficient σ^0, which is dimensionless (it can be thought of as the backscattering cross-section per unit surface area, so its units are $\mathrm{m}^2/\mathrm{m}^2$ although it is most commonly expressed in decibels), will in general depend on the incidence angle θ, and possibly also on the corresponding azimuth angle ϕ.

We have made no mention of polarisation in the foregoing discussion. In general, values of σ^0 can be defined for all possible combinations of incident and scattered polarisation states, so that, for example, σ^0_{HV} is the backscattering coefficient for horizontally polarised incident radiation and vertically polarised scattered radiation. In order to make the argument leading up to equation (9.4) as general as possible, all polarisation states should be considered. For example, a radar set to receive only horizontally polarised radiation will detect a power proportional to $P_H \sigma^0_{HH} + P_V \sigma^0_{VH}$, where P_H and P_V are the components of the transmitted power with horizontal and vertical polarisations, respectively. A complete description can be given by specifying a *scattering matrix*, showing how the Stokes vector (see section 2.2) of the radiation is changed.

9.4 Microwave scatterometry

A microwave scatterometer is a non-imaging radar system that provides a quantitative measure of the backscattering coefficient σ^0, often as a function of the incidence angle θ. It transmits a continuous signal or a series of pulses, the return signal is recorded, and its strength is used in conjunction with the radar equation (e.g. equation (9.4)) to determine the value of σ^0 for that part of the surface that is illuminated. It is especially useful if the scatterometer can be operated in such a way as to yield the value of σ^0 as a function of the incidence angle θ, since this function often allows the surface material to be identified or its physical properties to be deduced (see the discussion of microwave back-scattering in sections 3.3.4 and 3.5.4). There are three principal methods of achieving this. One is to use a narrow-beamwidth scatterometer that can be steered to point at the desired target area. As the platform (aircraft or satellite) carrying the scatterometer moves, the radar tracks the target area and the backscattering curve is built up. The second method is to use *Doppler processing* of the signal.

Let us consider a scatterometer with a power pattern that is broad in the along-track direction but narrow in the perpendicular, across-track, direction (figure 9.2). The scatterometer beam is inclined so that it looks forward. At any instant, the return signal is derived from a large range of angles $\Delta\theta$ (the beamwidth of the antenna), and hence from a long strip of the surface being sensed. The signal returned from the point X will be Doppler-shifted to a frequency $f_0 + \delta f$, where f_0 is the transmitted frequency and δf is given, following equation (2.20), by

$$\delta f = \frac{2f_0 v}{c} \sin \theta_0 \qquad (9.5)$$

In this equation, v is the platform velocity and c is the speed of light. This Doppler shift is unique to the incidence angle θ_0, so by feeding the return signal into a bank of filters tuned to select different Doppler shifts, data from a range of incidence angles can be extracted.

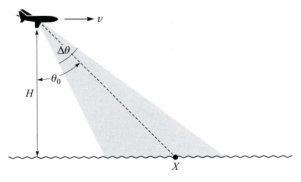

Figure 9.2. Principle of operation of a Doppler scatterometer. The radar emits a broad beam of angular width $\Delta\theta$, but radiation scattered from the point X (at incidence angle θ_0) can be identified by its Doppler shift.

The third method of scanning a range of incidence angles is to transmit very short pulses of radiation, and to analyse the time-structure of the returned signal. Unlike the Doppler method, this does not rely on motion of the sca- tterometer, and it simplifies the analysis if we assume that the platform is stationary. Again, we assume that the antenna power pattern is broad in one dimension (with a beamwidth $\Delta\theta$) and narrow in the perpendicular dimension, although, since we are assuming that the scatterometer is stationary, the orien- tation of the beam is unimportant as long as it is obliquely inclined to the surface (figure 9.3). The two-way propagation time from the scatterometer to the point X and back again is

$$\frac{2H}{c \cos \theta}$$

so by resolving the time-structure of the returned pulse we can uniquely iden- tify the contribution for the incidence angle θ. We may note that the ability to resolve the incidence angle is equivalent to spatial resolution, and that this approach to achieving spatial resolution is very similar to that of a pulse- limited radar altimeter (section 8.3).

9.4.1 *Applications of microwave scatterometry*

The useful output of a scatterometer, however it is realised, may be regarded as a plot of σ^0 as a function of the incidence angle θ, or, at least, one or more points representing this function. In chapter 3, and especially in sections 3.3.4 and 3.5.4, we discussed the principles that relate the function $\sigma^0(\theta)$ to the physical properties of the scattering medium. In general, we can state that the overall level of backscattering will be determined by the dielectric proper- ties of the scattering medium (and its internal structure, if volume scattering is

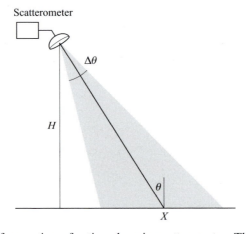

Figure 9.3. Principle of operation of a time-domain scatterometer. The radar emits a short pulse into a broad beam of angular width $\Delta\theta$, but radiation scattered from the point X (at incidence angle θ) can be identified by its time delay.

significant), while the dependence on incidence angle will be governed primarily by the surface geometry. A specularly smooth surface should, in principle, show a delta-function in a plot of $\sigma^0(\theta)$, centred at the value of θ that gives specular reflection into the radar. If the surface is horizontal, this will be $\theta = 0$. Real surfaces, however, are unlikely to be specularly smooth, especially at shorter wavelengths (see the discussion of the Rayleigh criterion in section 3.3.3), and in addition the plot of $\sigma^0(\theta)$ will be convolved with the antenna power pattern. Thus, a true delta function will never in practice be realised, and the plot of $\sigma^0(\theta)$ for a smooth surface will look more like figure 9.4.

At the opposite extreme, a Lambertian ('perfectly rough') surface has γ proportional to $\cos\theta$ (chapter 3), so that σ^0 will be proportional to $\cos^2\theta$. This has a characteristic shape if σ^0 is plotted logarithmically (i.e. as decibel values). Real materials lie somewhere between these extremes, unless volume scattering is also important, in which case the variation of σ^0 with incidence angle may be more complicated.

Even if a satisfactory physically based model of the backscattering is not available, a plot of $\sigma^0(\theta)$ may still be diagnostic of a particular surface material. Examples of such plots were given in section 3.5.4. Of course, more information can be obtained if observations can be made at more than one frequency or polarisation. Multiple-frequency observations are difficult to make, even from aircraft, because of the technical complexity of providing different 'front ends' for the radar (or the weight burden of carrying several radars), but multiple polarisations are easier to observe since little change needs to be made to the radar hardware.

9.4.1.1 *Microwave scatterometry over ocean surfaces*

The major application of microwave scatterometry to ocean surfaces is in determining wind velocity. This is somewhat similar in principle to the determination of wind speed from the significant wave height using radar altimetry (see chapter 8), although more information is available. The method relies on a model relating the roughness of the sea to the wind speed. The roughness is anisotropic (this fact was mentioned in section 3.5.4), as might be expected, since the crests and troughs of the surface wave field tend to align themselves

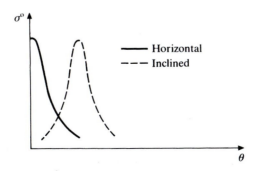

Figure 9.4. Variation of σ^0 with incidence angle for a smooth surface (schematic).

perpendicularly to the wind direction, and this is the key to determining the wind direction as well as the wind speed. This is shown in figure 9.5, which has been derived from a simple empirical model of microwave backscatter from ocean waves rather than directly from experimental data. It shows the effect of varying the azimuth angle ϕ while keeping the incidence angle, frequency and polarisation constant. The azimuth angle is defined such that $\phi = 0$ corresponds to the case when the horizontal component of the direction of the incident microwave radiation is opposite to the wind velocity vector; that is, when the scatterometer is looking upwind. The figure shows a strong contrast between the upwind and crosswind directions, and a much weaker contrast between the upwind and downwind directions.

The azimuthally dependent part of the model of sea-surface backscatter used to construct figure 9.5 can be written as

$$\sigma^0 = A + B \cos \phi + C \cos 2\phi \qquad (9.6)$$

where A, B and C are 'constants' that depend on the frequency, polarisation state, incidence angle and wind speed. By making at least three scatterometer observations in different azimuthal directions, the values of A, B and C, and hence the corresponding wind velocity, can be determined. This is illustrated schematically in figure 9.6 (in practice, a least-squares method, rather than the graphical approach suggested by the figure, is used). Here it is assumed that three observations have been made, all at the same frequency, polarisation and incidence angle, but with azimuth directions of 0 (i.e. looking north), 90° (east) and 180° (south). In each case, the observed value of σ^0 is consistent with a range of possibilities for the wind velocity. These are plotted as three curves in figure 9.6. For example, the curve a represents all the combinations of wind speed v and direction ψ consistent with the observed backscattering coefficient at azimuth 0. From the figure, we see that there is a mutual intersection of all three curves at a unique point, namely at a wind speed of $12.3\,\mathrm{m\,s^{-1}}$ in the

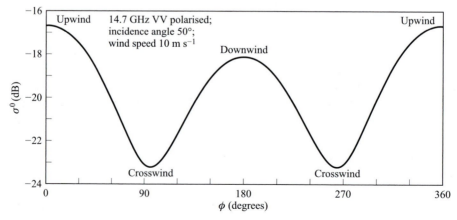

Figure 9.5. Typical variation of microwave backscattering coefficient with azimuth angle ϕ over a rough ocean surface.

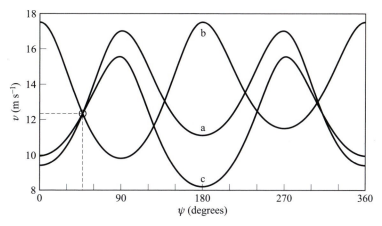

Figure 9.6. Determination of wind velocity over an ocean surface using three microwave scatterometer measurements. Curve (a) shows all values of the wind velocity (expressed as speed v and azimuth direction ψ) consistent with a certain value of backscattering coefficient measured by looking north. Curves (b) and (c) correspond to observations looking east and south, respectively. There is a unique intersection of all three curves, at $v = 12.3 \, \text{m s}^{-1}$, $\psi = 47°$.

direction $\psi = 47°$. We note, however, that there is also a 'near intersection' at about $12.5 \, \text{m s}^{-1}$, $310°$. Thus, if the scatterometer data are rather noisy, an unambiguous determination of the wind velocity may not always be possible.

Microwave scatterometry can give wind speeds, over the ocean, to an accuracy of about $2 \, \text{m s}^{-1}$, and wind directions to an accuracy of about $20°$, at least in the absence of rainfall. In the presence of rain, scattering from the falling droplets or from the rain-roughened sea can give rise to anomalous results. Nevertheless, scatterometry currently represents the most accurate technique for obtaining wind velocities over oceans.

A second oceanographic application of microwave scatterometry is the delineation and characterisation of sea ice. The technique can achieve much higher spatial resolutions than are possible with passive microwave radiometry, discussed in chapter 7.

9.4.1.2 *Microwave scatterometry over land surfaces*

Microwave scatterometry has been used extensively for characterising geological materials, using the variation of σ^0 with θ as a signature, in much the same way as materials are identified in the optical band by their spectral signatures. The technique has also been used for studying soil surfaces, where the principal retrievable parameters are moisture content, roughness and texture. It should be noted, however, that the greater spatial complexity of most land surfaces, when compared with oceans and sea ice, means that an imaging radar system is usually preferable to a simple scatterometer when interpreting surface types.

Microwave scatterometry has also found applications in the remote sensing of vegetation, particularly crops and forests. These can present a substantial theoretical problem because of their complicated geometries and comparatively open structures, with significant volume and surface scattering as well as, in some cases, scattering from the ground.

9.4.2   Example: ASCAT

In this section we will discuss the operation of a typical spaceborne microwave scatterometer. This is the *ASCAT* scatterometer to be carried on the *Metop* satellite. The ASCAT instrument is a C-band (5.3 GHz) VV-polarised scatterometer. Like most spaceborne scatterometers, ASCAT uses short pulses to obtain its spatial resolution. It emits radiation in six fan-beams (beams narrow in one dimension and broad in the perpendicular dimension), three on each side of the satellite, so that the different azimuths of these beams can be used to obtain wind velocities over the ocean as discussed in section 9.4.1.1. The geometry of the fan-beams is shown in figure 9.7. They are arranged so that they intersect the Earth's surface in swaths 500 km wide, extending from 150 to 650 km from the sub-satellite track. Two of these beams are perpendicular to the direction of the satellite's motion; these beams cover a range of incidence angles from 12° to 44°. The other four beams are arranged at 45° to the direction of motion, and these intersect the surface at incidence angles from 17° to 55°.

The instrument emits short pulses from each of its six antennas in turn. Time-resolution of the returned signal gives a spatial resolution of about 45 km. Since any point on the Earth's surface that falls within one of the two swaths of the instrument will be seen by all three of the fan-beams on that side, a 'triplet' of σ^0 values can be collected for input into the type of algorithm discussed in section 9.4.1.1.

9.5   Real-aperture imaging radar

Microwave scatterometers can be considered as imaging systems, although with a rather poor spatial resolution. In effect, they have sacrificed spatial resolution in order to achieve good radiometric resolution. In sections 9.5 and 9.6, we will consider radar systems in which the ability to generate images of reasonably high spatial resolution is a primary consideration. In this section

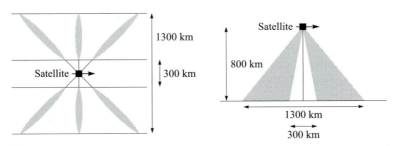

Figure 9.7. Geometry of the ASCAT scatterometer (schematic). Left: plan view; right: side view.

we discuss the real-aperture radars (RARs), usually referred to as *side-looking radars* (SLRs) or side-looking airborne radars (SLARs).

The SLR technique evolved in the 1950s, as a tool for military reconnaissance, from the plan-position indicator (PPI) radars developed during the Second World War. The SLAR is a pulsed radar system that looks to one side of the flight direction (hence 'side-looking') and is capable of producing a continuous strip image of the target area.

The basic idea of the SLR is very similar to that of the fan-beam scatterometer shown in figure 9.7. Just one of the fan-beams is used, and for a definite example we will assume that it is the middle beam on the right of the platform's motion. High spatial resolution in the along-track direction is achieved by ensuring that the narrower dimension of the fan-beam is as narrow as possible; that is, by arranging that the antenna is as long as possible in the along-track direction. Spatial resolution in the across-track direction is achieved by transmitting very short pulses (strong pulse compression, as discussed in section 8.3.6, is used) and time-resolving the returned signal. The main difference between an SLR and a scatterometer like ASCAT, apart from the fact that the latter will usually have several beams, is the use of very short pulses by the SLR.

Figure 9.8 shows the basic geometry of an SLR system. The upper part of the diagram shows the view from behind, so that the platform carrying the instrument is flying 'into the page'. The antenna, which has a width w, emits

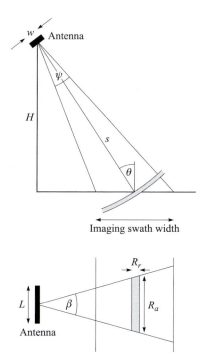

Figure 9.8. Geometry of an SLR system. Above: viewed from behind (the platform is flying 'into the page'); below: viewed from above (the platform is flying 'up the page').

radiation into a beam of angular width ψ, which will normally be set by the diffraction limit and therefore given by

$$\psi \approx \frac{\lambda}{w} \tag{9.7}$$

The intersection of this beam with the Earth's surface defines the width of the imaged swath. The transmitted radiation consists of a short pulse. We can use the same concept of a scattering region, introduced in section 8.3, that propagates away from the antenna at a speed of $c/2$. This region is shown shaded in figure 9.8, and its intersection with the Earth's surface is the region from which backscattered radiation is instantaneously detected.

The lower part of figure 9.8 shows a plan view of the same situation. The intersection of the scattering region with the surface is approximately rectangular. Its length in the along-track direction (also referred to as the *azimuth direction*) is governed by the beamwidth β of the antenna in this direction, and this is set by the length L of the antenna. The diffraction limit will normally mean that

$$\beta \approx \frac{\lambda}{L} \tag{9.8}$$

The length of the scattering strip in the azimuth direction is the *azimuth resolution R_a* of the system. Taking the azimuthal beamwidth as β (assumed to be much less than 1 radian) and the *slant range* from the antenna to the scattering region to be s, we must have $R_a \approx s\beta$, so we can see that the resolution will vary across the swath, being poorer at the farther edge and better at the nearer edge. If the effect of the Earth's curvature can be neglected, we can write the slant range in terms of the height H and the incidence angle θ:

$$s = \frac{H}{\cos \theta} \tag{9.9}$$

so, with the approximation of equation (9.8), the azimuth resolution becomes

$$R_a \approx \frac{H\lambda}{L \cos \theta} \tag{9.10}$$

If the length (duration) of the radar pulse is t_p, the *slant-range resolution* Δs is $ct_p/2$ – in other words, time-resolution of the returned signal allows two scatterers to be discriminated if their distances from the radar differ by at least this amount. By simple trigonometry, we can show that the *range resolution* (in the range direction, or across-track direction) is then given by

$$R_r = \frac{ct_p}{2 \sin \theta} \tag{9.11}$$

As an example, we consider an SLR system operating at $\lambda = 1\,\text{cm}$, with an antenna length of $5\,\text{m}$ and a pulse length of $30\,\text{ns}$, from an aircraft at an altitude of $6000\,\text{m}$. At a distance of $10\,\text{km}$ from the ground track the incidence angle $\theta = 59°$, so $R_a = 23\,\text{m}$ and $R_r = 5.2\,\text{m}$. At $25\,\text{km}$ from the ground track, the incidence angle is 77°, the azimuth resolution has coarsened to 42 m, but the range resolution is virtually unchanged at 4.7 m.

We note from equation (9.11) that the range resolution is independent of the platform height H, and can be made as small as 10 m, or less, provided that the incidence angle θ is not too small. The azimuth resolution R_a, on the other hand, is proportional to the platform height. Thus, although resolutions of the order of 10 m can be achieved from airborne systems, much poorer resolutions are available from spaceborne systems. This difficulty can be circumvented by the use of SAR systems, discussed in section 9.6.

9.5.1 *Image distortions*

The oblique imaging geometry, and the fact that the range (across-track) coordinate is determined from the slant range, introduces characteristic geometric and radiometric distortions into SLR images. The simplest of these is *slant-range distortion*. It arises only in the simplest form of signal processing, when the image is presented in such a way that it is the slant range, rather than the ground range, that increases uniformly across the image. It is illustrated schematically in figure 9.9a.

Slant-range distortion is relatively straightforward to correct, since it can be described by a small number of variables. Some SLR systems incorporate a correction within the radar's signal processing unit itself.

If the Earth's surface has appreciable relief, further geometric distortions will occur as a result of the imaging method. These are the phenomena *layover* and *shadowing*.

Layover arises from the fact that the pulse delay time is used to determine the across-track coordinate of a scatterer. Figure 9.9b shows a simple topography with five scatterers labelled A to E. Scatterers A, B, D and E are at the same altitude as one another, and this is equal to the altitude of the ground surface assumed for the slant-range to ground-range correction. However, scatterer C is located above the reference plane. The reduced slant-range is erroneously interpreted as a reduced across-track coordinate, and this is shown

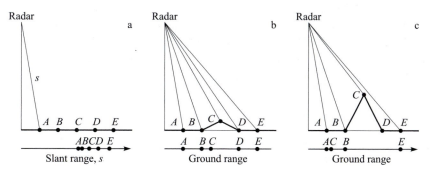

Figure 9.9. Image distortion phenomena in SLR imaging: (a) slant-range distortion; (b) layover; (c) shadowing.

schematically in figure 9.9b by a displacement of the point C to the left of its correct position.

Layover can be corrected if the surface topography is known, although the procedure is a laborious one. An uncorrected image of an area of high relief, for example a mountainous area, has a characteristically strange appearance in which the mountains appear to lean towards the radar. Figure 9.9b also demonstrates a radiometric consequence of layover. We can observe that, although the distances BC and CD, measured along the terrain surface, are similar to the distances AB and DE, the distance BC in the image will be shortened, and the distance CD lengthened, relative to flat terrain. As a result, the number of scatterers per unit length, measured on the image, is higher than normal in BC, with the consequence that this region will appear unusually bright. This phenomenon is known as *highlighting*, and is a purely geometrical effect (although it is likely to be enhanced by the fact that the backscattering coefficient σ^0 is usually significantly larger at small local incidence angles). Conversely, the density of scatterers within CD is lower than normal, so this region of the image will appear unusually dark. Although the term does not seem to be in widespread use, an obvious name for this phenomenon is *low-lighting*. Layover and highlighting effects can be clearly seen in figure 9.10,

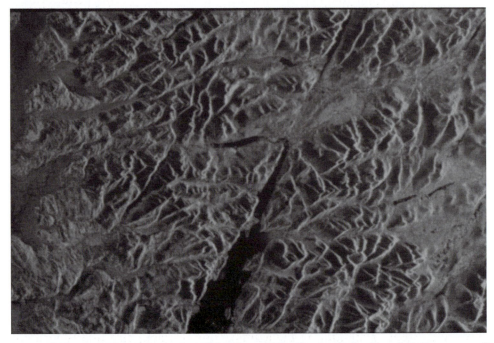

Figure 9.10. An ERS-1 radar image of the Great Glen, Scotland, showing the phenomena of layover (the mountains appear to lean to the right) and highlighting (the right-hand sides of the mountains are unusually bright). Although this image is presented in the conventional geographical orientation, with north approximately at the top, the reader may find it more intelligible to view it with the right-hand edge at the top. (Image © ESA 1991.)

which is in fact a Synthetic Aperture Radar (SAR) image rather than an SLR image.

Shadowing is the phenomenon in which one part of the surface is hidden from the radar's view by another. The shadowed region receives no radar illumination, and consequently no signal is returned from it. Unlike optical shadows, radar shadows are completely dark, because scattering of microwave radiation from the atmosphere into the shadowed region is entirely negligible. Shadowing is illustrated schematically in figure 9.9c, where the scatterer at D is obscured by the terrain in the vicinity of C and so does not appear in the image. It is clear from the figure that shadowing is a phenomenon associated with steep slopes away from the radar. However, the figure also shows that steep slopes *towards* the radar will introduce image distortions. Here, the layover of the scatterer at C is so large that the points B and C have been imaged in the reverse order. This can be thought of as an extreme example of the highlighting phenomenon.

9.5.2 *Instruments and applications*

As we remarked earlier, SLR systems are generally suitable for use only from airborne platforms, since the azimuth (along-track) resolution of a spaceborne system is normally unacceptably poor. Some SLRs have, however, been considered for deployment in space, for example the Ukrainian *RLSBO* instrument carried on the *Okean-O-1* satellite (at 650 km altitude) and planned for inclusion on the *Okean-O* series, and the Russian *SLR-3* instrument planned for inclusion on the *Almaz-1B* satellite (at 400 km altitude). The RLSBO is an X-band (9.7 GHz) VV-polarised instrument with a spatial resolution of the order of 2 km in azimuth and 1 km in range. The SLR-3 is also an X-band VV-polarised instrument, but the spatial resolution is improved to typically 1.5 km in azimuth and 0.2 km in range.

The applications of SLR systems are the same as those of SAR systems, and we therefore defer their discussion to section 9.6.7.

9.6   Synthetic aperture radar

The synthetic aperture radar (SAR) technique overcomes the problem of the altitude-dependence of the azimuth resolution of a SLR system (equation (9.10)). In external appearance, a SAR system is indistinguishable from a SLR. The same geometry applies, so that figure 9.8 describes both systems, and the same technique of emitting a very short pulse and analysing the temporal structure of the return signal is used to obtain resolution in the range direction. Higher resolution in the azimuth (along-track) direction is achieved by sophisticated signal-processing.

Equation (9.10) shows that improved azimuth resolution can be obtained from a longer antenna. Instead of making an antenna that is physically longer, the SAR technique relies on the motion of the platform. During some time interval T, the antenna is carried through a distance vT (where v is the plat-form velocity), so if we record the signal collected at the antenna during this interval, it ought to be possible to use it to reconstruct the signal that would have been collected from an antenna of length vT. This is the idea of the 'synthetic aperture'.

In order to discuss how this is achieved in practice, we consider the simple imaging geometry shown in figure 9.11. In this somewhat unrealistic case, the radar is looking vertically downwards (i.e. it is not side-looking), and we ignore the Earth's curvature. The radar is moving at a constant velocity v parallel to the x-axis of a Cartesian coordinate system, such that its position at time t is given by $(vt, H, 0)$. The radar's beam is broad in the azimuth direction, so that the radar can 'see' a large range of values of x. Somewhere within this range there is a scatterer with coordinates $(x, 0, 0)$.

When the time t is less than x/v, the radar is approaching the scatterer and the return signal will therefore be Doppler-shifted upwards. At $t = x/v$ the Doppler shift is zero, and at later times it is negative. Thus, by extract-ing the component of the return signal that has just the right variation of frequency with time, we can resolve any desired value of x. (Note that this is very similar to the Doppler processing for microwave scatterometry that was discussed in section 9.2). Clearly, the ability to extract different fre-quency components from the return signal requires that both its amplitude and phase should be stored, not just its intensity. This means that the transmitted radiation should be *coherent* (i.e. it should have a definite phase) and that it should be detected coherently. The path length from the radar to the scatterer at time t is $(H^2 + (vt - x)^2)^{1/2}$, so the phase delay for the two-way path from the radar to the scatterer and back to the radar is

$$\Delta\phi = 2k(H^2 + (vt - x)^2)^{1/2} \tag{9.12}$$

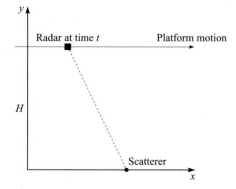

Figure 9.11. Simple geometry for considering SAR imaging. At time t, the radar is at $(vt, H, 0)$ and a scatterer has the fixed coordinates $(x, 0, 0)$.

where k is the wavenumber of the radiation (figure 9.12). To 'focus' the data on a particular value of x, the phase variation given by equation (9.12) is subtracted from the data.[3]

The length of the synthetic aperture is vT, where v is the platform velocity and T the time during which data are coherently collected for subsequent processing to generate the image. Since we expect the azimuth resolution to improve as the length of the synthetic aperture is increased, it appears that there should be no limit to the azimuth resolution. This is, however, not true, and we can estimate the best (finest) resolution as follows. If the length of the real antenna is L, it will have an angular beamwidth in the azimuth direction of roughly λ/L. Referring to figure 9.11, we see that this implies that a scatterer located at x will be within the radar beam, and hence visible to the radar, only for times t between

$$\frac{x}{v} - \frac{H\lambda}{2Lv}$$

and

$$\frac{x}{v} + \frac{H\lambda}{2Lv}$$

Thus, the maximum useful length of the synthetic aperture is $H\lambda/L$. The angular beamwidth, again in the azimuth direction, of *this* aperture is therefore roughly $\lambda/(H\lambda/L) = L/H$, giving a linear resolution at the surface of roughly L.

This is only an approximate calculation, although the result is almost correct (it should actually be $L/2$). A more satisfactory calculation is given in section 9.6.1. However, our simple derivation shows that the best possible resolution is

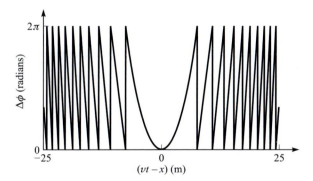

Figure 9.12. Phase delay $\Delta\phi$ given by equation (9.12) for the case when $H = 1000\,\text{m}$, at a frequency of $5\,\text{GHz}$. Values of $\Delta\phi$ greater than 2π have been reduced to the range 0 to 2π by subtracting integer multiples of 2π.

---

[3] This simple statement obscures the fact that the processing burden for a large image, which can easily consist of 10^7 pixels or more, is very large indeed. Until the 1980s, most SAR processing was in fact carried out *optically*, recording the data on film and using this as a diffraction mask from which the image could be reconstituted. However, the advent of powerful, extremely fast digital computers has meant that nearly all SAR processing is now performed digitally.

independent of the distance from the antenna to the surface, and that finer resolutions are, at least in principle, achievable using smaller antennas. Both of these results are contrary to the results we have obtained for other imaging systems. We can now see that spatial resolutions of the order of 10 m are feasible, even from spaceborne systems.

We should note that, if the optimum resolution of $L/2$ is to be achieved, it is necessary to preserve the coherence of the transmitted radar signal for a time $2H\lambda/Lv$. In practice, the maximum useful length of the synthetic aperture may be limited by the *coherence time* of the radiation[4] rather than by the beamwidth of the antenna. For example, suppose we consider a spaceborne SAR system designed to achieve an azimuth resolution of 5 m. The beamwidth criterion merely requires that the antenna should not be longer than 10 m. However, if we assume that $H = 800$ km, $\lambda = 6$ cm and $v = 7$ km s^{-1}, we see that the coherence time of the transmitted radiation must be at least 1.4 s. This would require that the transmitted radiation should have a bandwidth of no more than about 1 Hz.

9.6.1 *More exact treatment of the azimuth resolution*

The argument presented in the previous section was somewhat approximate. In this section, we rederive the result that the best possible azimuth resolution is given by $L/2$, using a somewhat more rigorous argument. No new principles are introduced here, and the section may be omitted by readers who do not share my enthusiasm for Fourier transforms. It does, however, illustrate in greater detail the idea of 'phase unwrapping' discussed in the previous section.

Figure 9.13 shows the necessary geometry. The antenna moves such that its position A at time t is given by $(x_a, H, 0)$ where $x_a = vt$, and the data will be processed so as to focus on the origin O. We wish to calculate the response of

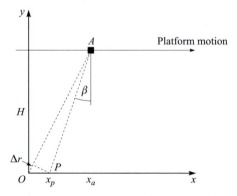

Figure 9.13. Geometric construction for calculating the azimuth resolution of a SAR. The antenna A has coordinates $(x_a, H, 0)$ and the data are processed to focus on the origin O. The text shows how the response to a scatterer at the point P can be calculated.

---

[4] The coherence time t_c is related to the coherence length l_c, introduced in equation (8.11), by $l_c = ct_c$, where c is the speed of light.

the system to a nearby point P with coordinates $(x_p, 0, 0)$. The signal $a(x_a, x_p)$ received from P when the antenna is at A is just

$$a(x_a, x_p) = f(\beta) \exp(2ik \, \Delta r) \tag{9.13}$$

where the distance AO exceeds the distance AP by Δr. (The reason that Δr, rather than the distance AP, appears in this expression is that the phase delay introduced by the path from A to O and back again is subtracted in the process of focussing the data.) $f(\beta)$ is the amplitude response of the antenna in the direction β, and k is the wavenumber of the radiation.

When the signals from different positions of the antenna are combined, equation (9.13) is integrated over x_a. For the present, we will suppose that the limits of the integration are given by $|x_a| \le X/2$, in other words that the length of the synthetic aperture is X. Using the approximations $\Delta r \approx \beta x_p$ and $x_a \approx \beta H$, we thus obtain the expression

$$a(x_p) = \int_{-X/2H}^{X/2H} f(\beta) \exp(2ik\beta x_p) \, d\beta \tag{9.14}$$

for the total amplitude $a(x_p)$ collected from the point P (we are neglecting constant factors that can be taken outside the integral).

Now the antenna's amplitude response $f(\beta)$ is obtained from the Fraunhofer diffraction pattern of its amplitude distribution. We can write this amplitude distribution as $A(y)$, where y is distance measured in the along-track direction from the centre of the antenna, so that, for example, a uniformly illuminated antenna of length L will have $A(y) = 1$ for $|y| \le L/2$, and zero otherwise. Whatever the form of $A(y)$, we have from equation (2.41) that

$$f(\beta) = \int_{-\infty}^{\infty} A(y) \exp(iky\beta) \, dy \tag{9.15}$$

provided that the antenna is long compared with the wavelength of the radiation.

Now we can substitute equation (9.15) into (9.14). We will also assume that X is infinite, so as to calculate the best possible azimuth resolution of the system. Thus,

$$a(x_p) = \int_{-\infty}^{\infty} A(y) \int_{-\infty}^{\infty} \exp(ik\beta[y + 2x_p]) \, dy \, d\beta$$

From the definition of the Dirac delta-function given in section 2.3, we can see that this simplifies to

$$a(x_p) = \int_{-\infty}^{\infty} A(y) \, \delta(k\beta[y + 2x_p]) \, dy$$

(we are again ignoring constant factors that can be taken outside the integral), and hence to

$$a(x_p) = A(-2x_p) \tag{9.16}$$

We have arrived at the desired result. Equation (9.16) shows that the response of the SAR system has exactly the same shape as the antenna's aperture function, but is half as wide. For example, the case of a uniformly illuminated antenna of length L gives $a(x_p) = 1$ for $|x_p| \leq L/4$, and zero otherwise. Thus, we have located the factor of 2 missing from the treatment in section 9.6, and we have also discovered the shape of the SAR response pattern.

9.6.2 *Speckle*

We noted, in section 9.6, that SAR is necessarily a *coherent* imaging technique, meaning that both the amplitude and phase of the received signal, and not just the intensity, are significant. An important consequence of this fact is that SAR images contain a characteristic type of granularity or image noise termed *speckle*. This adds to the uncertainty with which the backscattering coefficient σ^0 can be determined from a SAR observation. In this section, we will develop a simple one-dimensional model of this image speckle.

We begin by assuming that the radar system observes a surface that is nominally flat and uniform, consisting of isotropic point scatterers (this is the same starting point that we used to consider the behaviour of a radar altimeter in section 8.3). These scatterers are at various heights z above some datum, in order to model a rough surface. Figure 9.14 shows the necessary geometry.

The radar is located in the direction θ. The scattering surface does not necessarily pass through the point O, the origin of the coordinate system; this is just a convenient reference point. However, there is a scatterer at the point P, with coordinates (x, z). The ray that travels from the radar to P and back again is shorter than the ray from the radar to O and back again by twice the distance OA. Since OA is just

$$x \sin \theta + z \cos \theta \qquad ,$$

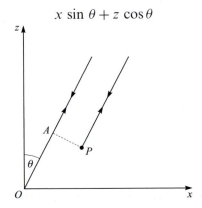

Figure 9.14. Geometry for calculating the effects of image speckle.

the phase $\phi(x)$ of the ray returning to the radar from P, relative to that from O, is

$$\phi(x) = -2k(x \sin \theta + z \cos \theta)$$

where k is the wavenumber of the radiation. The (complex) amplitude received in the direction θ is found by integrating over the whole surface:

$$a(\theta) = \int \exp(i\phi(x)) \, dx$$

To simplify this, we will find the speckle pattern near $\theta = 0$, so that we may put $\sin \theta \approx \theta$ and $\cos \theta \approx 1$. Thus,

$$a(\theta) = \int \exp(-2ik(x\theta + z)) \, dx \tag{9.17}$$

We recognise this as the Fourier transform of the function $\exp(-2ikz)$, where z is a function of x.

The exact nature of the speckle pattern will depend on the properties of the function $z(x)$ (in two dimensions, the function $z(x, y)$), which will in general be defined only statistically. Even if the r.m.s. value of $2kz$ is much less than 1 (i.e. the surface is very smooth), any realistic function $z(x)$ will generate a function $a(\theta)$ that changes sign on a small angular scale. Thus, the 'expected' image will be multiplied by a spatial intensity variation whose statistical properties will depend on the nature of $z(x)$. We can describe the image statistics by equation (9.18):

$$Q(j) = R(j)S(j) \tag{9.18}$$

Here, $Q(j)$ is the amplitude (or intensity) of pixel j in the image, $R(j)$ is the amplitude (or intensity) that it would have had in the absence of speckle, and $S(j)$ is the multiplicative speckle noise term, drawn at random from an appropriate probability distribution.

When the surface is rough (so that the r.m.s. value of $2kz$ is much greater than 1), equation (9.17) shows that the signal in a given direction is found by adding together a large number of components, each of which has the same amplitude but a phase drawn at random from a uniform distribution from 0 to 2π. In this case, which is called *fully developed speckle*, the probability distribution $p(S)$ for an intensity image is a negative exponential function:

$$p(S) = \exp(-S) \tag{9.19}$$

For an amplitude image, the corresponding form of $p(S)$ is a Rayleigh distribution:

$$p(S) = 2S \exp(-S^2) \tag{9.20}$$

Figure 9.15 shows an example of 'pure' speckle. It is a simulated amplitude image corresponding to the case where $R(j)$ in equation (9.18) is constant; that is, corresponding to a completely homogeneous region. The characteristic

Figure 9.15. 'Pure' fully developed speckle in an amplitude image.

granularity can clearly be seen. Real, as opposed to simulated, speckle can be seen in figures 9.17 and 9.21, for example.

Image speckle is an undesirable consequence of the coherent imaging mechanism that is needed to obtain the high azimuth resolution of a SAR. The effects of speckle can be reduced by some form of spatial averaging of the data (e.g. Rees and Satchell, 1997). Often, this is performed at the stage of generating the radar image from the raw amplitude and phase data. This is referred to as a multi-look image, where the number of 'looks' refers to the number of samples that are combined incoherently to process a given pixel.

9.6.3   *Distortions of SAR images*

Synthetic aperture radar images are subject to the same geometrically nduced distortions as SLR, discussed in section 9.5.1. However, if the 'target' is moving, a further source of distortion is introduced. This arises from the complicated way in which the data are processed to generate the image, and can most easily be thought of in terms of the Doppler frequency analysis presented briefly in section 9.6. We noted there that the Doppler shift of the signal received from a given scatterer is first positive, falls to zero at the instant when the radar has the same along-track coordinate as the scatterer, then becomes negative. If the scatterer is in motion, an extra Doppler shift will be added to that due to the platform motion. The total Doppler shift will thus fall to zero at a different value of the along-track coordinate, and the processor will assign this wrong value to the along-track coordinate of the scatterer.

We can model this phenomenon quite simply. Figure 9.16 shows a 'target' scatterer located at the origin of a Cartesian coordinate system, moving at speed u in a direction that makes an angle ψ with the x-axis. The radar moves parallel to the y-axis such that its position is (x, y, H), where x and H are constants, and y increases uniformly with time at the rate v. At the instant illustrated in the figure, the velocity of the target with respect to the radar is

$$\mathbf{v}' = (u \cos \psi, u \sin \psi - v, 0)$$

and the position of the target with respect to the radar is

$$\mathbf{r}' = -(x, y, H)$$

The Doppler shift falls to zero when these two vectors are perpendicular (this follows from equation (2.20)); that is, when $\mathbf{v}' \cdot \mathbf{r}' = 0$. This gives

$$y = \frac{ux \cos \psi}{v - u \sin \psi} \tag{9.21}$$

This is the value of the y-coordinate (i.e. the along-track or azimuth coordinate) that will be assigned to the target, instead of its correct value of zero.

Equation (9.21) shows that the azimuth shift is zero when $\cos \psi = 0$, that is, when the target is moving parallel or antiparallel to the radar track. If the target is moving perpendicularly to the radar track and towards it, $\cos \psi = 1$ and the displacement is in the direction in which the radar moves. Conversely, when the target moves away from the radar track, the displacement is opposite to the direction of the radar. These effects can be quite significant. For a spaceborne SAR, for which $v \gg u$, equation (9.21) can be approximated as $y \approx ux \cos \psi / v$. Taking $u = 10 \, \mathrm{m \, s^{-1}}$, $x = 300 \, \mathrm{km}$, $\cos \psi = 1$ and $v = 7 \, \mathrm{km \, s^{-1}}$ gives

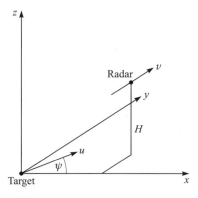

Figure 9.16. Geometry for discussing azimuth shift in a SAR image. The target scatterer is located at the origin with velocity coordinates $(u \cos \psi, u \sin \psi, 0)$; the radar is at (x, y, H) with velocity coordinates $(0, v, 0)$.

an azimuth shift y of over 400 m. Figure 9.17 shows an example of azimuth shift in a SAR image.

The azimuth shift will be fully developed only if the motion of the target is maintained throughout the coherence time of the SAR. If the period during which the target's motion can be considered steady is shorter than the coherence time, the image of the target will be blurred in the azimuth direction.

Other forms of motion-induced distortion and blurring can occur in SAR images. *Range walk* occurs when the target's range-direction coordinate changes by more than the range resolution during the time taken to acquire the image. This will obviously cause blurring in the range direction, but in fact also in the azimuth direction (see Robinson, 1994). For a platform velocity v, the time t taken to acquire an image with an azimuth resolution R_a is of the order of

$$t = \frac{S\lambda}{2R_a v} \tag{9.22}$$

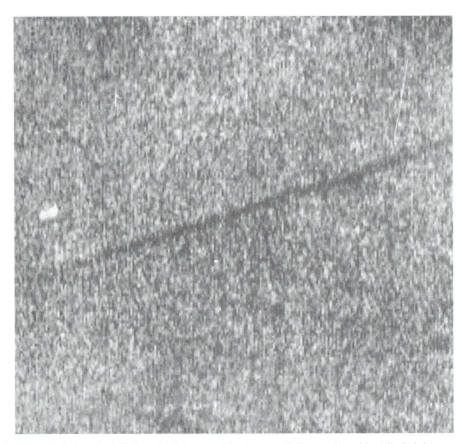

Figure 9.17 Azimuth shift in a SAR image. The image of the moving ship (the bright rectangular region, left centre) is displaced from the image of its wake (the dark diagonal stripe). (Image reproduced from Wahl et al. (1986) by courtesy of the European Space Agency.)

where λ is the wavelength and S the slant range to the target. The condition for range walk to occur is thus

$$|u_r| > \frac{2 R_r R_a v}{S\lambda} \qquad (9.23)$$

where u_r is the range component of the target's velocity and R_r is the range resolution. For a typical spaceborne system, one might have $R_a = R_r = 10\,\mathrm{m}$, $S = 350\,\mathrm{km}$, $\lambda = 6\,\mathrm{cm}$ and $v \approx 7\,\mathrm{km\,s^{-1}}$ giving an upper limit on the range velocity (to avoid range walk) of about $70\,\mathrm{m\,s^{-1}}$. At similar target velocities to those given by equation (9.23), the phenomenon of *azimuth defocussing* can also occur. This happens when the rate of change of the Doppler shift of the returned signal is significantly different from the rate of change expected from a stationary target.

9.6.4 Limitations imposed by ambiguity

In section 8.2 we introduced the concept of range ambiguity, and the desirability of avoiding it, for a pulsed system. In the case of a SAR system, this produces some rather unexpected limitations on its performance.

Figure 9.18 shows a rear view (the radar is flying 'into the page') of a SAR imaging a swath of width W from a height H. For simplicity, we assume that H is sufficiently small that the Earth's curvature can be ignored. The time taken by electromagnetic radiation for a round-trip journey from the radar to the near edge of the imaged swath and back again is just

$$\frac{2H}{c} \sec \theta_1$$

and similarly for the far edge. To avoid ambiguity, the time interval between successive pulses must be greater than the difference between these two times, so the pulse repetition frequency p must satisfy

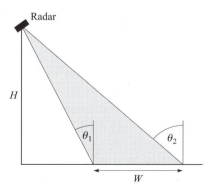

Figure 9.18. Relationship between the near-swath and far-swath incidence angles θ_1 and θ_2, and the swath width W, for a flat-Earth geometry.

$$\frac{1}{p} > \frac{2H}{c}(\sec \theta_2 - \sec \theta_1) \tag{9.24}$$

On the other hand, a pulse repetition frequency that is too low will degrade the azimuth resolution of the system. If the platform velocity is v and the azimuth resolution is R_a, the condition that the azimuth resolution is not degraded by excessively slow sampling is

$$\frac{1}{p} < \frac{R_a}{v} \tag{9.25}$$

Combining the two inequalities (9.24) and (9.25) gives

$$\frac{\sec \theta_2 - \sec \theta_1}{R_a} < \frac{c}{2Hv} \tag{9.26}$$

Since $\sec \theta_2 - \sec \theta_1$ is related to the swath width W, we see that the swath width and the azimuth resolution cannot be varied independently of one another. To make this more definite, we will suppose that $\theta_2 - \theta_1$ is small. In this case, the inequality (9.26) can be expressed as

$$\frac{W}{R_a} < \frac{c}{2v \sin \theta} \tag{9.27}$$

Thus, for example, a spaceborne SAR system (with $v \approx 7\,\mathrm{km\,s^{-1}}$) designed to operate at an incidence angle θ near $30°$ cannot have a swath width greater than approximately 4×10^4 times the azimuth resolution.

9.6.5 *SAR interferometry*

A SAR determines the across-track position of a target from the slant range, which depends on both the ground-range coordinate and the height of the target. As we have already seen in discussing the geometric distortions in SLR images (section 9.5.1), the contributions from these two components cannot be disentangled from a single observation. However, if *two* SAR images are available, acquired from positions separated by a very small distance, it is possible to separate the ground-range and height effects, and hence to derive information about the surface topography. This is the idea of SAR interferometry (Graham, 1974; Zebker and Goldstein, 1986; Massonnet, 1993), which shares some similarities with stereophotography (section 5.5.2).

The basic idea behind SAR interferometry can be understood from figure 9.19. Two SAR images are acquired, from the positions M and S, respectively (these denote 'master' and 'slave'). We consider a particular target scatterer located at A, such that its slant range from M is r_M. If no more information than this is available, all we can say about the position of the scatterer is that it lies somewhere on BB, the locus of points that are distance r_M from M. Now we also consider the observation from S. The slant range to A is r_S, and the locus of points that are this distance from S is CC. In principle, this is enough to distinguish between different positions, but we can see from figure 9.19 that

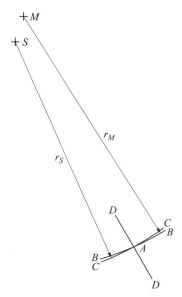

Figure 9.19. Basic idea of SAR interferometry. Two SAR observations are made of the region near the target A, from positions M ('master') and S ('slave') respectively. M and S are very close to one another. BB is the locus of points that are the same distance r_M from M as A; CC is the locus of points that are the same distance r_S from S as A; and DD is the locus of points for which the difference between the distance to M and the distance to S is equal to $r_M - r_S$.

if M and S are very close together the loci BB and CC will be almost coincident. Once we take into account the finite thickness of BB and CC (as a result of the finite range resolution of the SAR), we can see that in practice the two curves are coincident for a significant part of their lengths, centred at A. However, if we now consider the locus DD of points for which the *difference* between the distance to M and the distance to S has the same value as for the point A, we see that it is *perpendicular* to the locus BB (or CC). Thus, we can see that a measurement of the slant range from M, and the difference between the slant ranges from M and from S, will in general be enough to identify the location of a particular scatterer.

The slant-range difference is in practice determined by comparing the phases of the signal in the two images, which of course means that both the amplitudes and phases of the backscattered signal must be available (in which case, the images are referred to as 'complex'). In general, we can write the complex amplitude detected from a given pixel in the 'master' image as

$$a_M = a_1 \exp(i\phi_1) \exp(2ikr_M)$$

where a_1 and ϕ_1 are real numbers denoting the backscattered amplitude, and k is the wavenumber of the radiation. The same pixel observed in the 'slave' image has

$$a_S = a_2 \exp(i\phi_2) \exp(2ikr_S)$$

If the local observing geometry at the pixel, and its backscattering properties, are practically identical between the two observations, we will have

$$a_1 = a_2$$

and

$$\phi_1 = \phi_2$$

which means that if we multiply one image by the complex conjugate of the other we will obtain

$$a_M a_S^* = |a_1^2| \exp(2ik[r_M - r_S])$$

This is an intensity image with interference fringes superimposed. The fringes contain the information about how $r_M - r_S$ changes between the two images, with one complete fringe corresponding to a change of half a wavelength. Since the wavelength of a SAR system is typically only a few centimetres, we can see that the technique has the potential to achieve extremely high resolutions.

A convenient way of characterising the potential offered by a particular imaging geometry for the generation of interference fringes is to calculate the ambiguity distance e, as shown in figure 9.20. This is defined by the point B, which has the same distance from M as the reference point A, but for which the difference between the distances to M and S has changed by half a wavelength. Since this changes the difference in the round-trip distances by one wavelength, it means that the use of interferometry cannot distinguish between the positions A and B. Simple flat-Earth geometry shows that

$$e = \frac{R^2 \lambda}{2(Hd - hD)} \tag{9.28}$$

For example, if we take $H = 800\,\text{km}$, $D = 350\,\text{km}$ (and hence $R = 873.2\,\text{km}$), $h = 0$ and $d = 1\,\text{km}$ for a SAR with $\lambda = 6\,\text{cm}$, we obtain $e \approx 30\,\text{m}$. The ambi-

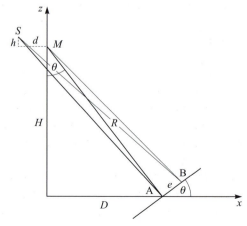

Figure 9.20. The ambiguity distance e is defined such that $(SB - MB)$ differs from $(SA - MA)$ by $\lambda/2$, where λ is the wavelength.

guity height, $e \tan \theta$, is about 11 m in this case, which means that if the signal-to-noise ratio in the complex image is high enough to detect, say, one tenth of an interference fringe, altitudes can be determined from the interferogram with an accuracy of about 1 m.

Equation (9.28) shows that when $Hd = hD$, the ambiguity distance e is infinite. In this case, no interference fringes are formed, and it arises when the target, master and slave are collinear. Values of e that are too small also pose a problem, since if e is much smaller than the slant-range resolution of the SAR system a single pixel will contain many fringes and it will be impossible to count the number of fringes. Equation (9.28) shows that small values of e arise from large values of the baseline (h or d) between the master and slave, so we can see that this baseline must not be too large. For typical spaceborne SAR systems this maximum baseline is of the order of 1 km.

The SAR interferometry technique has an obvious application to measuring surface topography. It can also be used to measure bulk translation of parts of the surface between images obtained at different times (e.g. to monitor the motion of a glacier or the bulging of a volcano or earthquake zone). Plate 7 shows an example of this. For this application, it is necessary to resolve the ambiguity between vertical and horizontal displacements. In general, SAR interferometry is a difficult technique that can yield excellent results when it works. Because it is very unlikely that one will have sufficiently accurate *a priori* information about the locations from which the master and slave images were obtained, it is usually necessary to use a cross-correlation method on the images first, in order to determine which pixels in the slave image correspond to those in the master image. Only if the correlation coefficient is high enough will further processing be worth undertaking. Similarly, the baseline vector (d, h) cannot usually be predicted and must be found from ground control points.

9.6.6   *Major applications of radar imaging*

In considering the applications of SLR and SAR images, it is helpful to compare them with the optical systems discussed in chapter 6. The SLR and SAR images are generally produced in a format similar to that of a black-and-white aerial photograph, the brightness of the displayed image being dependent on the value of σ^0. Such images may be visually interpreted singly or as stereo pairs (although the side-looking geometry complicates the stereo effect). Even simple visual interpretation can often reveal important spatial relationships in the data, although increasing use is now made of digital imagery which can be processed and analysed using a computer, making the maximum use of the quantitative nature of the data.

The spatial resolution achievable by imaging radar systems is often comparable to that obtained from optical imaging systems. Since SLR and SAR are active remote sensing techniques, in which the illumination of the target is supplied by the instrument itself and not by solar radiation, they can be applied at night as well as during the daytime. Furthermore, since the propagation of microwave radiation is little affected by the presence of cloud or even rainfall (unless it is very intense), radar images can be acquired during most weather conditions. These are obviously major advantages.

On the other hand, the physical processes that modulate the 'brightness' (i.e. the value of the backscattering coefficient) of a radar image are things for which most image analysts do not have a strong intuitive feeling. The relevant factors include the microwave dielectric constant of the surface material, its surface roughness and internal structure, as well as parameters of the observing system itself, such as the frequency, polarisation and incidence angle. In some cases, the counter-intuitive nature of the relationship between a surface material and its backscattering coefficient can be quite strong. An example of this is a situation in which a medium that is optically thick at visible wavelengths is practically transparent to microwave radiation, such as dry snow. It is clear that in such cases, *modelling* of the interaction between the microwave radiation and the target material will be especially important. Further difficulties are introduced by geometrical and topographic effects, such as highlighting, layover and shadowing, and by speckle.

Despite these complications, imaging radar data are being applied to an increasingly wide range of tasks. A number of these are suggested by figure 9.21. Over land surfaces, applications include cartography (where the main task is to recognise and delineate natural and man-made features), the detection and characterisation of geomorphological lineaments, vegetation mapping (including the identification and monitoring of different agricultural crops, change and damage detection, and the monitoring of soil moisture content), and highly 'applied' tasks such as monitoring flooding events and (using radar interferometry) subsidence and landslides.

Imaging radar also finds many applications in the marine environment. Surface wave fields can be imaged and their power spectra deduced, although with some difficulty as a result of the motion-dependent distortions discussed in section 9.6.3. The azimuth shift phenomenon can be used positively to identify ship wakes and hence to monitor shipping, for example in an area where fishing restrictions apply. Diffraction of waves by coastal features, and refraction by variations in bottom topography, are often clearly visible, and the latter phenomenon has been used as a bathymetric technique. There is also some evidence for the imaging of internal waves. Small-scale surface roughness is reduced by the presence of natural and artificial slicks, and these have been detected and monitored in radar imagery. The natural slicks can sometimes be indicators of unexploited oil reserves.

Imaging radar finds extensive applications in the delineation and monitoring of snow and ice. The presence of a snow cover can be detected provided the snow is not completely dry (in which case the microwave penetration depth can be very large, thus rendering the snowpack effectively transparent). The boundaries of ice sheets and glaciers can be delineated, and different surface zones, corresponding to different thermodynamic regimes, can often be located. Year-to-year monitoring of these zones, and of the glacier's total

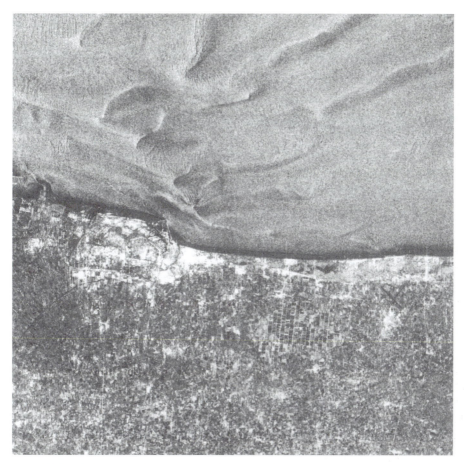

Figure 9.21 An L-band SAR image showing an area approximately 37 km × 40 km near Dunkirk, France. The image was obtained in 1978 by the SEASAT satellite from an altitude of approximately 800 km. Note the features visible in the water, which are probably a manifestation on the surface of the bottom topography. Note also the many features of the land surface that are visible, including roads, rivers, canals, built-up areas and agricultural fields. (Reproduced by courtesy of the National Remote Sensing Centre, UK.)

area, can be used to infer its *mass balance*, namely whether it is growing, shrinking or in dynamic equilibrium. If the glacier or ice sheet calves icebergs, these can often be detected and tracked using radar imagery. Imaging radar is also applied extensively to the study of sea ice. The delineation of boundaries between ice floes and open water is usually straightforward, and comparison of consecutive images of the same area of ocean allows the motion of ice floes to be tracked. A rather more difficult problem is the determination of ice type from radar images.

Although imaging radar already finds a wide range of applications, the extraction of quantitative parameters from the images is often difficult. One of the main reasons for this difficulty, at least for spaceborne radar imagery, is that the current generation of imaging radars measures only a single variable (the backscattering coefficient σ^0 for a given polarisation state, frequency and incidence angle) from each pixel, yet this variable is likely to depend on more than one physical property of the target material. There is thus not enough information to disentangle the contributions of the various physical properties. Very promising work has been conducted with airborne multipolarisation and multifrequency SAR systems (see plate 8), and it seems likely that the next major advance in the development of imaging radar applications will occur when systems like these are operated from space.

9.6.7 Example: Radarsat

We conclude this chapter by briefly describing a spaceborne SAR system, namely the instrument carried on the Radarsat satellite. This is a Canadian satellite, launched in November 1995 into an orbit with an altitude of 800 km, with a mission entirely dedicated to SAR imaging. The SAR system operates at C-band (5.3 GHz), HH-polarisation. The imaged swath lies to the right of the spacecraft (i.e. to its east when it is travelling north). Unlike earlier spaceborne SARs, the Radarsat instrument has substantial flexibility in observing modes, illustrated schematically in figure 9.22. For example, the wide-swath 'ScanSAR' mode has a swath width of 510 km, with incidence angles ranging from 20° at near swath to 49° at far swath and a spatial resolution of 100 m × 100 m, whereas the 'fine beam' mode observes a region 45 km × 45 km

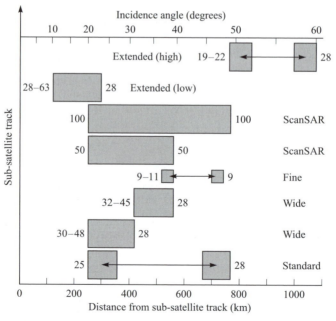

Figure 9.22. The different observing modes of the Radarsat SAR. The figures to the left of each box are the range resolutions, and the figures to the right are the azimuth resolutions, both expressed in metres. Double-headed arrows denote that the beam position can be selected within the range shown.

with a resolution of about $10\,\text{m} \times 10\,\text{m}$. The figure illustrates the poorer range resolution expected at small incidence angles on the basis of equation (9.11).

PROBLEMS

1. A radar transmits a power of 4 kW at a wavelength of 5 cm and over a range of 800 km. If the radar can detect a received signal of 10^{-16} W, calculate the required gain of the antenna, and hence estimate the area of the antenna, if it is to be able to detect values of σ^0 down to -25 dB from an area of $10^3\,\text{m}^2$. Assume that the antenna has unit efficiency.

2. A very simplified model of the radar backscattering coefficient of a sea surface is

$$\sigma^0 = A + B \cos 2\phi$$

where ϕ is the angle between the wind direction and the radar look azimuth, and A and B are as follows for Ku-band scattering at $40°$ incidence angle, HH-polarisation:

$$A = 0.8v - 30$$
$$B = 3.5 - 0.1v.$$

where v is the wind speed in m s^{-1} and σ^0 is given by these expressions in decibels.

A scatterometer observation measures σ^0 values of $-22.9\,\text{dB}$ and $-21.1\,\text{dB}$ looking north and east, respectively. Find the wind velocity. Is there any ambiguity in your answer? If so, could it be removed by a third observation?

3. Show that the amplitude of fully developed speckle has a Rayleigh distribution. Assume that the signal is composed of a very large number of components, each of which has the same amplitude but a phase randomly selected from the range 0 to 2π.

4. Two SAR tracks are used to obtain a SAR interferogram at a wavelength of 5 cm. The (x, z) coordinates, in metres, of the SAR when it acquires the 'master' image are $(-350\,000, 800\,000)$, and the coordinates when the 'slave' image is acquired are $(-350\,100, 800\,000)$. A strip of four pixels is imaged in the cross-track direction. The first (nearest) pixel has coordinates $(0, 0)$, and adjacent pixels are separated by 15 m in slant range. The table below gives the measured real and imaginary parts of the signal detected in each image and from each position. Find the coordinates of the pixels. [It is known that the surface slope does not exceed 0.2 in any direction.]

Pixel	Master image		Slave image	
	Re	Im	Re	Im
1	−13	−99	59	81
2	−57	82	81	−59
3	84	54	37	−93
4	0	−100	−80	−60

10

Platforms for remote sensing

10.1 Introduction

In this chapter we consider aircraft and satellites as platforms for remote sensing. There are other, less commonly used, means of holding a sensor aloft, for examples towers, balloons, model aircraft and kites, but we will not discuss these. The reason for this, apart from their comparative infrequency of use, is that most remote sensing systems make direct or indirect use of the relative motion of the sensor and the target, and this is more easily controllable or predictable in the case of aircraft and spacecraft. Figure 10.1 shows schematically the range of platforms, and their corresponding altitudes above the Earth's surface.

The spatial and temporal scales of the phenomenon to be studied will influence the observing strategy to be employed, and this in turn will affect the choice of operational parameters in the case of an airborne observation or of the orbital parameters in the case of a spaceborne observation.

10.2 Aircraft

Aircraft of various types provide exceptionally convenient and operationally flexible platforms for remote sensing, carrying payloads ranging from a few tens of kilograms to many tonnes. With a suitable choice of vehicle, a range of altitudes can be covered from a few tens of metres, where atmospheric propagation effects are negligible, to many thousands of metres, above most of the Earth's atmosphere. The choice of flying altitude will obviously have an impact on the scale, spatial coverage and spatial resolution of the data collected. It is also important in the case of pulsed systems, such as laser profilers, where the question of range ambiguity arises. The range of available platform speeds is more or less continuous from zero (in the case of a hovering helicopter) to several hundred metres per second. It is particularly important to match the flying speed to the characteristics of a scanner or a SAR system, and flying speed also has an obvious role in determining the total area from which data can be collected. Flying routes and times can also be chosen with great flex-

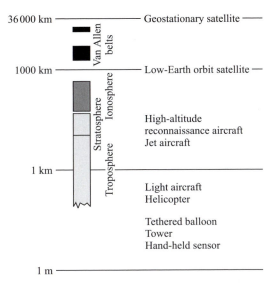

Figure 10.1. Remote sensing platforms arranged by typical altitude above the Earth's surface.

ibility, subject of course to any restrictions on the use of air space or those imposed by weather.

The main disadvantages of aircraft as platforms for remote sensing, when compared with spacecraft, are as follows. Firstly, a typical airborne observing mission has a duration of only a few hours, as compared with a few years for a spaceborne mission. This means that it is much more difficult to provide continuity of data, for example for a 10-year monitoring programme. Secondly, since airborne observations are acquired from much lower altitudes than spaceborne observations, the spatial coverage of the data is smaller, and airborne observations are obviously unsuitable for studying very large areas. (On the other hand, of course, they are much more suitable than spaceborne observations for detailed investigations of smaller area.) Finally, since aircraft necessarily operate within the Earth's atmosphere, and the atmosphere is in motion, neither the position nor the motion of the aircraft may be exactly what was intended. (These are problems for spacecraft too, just less severe.)

An aircraft's true position can be determined in two ways, often in combination. The recent enormous improvements in radiolocation methods, notably the Global Positioning System (GPS) discussed in appendix 1, mean that it is now easily possible to determine the position of an aircraft, with an accuracy of the order of 1 m or better, at the time. The second approach is through the use of ground control points (GCPs). These are features whose exact locations on the ground are known and which can also be observed and precisely located in the image collected by the airborne sensor. They can thus be used to correct the position, orientation and scale of the image, and also to correct any distortions that may be present. This procedure is discussed in section 11.3.1.2. Suitable GCPs can be 'naturally' occurring – for example, coastal features or road intersections – or can be emplaced especially for the purpose. In the latter case, suitable GCPs might be reflective markers fixed to the ground or, for active microwave sensors, passive or active radar transponders.

In addition to uncertainty in its position, an aircraft may also be subject to uncertainty or variation in its motion. The most important of these are roll, pitch and yaw, which are oscillations about three mutually perpendicular axes. These motions, and the distorting effect they have on uncorrected scanner imagery, are illustrated in figure 10.2. Many scanners for airborne use now include sensors and at least partial correction for roll. Pitch and yaw can lead to over- or under-sampling in the along-track direction, but provided the effects are small enough they can be corrected for using GCPs.

10.3 Satellites

Placing a satellite in orbit about the Earth is clearly more expensive than mounting an airborne remote sensing campaign, but the advantages, in terms of the increased platform speed and potential swath width, as well as continuity of observations, are substantial. In general, the spatial data coverage from a satellite mission is better than that obtainable from an airborne mission, and the fact that a spaceborne sensor may continue to function for three years or more (sometimes much more) means that temporally homogeneous datasets can be obtained.

The cost-effectiveness of satellite remote sensing was discussed briefly in chapter 1, where it was suggested that the economic benefits[1] of the data generated by an operational satellite (as opposed to one launched purely for the purpose of research) normally more than justify the cost of launching and operating it. An obvious advantage of using satellites for remote sensing, but one that poses interesting legal, security and moral questions, is the fact that

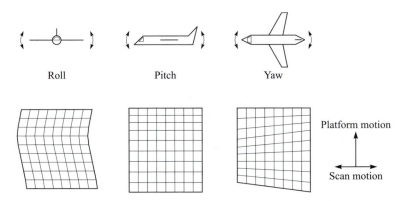

Figure 10.2. Roll, pitch and yaw as oscillatory motions of an aircraft, and the corresponding distortions of a scanned image of a uniform square grid.

---

[1] That is, the economic benefits to a nation or a consortium of nations, where economies of scale will apply. Most of the world's remote sensing satellites are still operated on this basis, although there is now an increasing trend towards commercial operation.

the laws of orbital dynamics do not respect political boundaries. A recent discussion of these questions is presented in the article 'Legal and international aspects [of remote sensing]' in Rees (1999).

10.3.1   Launch of satellites

This section provides a brief introduction to the considerations that apply to the placing of a satellite in orbit about the Earth. A more detailed treatment is provided by Chetty (1988).

To place a satellite in a stable[2] orbit, it is necessary to overcome the Earth's gravitational attraction and, to a lesser extent, the resistance of the lower atmosphere. This is achieved using a *rocket*, which is a vehicle that carries all its own fuel, including the oxidising agent, deriving a forward thrust from the expulsion backwards of the combustion products. Elementary classical mechanics shows that a rocket of total initial mass M_i burning a mass M_f of fuel will increase its speed, in the absence of gravitational and friction forces, by

$$\Delta v = u \ln \frac{M_i}{M_i - M_f} \tag{10.1}$$

where u is the speed of the exhaust gases relative to the rocket.[3]

Figure 10.3 illustrates a very simple model of the insertion of a satellite into a circular orbit of radius R around the Earth. The Earth is assumed to be a non-rotating uniform sphere of mass M and radius R_E. The first part of the process is to launch the rocket vertically upwards (i.e. radially away from the

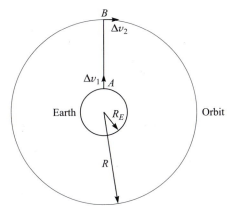

Figure 10.3. Schematic illustration of one method of placing a satellite in orbit around the Earth.

---

[2] Apart from the effects of atmospheric friction and the influence of the solar wind.

[3] In fact, the speed of the exhaust gases is often specified through the *specific impulse*, I, defined by $I = u/g$ where g is the gravitational field strength at the Earth's surface. Typical values of the specific impulse range from 200 to 300 s.

Earth's surface), with an essentially instantaneous 'burn' of the rocket motor at
A. This gives the rocket a speed Δv_1. The rocket then travels 'ballistically', that
is, under the influence of gravity alone (we ignore air resistance), until it comes
to rest at *B*. It is straightforward to show that the velocity increment Δv_1
needed to achieve this is given by

$$\Delta v_1^2 = 2\,GM\left(\frac{1}{R_E} - \frac{1}{R}\right) \tag{10.2}$$

where *G* is the gravitational constant. For example, if we take $R_E = 6400\,\text{km}$
and $R = 7200\,\text{km}$ (corresponding to an orbital altitude of 800 km), equation
(10.2) gives $\Delta v_1 = 3.7\,\text{km}\,\text{s}^{-1}$.

The final part of the process is to give the satellite the velocity necessary for a
circular orbit of radius *R*. Since the rocket starts from rest at *B*, the required
velocity increment Δv_2 is just the orbital velocity and hence is given by

$$\Delta v_2^2 = \frac{GM}{R} \tag{10.3}$$

For the example of $R = 7200\,\text{km}$, this gives $\Delta v_2 = 7.5\,\text{km}\,\text{s}^{-1}$, so the total
velocity increment needed for this type of launch is $11.2\,\text{km}\,\text{s}^{-1}$.

Figure 10.4 illustrates a more efficient procedure. In this case, the rocket is
launched tangentially to the Earth's surface at *A*, with a velocity Δv_1 sufficient
to put it into an elliptical orbit that just grazes the desired orbit at *B*. At *B*, a
further velocity increment Δv_2 is applied to inject the satellite into its circular
orbit. It can be shown that the required velocity increments in this case are

$$\Delta v_1 = \sqrt{\frac{2RGM}{R_E(R + R_E)}} \tag{10.4}$$

and

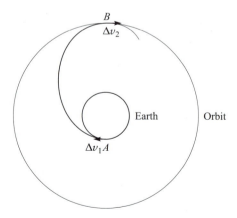

Figure 10.4. Injection of a satellite into a circular orbit using an elliptical Hohmann transfer
orbit.

$$\Delta v_2 = \sqrt{\frac{GM}{R}} - \sqrt{\frac{2R_E GM}{R(R + R_E)}} \qquad (10.5)$$

For our example of $R = 7200$ km, Δv_1 is 8.2 km s^{-1} and Δv_2 is 0.2 km s^{-1}, giving a total velocity increment of 8.4 km s^{-1}. However, if the satellite revolves in its orbit in the same sense as the Earth rotates (this is called a *prograde* orbit – see section 10.3.2), some of the required velocity is imparted by the Earth's rotation. Since the tangential speed of the Earth's surface is roughly 0.5 km s^{-1}, this can reduce the total velocity increment to about 7.9 km s^{-1}.

Since a typical value of u (the velocity of the exhaust relative to the rocket) is about 2.4 km s^{-1}, and we have seen that the rocket needs to accelerate the satellite to a speed of at least (roughly) 8 km s^{-1}, we can estimate from equation (10.1) that a rocket capable of inserting the satellite into our chosen orbit must initially consist of at least 96% fuel. Naturally, the payload (the satellite) can account for only a fraction of the remaining 4%, reduced still further when the effect of atmospheric friction is taken into account. For this reason, single-stage rockets are capable of placing only rather small masses into orbit. Instead, multiple-stage rockets, with three or four stages, are used, and these are capable of putting payloads of a few tonnes into low-Earth orbit and somewhat smaller masses into geostationary orbit.

10.3.2   Description of the orbit

If the Earth were a spherically symmetric mass with no atmosphere, and there were no perturbing influences on the motion of a satellite from the Sun, Moon, other planets, solar wind, and so on, the motion of a small satellite would in general follow an elliptical path with the Earth's centre as one focus of this ellipse. It will be useful to begin our investigation of satellite orbits by considering this idealised case, and figure 10.5 illustrates some of the terms we shall need to describe it.

The points P (the *perigee*) and A (*apogee*) are, respectively, the nearest and furthest points of the orbit from the Earth's centre E. The line AP is called the major axis, and its perpendicular bisector is the minor axis. The terms semi-major and semi-minor axes are normally used to refer to half the lengths of

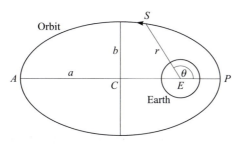

Figure 10.5. Description of a satellite S in orbit about the Earth.

these axes, namely a and b, respectively. The *eccentricity e* of the ellipse is the ratio of the lengths CE to CA; it is also related to a and b through

$$b^2 = a^2(1 - e^2) \tag{10.6}$$

According to Newton's law of gravitation, the period of the orbital motion (i.e. the time interval between successive passes through the same point in the orbit) is given by

$$P_0 = 2\pi\sqrt{\frac{a^3}{GM}} \tag{10.7}$$

where, as before, G is the gravitational constant and M is the Earth's mass. Although neither G nor M has yet been measured particularly accurately, their product has been measured to very high precision by observing the orbits of artificial satellites about the Earth. A recent value (Smith et al., 1985) is

$$GM = (3.98600434 \pm 0.00000002) \times 10^{14}\,\mathrm{m}^3\,\mathrm{s}^{-2}$$

The position of the satellite S in the orbital plane can be specified by the angle θ that it has rotated from the perigee, and its distance r from the Earth's centre E. These variables are related by

$$r = \frac{a(1 - e^2)}{1 + e\cos\theta} \tag{10.8}$$

which tells us the shape of the ellipse but not the rate at which the satellite travels along it. The position of the satellite in the orbit, specified by the angle θ (between $-\pi/2$ and $+\pi/2$), is related to the time t since it passed through the perigee by the following equation:

$$\frac{t}{P_0} = \frac{1}{\pi}\arctan\frac{(1 - e)\tan(\theta/2)}{\sqrt{1 - e^2}} - \frac{e}{2\pi}\frac{\sqrt{1 - e^2}\sin\theta}{1 + e\cos\theta} \tag{10.9}$$

Equation (10.9) is actually rather inconvenient since it gives the time in terms of the position, and we will more commonly want to know the position in terms of the time. Unfortunately, the equation is not analytically invertible, in other words it cannot be rewritten as a closed-form expression for θ in terms of t. It can, however, be inverted into a series expansion that is useful for small values of the eccentricity e, as shown in equation (10.10):

$$\theta = \frac{2\pi t}{P_0} + 2e\,\sin\!\left(\frac{2\pi t}{P_0}\right) + \frac{5e^2}{4}\sin\!\left(\frac{4\pi t}{P_0}\right) + \cdots \tag{10.10}$$

Use of the first three terms of this expansion, as shown, results in a maximum error in θ of approximately $4e^3/3$ radians.

In fact, most artificial satellites used for remote sensing are placed in orbits that are nominally circular and in practice have very small eccentricities, typically less than 0.01, and we may continue for the time being to develop our description of orbital motion on the assumption that $e = 0$. In this case, equations (10.9) or (10.10) show that θ increases uniformly with time. The orbit must be concentric with the Earth (which we are still assuming to be spherically symmetric) but need not be parallel to the equator. The angle between the plane of the orbit and the plane of the equator is called the *inclination* of the orbit (see figure 10.6). The inclination i is by convention always positive, and is less than 90° if the orbit is *prograde* (i.e. in the same sense as the Earth's rotation about its axis), and greater than 90° if the orbit is *retrograde*. An exactly *polar* orbit, in which the satellite passes directly over the Earth's poles, has an inclination of 90°. Near-polar orbits give the greatest coverage of the Earth's surface, and are widely used for low-orbit meteorological and other remote sensing satellites. It is, however, in general more expensive to inject a satellite into a near-polar orbit than into a prograde orbit, because of the advantage that can be taken of the Earth's rotation during the launch phase in the latter case.

The position of the satellite in space is in general specified by six variables, although there are various equivalent sets of variables. For example, we may specify the direction of Ω, the values of a, e and i, the instantaneous value of ϕ and the value of ϕ at perigee. These variables are usually called the *Kepler elements* of the orbit. For a circular orbit, the number of useful variables is four, since the eccentricity is zero and the orbit has no perigee.

In many cases, we will want to define the position of the *sub-satellite point*, namely the point on the Earth's surface that is directly below the satellite. If we continue to assume that the Earth is spherical, the latitude b and longitude l of this point[4] can be calculated using spherical trigonometry. The point Ω in figure 10.6 is the *ascending node* of the orbit, and is the point where the satellite crosses the equatorial plane from south to north. If the angle $SE\Omega$ is ϕ, as

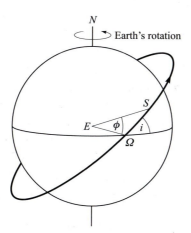

Figure 10.6. Circular satellite orbit in relation to the Earth's surface. E is the centre of the Earth, N the north pole, S the instantaneous position of the satellite, and Ω is the ascending node. i is the inclination of the orbit and ϕ is the angle $SE\Omega$.

---

[4] We adopt the convention that north latitudes and east longitudes are positive.

shown in figure 10.6, and the instantaneous longitude of Ω is l_0, the position of the sub-satellite point is given by

$$\sin b = \sin \phi \sin i \qquad (10.11.1)$$

$$l = l_0 + \mathrm{atan2}\left(\frac{\cos \phi}{\cos b}, \frac{\tan b}{\tan i}\right) \qquad (10.11.2)$$

(atan2 is a convenient inverse trigonometrical function that avoids the ambiguity involved in taking inverse sines, cosines or tangents. It is defined such that atan2(x, y) is the angle whose cosine is $x/(x^2 + y^2)^{1/2}$ and whose sine is $y/(x^2 + y^2)^{1/2}$.)

Even if the plane of the satellite's orbit were fixed in space (it is not, in general, as we shall in see in the next section), the sub-satellite track would not describe a great circle. This is because of the Earth's rotation. When the satellite has completed one orbit, the Earth will have turned to the east and so the orbit will appear to drift to the west. This is true of both prograde and retrograde orbits, and of course also polar orbits. The rotation of the Earth may be taken into account in equation (10.11.2) by noting that it is equivalent to a uniform rate of change of l_0. Typical sub-satellite tracks are illustrated in figure 10.7. This figure clearly shows the westward drift of the orbits. It also shows the fact, which can be deduced from equation (10.11.1), that a prograde orbit of inclination i reaches maximum north and south latitudes of i, and a retrograde orbit of inclination i reaches maximum latitudes of $180° - i$.

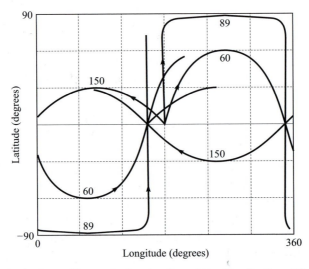

Figure 10.7. Typical sub-satellite tracks for circular orbits of inclination 60°, 89° and 150°. All the tracks begin at the equator and at longitude 180°. The period of the orbits is 100 min, and just over one complete orbit (120 min) has been plotted.

10.3.3 Effects of the Earth's asphericity

Thus far, we have assumed the Earth's mass to be distributed with spherical symmetry. It is not; roughly speaking, it is an oblate spheroid (i.e. the equator bulges outwards). The most convenient way to describe mathematically the effect of this non-spherical Earth on the motion of a satellite is to write the gravitational potential as a (possibly infinite) sum of spherical harmonics. As we might expect, the longitudinal variations are quite small when compared with the latitudinal variations, and in fact we normally need to consider only the latitudinal terms. In this case, the gravitational potential V per unit mass at latitude b and distance r from the Earth's centre can be written as

$$V = -\frac{GM}{r}\left(1 - \frac{a_e^2 J_2}{2r^2}(3\sin^2 b - 1) + \cdots\right) \tag{10.12}$$

where a_e is the Earth's equatorial radius,

$$a_e \approx 6\,378\,135\,\text{m}$$

The dimensionless term J_2, often called the *dynamical form factor*, expresses the equatorial bulge and has a value of

$$J_2 \approx 0.00108263$$

It has three important effects on the orbit of a satellite. Firstly, it increases the orbital period relative to the value P_0 given by equation (10.7). The nodal period (i.e. the time between successive ascending or descending nodes) is given by

$$P_n = 2\pi\sqrt{\frac{a^3}{GM}}\left(1 + \frac{3J_2 a_e^2}{4a^2}\left\{1 - 3\cos^2 i + \frac{1 - 5\cos^2 i}{(1 - e^2)^2}\right\}\right) \tag{10.13}$$

and the rate at which even a circular orbit is described is no longer exactly uniform.

Secondly, it causes the orbital plane to rotate (precess) about the Earth's polar axis, so that the plane is not fixed in space. This is illustrated in figure 10.8. The precession occurs at an angular velocity of

$$\Omega_p = -\frac{3J_2\sqrt{GM}\,a_e^2 a^{-7/2}\cos i}{2(1 - e^2)^2} \tag{10.14}$$

where a positive value indicates prograde precession. Because of the negative sign in equation (10.14), prograde precession can only occur when $\cos i$ is negative; that is, when the orbit itself is retrograde.

Finally, if the orbit is elliptical, the Earth's asphericity will cause the ellipse to rotate (precess) in its own plane, as illustrated in figure 10.9. The angular velocity of this precession is given by

$$\omega_p = -\frac{3J_2\sqrt{GM}\,a_e^2 a^{-7/2}(1 - 5\cos^2 i)}{4(1 - e^2)^2} \tag{10.15}$$

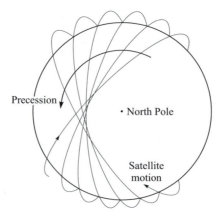

Figure 10.8. Precession of a satellite orbit around the Earth's polar axis, viewed from a fixed location in space looking down onto the North Pole. The rate of precession has been greatly exaggerated.

As before, a positive sign indicates prograde precession.

10.3.4 Special orbits

The dependence of the motion of a satellite on several parameters introduces the possibility of 'tuning' these parameters to give orbits with especially useful characteristics. In this section, we discuss the most important of these special orbits.

10.3.4.1 *Geostationary orbits*

A satellite in a geostationary orbit is, as its name suggests, at rest with respect to the rotating Earth. This is achieved by putting the satellite into a circular orbit above the equator, with a nodal period P_n equal to the Earth's rotational period P_E. The period P_E is not equal to 24 h. This is because, in a period of 24 h, the Earth does indeed rotate once with respect to the Sun, but since it is also orbiting the Sun in the same direction as it rotates on its axis, it has in fact

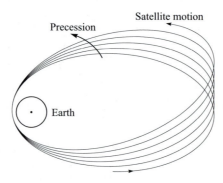

Figure 10.9. Precession of an elliptical orbit in its own plane (exaggerated for clarity).

rotated by slightly more than one complete turn with respect to a fixed reference system. The Earth takes approximately 365.24 days to orbit the Sun, so in 24 h it makes $1 + 1/365.24$ complete turns. Thus we can see that P_E must be about 23.9345 h or 86 164 s. This is called a *sidereal day* (from the Latin *sidus*, a star).

The requirements for a geostationary orbit are therefore that its inclination and eccentricity should be zero, and its nodal period should be 86 164 s. The value of the semi-major axis a that satisfies these requirements in equation (10.13) is about 42 170 km (about 6.6 times the Earth's radius), so geostationary satellites are located approximately 35 800 km above the equator. Such orbits are used for geostationary meteorological satellites such as GOES and METEOSAT, as well as for telephone and television relay satellites.

Figure 10.10 shows a typical view of the Earth from a geostationary satellite. The part of the Earth's surface that is visible from a geostationary satellite is a

Figure 10.10. View of the Earth from a geostationary satellite. The image was recorded by a METEOSAT satellite located at longitude 0° and above the equator, and it shows radiation reflected from the Earth in the band 0.4 to 1.1 μm. (Reproduced by courtesy of the National Remote Sensing Centre, UK.)

small circle centred on the sub-satellite point and having a radius of just over 81°, but in practice the useful coverage for quantitative analysis is assumed to be within a radius of 55°, and 65° for qualitative analysis. Figure 10.10 shows clearly that, while the coverage of latitudes reasonably close to the equator is excellent, the imaging of higher latitudes is impaired by the oblique viewing angle.

It would be convenient, but of course impossible, to place a satellite in a geostationary orbit above a point that is not located on the equator. If a satellite is placed in a circular orbit that has a nodal period of one sidereal day and a non-zero inclination i, the sub-satellite path traces a figure-of-eight pattern, crossing at a fixed point on the equator and reaching maximum north and south latitudes of i, as shown in figure 10.11. This can be called a *geosynchronous orbit*. The sub-satellite track is traced at an approximately uniform rate, so that for roughly half a day the satellite is above the 'wrong' hemisphere. Consequently, such orbits are not used in remote sensing.[5] An ingenious partial solution to this problem is the *Molniya orbit*, discussed in the next section.

10.3.4.2   *Molniya orbits*

The Molniya (Russian for 'lightning') orbit provides a partial solution to the problem of placing a satellite above a fixed point on the Earth's surface that is not on the equator. The orbit is highly eccentric, with the apogee (furthest distance) positioned above the desired point. Since a satellite's angular velocity is much smaller when it is far from the Earth than when it is near to it (this follows from equation (10.9)), it can, with a suitable choice of orbital para-

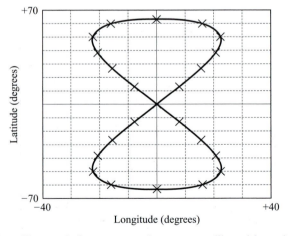

Figure 10.11. Subsatellite track for a geosynchronous satellite with an inclination of 63.4°. The crosses are at intervals of one sidereal hour.

---

[5] Except, of course, that a real geostationary satellite is unlikely to have an inclination that is exactly equal to zero, so that in practice the sub-satellite point will make small figure-of-eight oscillations.

meters, be arranged to spend much longer 'on station' than in the wrong hemisphere.

It is clear that a Molniya orbit must not rotate in its own plane, otherwise the position of the apogee would change over time. From equation (10.15) we see that this requires that the inclination i must satisfy the equation

$$1 - 5\cos^2 i = 0$$

so $i = 63.4°$ or $116.6°$. In practice, therefore, the latitude of the apogee must be fixed at $63.4°$ north or south. The nodal period is chosen to be *half* a sidereal day, so from equation (10.13) we see that the semi-major axis is about 26 560 km. (If the nodal period were one sidereal day, the semi-major axis would be of the order of 42 000 km, similar to that for a geostationary orbit, and the large eccentricity of the orbit would result in an unhelpfully large apogee distance.) The eccentricity is chosen to give a minimum altitude above the Earth's surface of the order of 500 km. For example, if we set $e = 0.740$, the satellite will have a perigee distance of about 6900 km, giving a minimum altitude of about 500 km, and an apogee distance of about 46 200 km. Figure 10.12 shows the sub-satellite track of an orbit with these parameters. It can be seen that the satellite spends a large proportion of its time near the perigee point; for practical purposes, it can be assumed to be 'on station' for 8 h, so that three satellites in the same orbit can provide continuous coverage.

Molniya orbits are used for telephone relay satellites by the former Soviet Union. They have not, as far as the author is aware, so far been used for remote sensing satellites, although they clearly have a useful potential in this role.

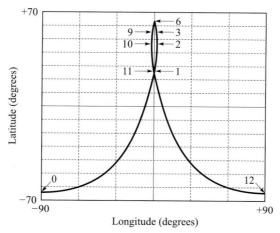

Figure 10.12. Subsatellite track of a satellite in a Molniya orbit with an inclination of $63.4°$ and an eccentricity of 0.74. The track is labelled with the elapsed time in sidereal hours. It repeats itself at intervals of 24 sidereal hours; after 12 sidereal hours, the track advances exactly $180°$ in longitude.

10.3.4.3 Low Earth orbits

Placing a remote sensing satellite in low Earth orbit (LEO) has the obvious advantage of improving the potential spatial resolution with respect to what can be achieved from a geostationary orbit, at the expense of reduced coverage. Such orbits are very widely used in spaceborne remote sensing.

The useful range of orbital altitudes for LEO satellites is constrained by the Earth's atmosphere and by the van Allen belts. If a satellite orbits too low above the Earth's surface, it will experience an unacceptably high degree of atmospheric friction and will spiral out of its orbit and towards the Earth (see section 10.3.5). Except for short satellite missions and very high resolution military reconnaissance satellites, the minimum useful altitude is effectively about 500 km.

The *van Allen belts* are rings of high-energy charged particles (mostly electrons and protons) in the Earth's equatorial plane, at altitudes between about 2000 and 5000 km and between 13 000 and 19 000 km. They have probably been 'captured' from the solar wind by the Earth's magnetic field. They represent a particularly difficult environment in which to operate a remote sensing satellite, so in practice there is an upper limit of about 2000 km to the altitude of a LEO satellite.

Careful choice of the orbital parameters of a satellite in LEO can give a range of useful properties. One particularly important type of LEO orbit is the *sun-synchronous orbit*. Such an orbit has the property that it precesses about the Earth's polar axis at the same rate (one revolution per year) that the Earth orbits the Sun.[6] This is illustrated in figure 10.13. The figure shows that the angle between the normal to the orbital plane and the line joining the centres of the Earth and the Sun is constant for a sun-synchronous orbit, whereas it increases at the rate of 360 degrees per year for a non-precessing orbit.

The mean angular speed Ω_s at which the Earth orbits the Sun is 2π radians per year or equivalently

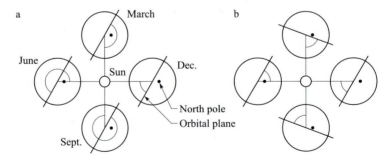

Figure 10.13. Schematic illustrations of (a) a non-precessing orbit and (b) a sun-synchronous orbit. In each case, the view (not to scale) is normal to the plane in which the Earth rotates around the Sun.

---

[6] That is, the precession rate is chosen to be the same as the *average* angular speed at which the Earth orbits the Sun.

$$\Omega_s = 1.991 \times 10^{-7}\mathrm{s}^{-1}$$

Substituting $\Omega_p = \Omega_s$ into equation (10.14), and setting $e = 0$ for simplicity, we can thus find the relationship between the inclination i and the semi-major axis a for a sun-synchronous orbit. This is plotted in figure 10.14, along with the corresponding nodal period calculated from equation (10.13). For example, the figure shows that a satellite in a circular sun-synchronous orbit with an altitude of 800 km will have inclination of about 99° and a nodal period of about 101 min. The fact that sun-synchronous low Earth orbits have inclinations close to 90°, and are hence near-polar orbits, is convenient since it means that a satellite in such an orbit has the potential to view a large fraction, perhaps all,[7] of the Earth's surface.

The most useful consequence of putting a satellite in a sun-synchronous orbit is that that it will cross the same latitude, in a given sense (i.e. north-bound or southbound), at the same local solar time, regardless of the long-itude or the date. This can be seen by considering the northbound crossing of the equator, namely the ascending node, although the argument can be gen-eralised to any latitude. Let us suppose that the satellite has a nodal period P_n and that it passes through the ascending node, at longitude zero, at time zero. This is the *Universal Time* (UT), which is equivalent to the Greenwich Mean Time to sufficient accuracy for the purpose of this discussion. At UT P_n the satellite has made one orbit, so it is again at its ascending node. During this time, the Earth has rotated through an angle (expressed in radians) of $P_n\Omega_E$ to the east, where Ω_E is the Earth's angular velocity of rotation, given by

$$\Omega_E = \frac{2\pi}{P_E}$$

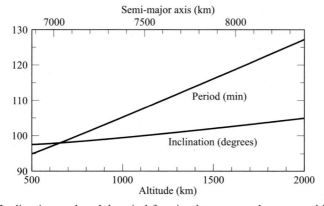

Figure 10.14. Inclination and nodal period for circular sun-synchronous orbits.

---

[7] This depends on the swath width of the sensor.

and P_E is one sidereal day. However, the satellite orbit has precessed through an angle $P_n\Omega_s$ to the east, and thus the longitude of this ascending node is, in radians,

$$-P_n(\Omega_E - \Omega_s)$$

where the minus sign indicates a longitude that is west of the Greenwich meridian. Now although the UT is P_n, the local solar time must be corrected for the longitude. The correction is 24 hours per 2π radians of longitude, so we can write the local solar time as

$$P_n - P_n(\Omega_E - \Omega_s)\frac{P'_E}{2\pi}$$

where P'_E is one solar day of 24 hours. Finally, we observe from the definition of the sidereal day given in section 10.3.4.1 that

$$\Omega_E - \Omega_s = \frac{2\pi}{P'_E} \qquad (10.16)$$

so the local solar time at the new ascending node is zero, and our demonstration is complete.

The fact that a sun-synchronous orbit crosses a given latitude at the same solar time is particularly useful for satellites carrying passive optical or infrared sensors, and is widely used. For example, the *Landsat-7* satellite is in a sun-synchronous orbit with a descending node at a local solar time of 10:00; that is, in the late morning. Thus, depending on the time of year, data can be collected from most of the descending (southbound) pass of the satellite while it is above the sunlit side of the Earth. This is illustrated in figure 10.15, which shows how the local solar time varies with the latitude for the descending pass.

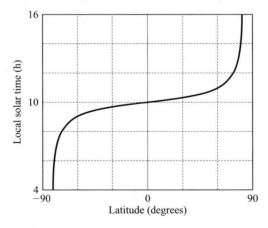

Figure 10.15. Local crossing time as a function of latitude for the descending pass of the Landsat-7 orbit, assuming a crossing time at the descending node of 10:00.

10.3.4.4 Exactly repeating orbits

It is often convenient to arrange that the sub-satellite track will form a closed curve on the Earth's surface, in other words that it will repeat itself exactly after a certain interval of time. This allows images having the same viewing geometry to be acquired on many occasions during the satellite's lifetime, and makes available a particularly simple method of referring to the location of images, for example the 'path and row' system used for the Landsat World Reference System (WRS).

In order for the sub-satellite track to repeat itself, it is clear that the Earth must make an integral number of rotations, say n_1, in the time required for the satellite to make an integral number of orbits, say n_2. However, we must also allow for the precession of the satellite orbit. Thus, we can write the condition for an exactly repeating orbit as

$$P_n(\Omega_E - \Omega_p) = 2\pi \frac{n_1}{n_2} \qquad (10.17)$$

Provided that the fraction n_1/n_2 is expressed in its lowest terms (i.e. that n_1 and n_2 have no common factors), this can be described as an n_1-day repeating orbit. If the orbit is also sun-synchronous, substituting equation (10.16) into equation (10.17) yields the simpler condition

$$\frac{P_n}{P'_E} = \frac{n_1}{n_2} \qquad (10.18)$$

where P'_E is one solar day of 24 hours.

It is desirable that n_1 should be as small as possible, since this determines the time interval between successive opportunities to observe a given location. On the other hand, n_2 governs the density of the sub-satellite tracks on the Earth's surface since there are n_2 ascending and n_2 descending passes. Thus, we require a large value of n_2. Because both the satellite's nodal period P_n and precession rate Ω_p depend on its semi-major axis a and inclination i,[8] there is some scope for 'fine-tuning' of the ratio n_1/n_2 by adjusting these orbital parameters. However, the ratio is mainly governed by the nodal period, and as figure 10.14 shows, this is generally between about 90 min and 120 min for satellites in LEO. Thus, in practice the ratio is constrained to roughly 0.07 ± 0.01. For example, an orbit suitable for observing highly dynamic phenomena, for which a revisit interval of no more than one day would be acceptable, would have a maximum value of n_2 of only about 16. This would give rather coarse spatial sampling since spatially adjacent sub-satellite tracks would be about 2500 km apart. On the other hand, if we required that the spacing of the sub-satellite tracks should be at most 100 km (e.g. because we wish to obtain complete coverage of the Earth's surface using a narrow-swath sensor), n_2 would have

---

[8] And also on the eccentricity, of course. However, practically all exactly repeating orbits are at least nominally circular.

to be at least 400, implying that successive revisits of a given location could not occur more frequently than once in 24 days.

Figure 10.16 shows the pattern of sub-satellite tracks for the orbit of Landsat-7. This is an exactly repeating sun-synchronous orbit with $n_1 = 16$ and $n_2 = 233$. From the figure, it can be seen that spatially adjacent tracks are separated by $360/233 = 1.55°$, as expected. It can also be noted that spatially adjacent tracks are followed at intervals of seven days, in the sense that the sub-satellite track moves $1.55°$ westward every seven days. This can be described as a seven-day *subcycle*. If the swath width of the sensor is sufficiently broad,[9] a given location can be observed more than once per repeat cycle. The time interval between these opportunities will be equal to the period of the subcycle, and this can provide a partial solution to the conflict between the requirements for frequent observation and for a dense network of sub-satellite tracks. Rees (1992) provides a detailed discussion of this technique.

10.3.4.5 Altimetric orbits

The last kind of special orbit that we shall consider is an orbit suitable for spaceborne altimetry. This has two desirable features. The first is that the ascending and descending sub-satellite tracks should intersect at roughly $90°$ on the Earth's surface, so that orthogonal components of the surface slope can be determined with equal accuracy. The second feature is important when considering altimetry of the ocean surface. As discussed below, the existence of ocean tides has implications for the temporal frequency with which measurements should be made.

In order to address the first requirement, we will calculate the orientation of the sub-satellite track on the Earth's surface. We will assume the Earth to be spherically symmetric. For a circular orbit of inclination i, the latitude b of the subsatellite point is given by equation (10.11.1) as

$$\sin b = \sin \phi \sin i \qquad\qquad (10.11.1)$$

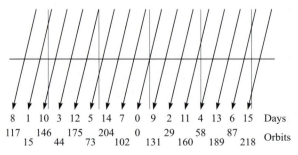

| 8 | 1 | 10 | 3 | 12 | 5 | 14 | 7 | 0 | 9 | 2 | 11 | 4 | 13 | 6 | 15 | Days |

| 117 | 146 | 175 | 204 | 0 | 29 | 58 | 87 | | Orbits |
| 15 | 44 | 73 | 102 | 131 | 160 | 189 | 218 | |

Figure 10.16. Repeat pattern for an orbit with $n_1 = 16$ and $n_2 = 233$. The arrows represent descending passes near the equator; the spacing of the graticule is $5°$ of longitude.

[9] Or if the instrument's field of view can be steered, as is possible with, for example, the SPOT satellites.

where ϕ is the angle defined in figure 10.6. Differentiating this expression with respect to ϕ gives

$$\frac{db}{d\phi} = \frac{\sin i}{\cos b} \sqrt{1 - \frac{\sin^2 b}{\sin^2 i}} \tag{10.19}$$

for the ascending (northbound) track, so the northward component of the velocity of the sub-satellite point is given by

$$v_N = \frac{2\pi r}{P_n} \frac{db}{d\phi} = \frac{2\pi r \sin i}{P_n \cos b} \sqrt{1 - \frac{\sin^2 b}{\sin^2 i}} \tag{10.20}$$

where r is the Earth's radius and P_n is the satellite's nodal period.

The simplest way to find the eastward component v_E of the velocity is first to consider the case of a non-precessing orbit around a non-rotating Earth. In this case, we can just apply Pythagoras's theorem to the components v_E and v_N to obtain

$$\pm \frac{2\pi r}{P_n} \sqrt{1 - \left(\frac{db}{d\phi}\right)^2}$$

for v_E, where a positive sign indicates a prograde orbit and a negative sign a retrograde orbit. Finally, including the effect of the Earth's rotation at Ω_E and the orbital precession at Ω_p gives the result

$$v_E = -(\Omega_E - \Omega_p) r \cos b \pm \frac{2\pi r}{P_n} \sqrt{1 - \left(\frac{db}{d\phi}\right)^2} \tag{10.21}$$

where, as before, the sign is chosen according to whether the orbit is prograde or retrograde.

Equations (10.19) to (10.21) define the velocity of the sub-satellite point, and hence its direction, relative to the Earth's surface. They have been derived only for the ascending track of the orbit; for the descending track, the sign of v_N is reversed. We can use these results to determine the suitability of an orbit for altimetry. Figure 10.17 shows how the crossing angle between the ascending and descending tracks depends on the latitude and inclination for satellites in LEO. It shows that the ideal inclination is near $42°$ or $132°$ for observations near the equator, although at latitude $60°$ it is nearer $69°$ or $110°$.

The second criterion for altimetric orbits, of particular relevance to studies of the ocean surface topography, is the choice of repeat interval, namely the frequency at which a given location is observed. If the altitude of this location does not change then the problem does not arise. However, the Earth's ocean and, to a lesser extent, its solid surface are subject to tidal action, which causes them to vary in height. These variations have strong periodicities: for example, the largest components of the deep-water tides have periods of 12.421, 23.934, 12.000 and 25.819 h.

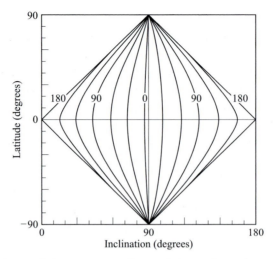

Figure 10.17. Crossing angle between ascending and descending sub-satellite tracks for satellite in low Earth orbit.

Suppose we try to observe the ocean surface with an altimeter that repeats its observation of a given location at intervals of exactly 3 days. Since exactly six cycles of the 12-hour tide will elapse between each measurement, the same point in the cycle will be measured on each occasion and we will not in fact observe any variation at all as a result of this tidal component. If our goal is to measure the mean ocean topography, rather than its mean instantaneous configuration at some point in the tidal cycle, we will fail to achieve it.

Now let us suppose that we try to observe the 12-hour tidal component by increasing the sampling interval very slightly to 3.01 days. After one sample interval, the 12-hour cycle will have occurred 6.02 times, so our measurement point will have advanced in effect by only 0.02 cycles. Thus, at least 50 such samples, requiring 150.5 days, will be necessary in order to observe this component of the tide at all phases of its cycle. This is the phenomenon of *aliasing*, which arises when a periodic phenomenon of frequency f_1 is observed by means of repeated measurements at frequency f_0. The frequency of the phenomenon is reduced to an apparent frequency f_a in the range $-f_0/2$ to $+f_0/2$ (the aliased frequency), which differs from the true frequency f_1 by an integral multiple of f_0. This can be expressed mathematically as

$$f_a = f_1 - f_0 \left[\frac{f_1}{f_0} + \frac{1}{2} \right] \tag{10.22}$$

where $[x]$ is a function of x defined as the largest integer not exceeding x. Returning to the first of our two examples above, we have $f_1 = 2$ day^{-1} and $f_0 = 1/3$ day^{-1}, so equation (10.22) shows that the aliased frequency is zero. In the second example, f_0 is decreased to 0.33223 day^{-1}, which gives an aliased frequency of 0.00664 day^{-1} and hence a period of 150.5 days.

10.3.4.6   *Station-keeping and orbital manoeuvres*

As a final remark on special orbits, we should note that some 'station-keeping' operations will probably be necessary unless the satellite mission is a particularly brief one. Some satellite missions include deliberate changes to the orbital parameters, for example to change the values of n_1 and n_2 of an exactly repeating orbit. Even if such alterations are not needed, the satellite will be perturbed in its orbit, for example by atmospheric friction and by the solar wind, and small adjustments may be necessary to bring it back into the desired orbit. For this reason, remote sensing (and other) satellites are often equipped with small rocket motors to enable them to make these adjustments. As discussed in section 10.3.1, the total amount of fuel needed is proportional to the mass of the satellite and to the total velocity increment that is required. For example, the velocity increment needed to change the semi-major axis of a circular orbit from a to $a + \Delta a$ is

$$\frac{v}{2}\frac{\Delta a}{a} \tag{10.23}$$

where v is the orbital velocity. Thus, for a satellite of mass M using a propellant of specific impulse I, the mass of fuel required for such an adjustment will be

$$\frac{Mv\Delta a}{2gIa}$$

where g is the strength of the Earth's gravitational field at the surface. A 5-km adjustment to a 1-tonne satellite in LEO might therefore require of the order of 1 kg of fuel.

10.3.5   *Decay of orbits and orbital lifetimes*

We have already mentioned the influence of atmospheric friction on launch vehicles and on orbiting satellites. Ultimately, uncompensated loss of energy through the action of atmospheric drag will cause a satellite to fall back to Earth, or to burn up in the atmosphere, and this provides an upper limit to the useful life of a satellite.

It can easily be shown that the reduction Δa in the semi-major axis of a satellite in a circular orbit, during one orbit of the Earth, is given roughly by

$$\Delta a \approx \frac{4\pi A \rho a^2}{M} \tag{10.24}$$

where A is the cross-sectional area of the satellite, M is its mass and ρ is the atmospheric density. Equation (10.24) is an approximation, since differences in drag coefficient can increase or decrease Δa by factors of up to about 2.

As an example of the application of equation (10.24), consider the Landsat-5 satellite. This has a mass of about 1700 kg and a cross-sectional area of about 10 m^2. It orbits at an altitude of 700 km, where the atmospheric density is of the order of 10^{-13} kg m^{-3}, so from equation (10.14) we expect the satellite to descend by about 0.4 metres per orbit or 5 metres per day. In fact, observations in changes of the orbital parameters of satellites have been used extensively for the determination of the properties of the Earth's outermost atmosphere.

The maximum useful lifetime would be expected, on the basis of equation (10.24), to be proportional to M/A for a given orbital configuration. To a good approximation, we may put

$$\tau = \frac{M}{A} f(h, e) \qquad (10.25)$$

where τ is the lifetime of the satellite in its orbit and f (which has SI units of m^2 s kg^{-1}) is a function only of the perigee altitude h and the eccentricity e. Figure 10.18 summarises the behaviour of the function $f(h, e)$, although it should be remembered that the properties of the atmosphere at large distances from the Earth are strongly affected by the Sun and hence are quite variable.[10] From the figure, we can estimate the lifetime of Landsat-5 to be about 500 years. Naturally, in this case, it is failure of the equipment carried by the satellite, rather than decay of the orbit, that is the limiting operational factor. However, the figure also shows that if the same satellite were put into a circular orbit with an altitude of only 100 km, its lifetime would be expected to be only a couple of days. Very low orbits, below about 150 km, are thus only used for military

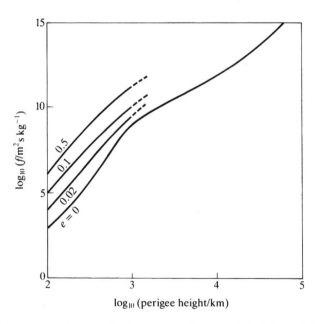

Figure 10.18. Approximate behaviour of the function $f(h, e)$ defined in the text.

---

[10] This variability also means that the position of a satellite in its orbit is not easy to predict accurately. In a typical LEO, an uncertainty of the order of 1 km in the along-track postiton can develop in a few days. For this reason it is normally necessary to have some means of accurately monitoring a satellite's position, and if necessary correcting its orbital parameters as discussed in section 10.3.4.6.

reconnaissance missions in which mission duration can be sacrificed in favour of very high spatial resolution.

PROBLEMS

1. Prove equations (10.2) to (10.5).

2. Consider a satellite in orbit about a spherically symmetric planet. If the orbit is circular, the sub-satellite point travels uniformly along its circular track. If the orbit is not circular, however, the sub-satellite track is still circular, but the sub-satellite point moves along it at a variable rate. Show that the maximum along-track error in calculating the position of the sub-satellite point on the assumption that it moves uniformly is about eD, where e is the eccentricity of the orbit and D is the planet's diameter.

3. Landsat-5 is in a circular sun-synchronous orbit with an inclination of 98.2°. It makes exactly 233 orbits of the earth in exactly 16 days, and crosses the equator southbound at 09.30 local time. Calculate the local time at which it crosses latitude 52° north in the southbound direction.

4. A certain satellite is placed in an exactly repeating sun-synchronous orbit in which 502 orbits are made every 35 days. Show that there is an orbital 'sub-cycle' with a period of 3 days, and calculate the rate and direction in which it drifts.

11

Data processing

11.1 Introduction

The general direction of this book has been to follow approximately the flow of information, from the thermal or other mechanism for the generation of electromagnetic radiation, to its interaction with the surface to be sensed, thence to its interaction with the atmosphere, and finally to its detection by the sensor. It is clear that that the information has not yet reached its final destination. Firstly, it is still at the sensor and not with the data user. Secondly, the 'raw' data will in general require a significant amount of processing before they can be applied to the task for which they were acquired.

In this chapter, we shall discuss the more important aspects of the processes to which the raw data are subjected. For the most part, it will be assumed that the data have been obtained from an imaging sensor so that the spatial form of the data is significant. The principal processes are transmission and storage of the data, preprocessing, enhancement and classification. The last three processes are generally regarded as aspects of *image processing*, a major field of study in its own right, and we shall not be able to do much more here than outline its general features. There are many books on the subject to which the interested reader may be referred, for example Mather (1987), Richards (1993) and Schowengerdt (1997).

11.2 Transmission and storage of data

It is clear that the data must be brought from the sensor to the place where they are to be analysed. In the case of airborne remote sensing this presents no fundamental difficulty, since missions are comparatively short and it is easy to transport the data, whether recorded on photographic film or digitally on some medium. Similar remarks apply to short-duration missions on low-altitude reusable platforms such as the Space Shuttle,[1] but when we consider an unmanned satellite in a long-duration mission (perhaps five years or more)

---

[1] Or even non-reusable platforms. See the discussion of the Corona system in section 5.7.

from which the satellite is unlikely to be recovered, the situation is rather different. Here, there are essentially three possibilities.

The simplest and most usual method of transmitting data from a satellite to the ground is to broadcast them continuously from the satellite, as they are received, to a network of receiving stations on the Earth's surface. Successful reception of the data requires that the line of sight from the receiving station to the satellite is not obscured, so that siting of the receiving station is important, and also that the elevation of the line of sight above the horizon is sufficiently large, so that atmospheric degradation of the signal is not significant. Together, these requirements define a *station mask* for a given receiving station. This is the region on the Earth's surface within which the sub-satellite point must lie in order that data can be received from the satellite. If the sensor on board the satellite is a narrow-swath nadir-viewing instrument, it is clear that the station mask also corresponds to the region from which data can be collected.

We can calculate the station mask quite simply if we assume that the Earth is spherical with radius R, the satellite is in orbit at an altitude h above the Earth's surface, and that there are no obstructions on the horizon as seen from the receiving station. This is shown in figure 11.1. The angular distance ϕ subtended at the Earth's centre between the receiving station and the satellite is given by

$$\cos(\theta + \phi) = \frac{R}{R+h}\cos\theta \qquad (11.1)$$

where θ is the elevation angle of the line of sight. For example, if we set $\theta = 5°$ (a reasonable minimum value in order to avoid an excessively long atmospheric path length), equation (11.1) shows that for a satellite at an altitude of 700 km, $\phi \approx 21°$. In this case, therefore, the station mask will be a circle of radius 2400 km centred on the receiving station.

Figure 11.2 shows the typical area covered by a single station mask. Although there are now many receiving stations throughout the world, they

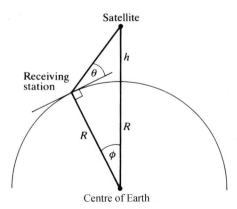

Figure 11.1 Relationship between the elevation angle θ of the line of sight to a satellite and the angle ϕ subtended at the Earth's centre between the receiving station and the satellite.

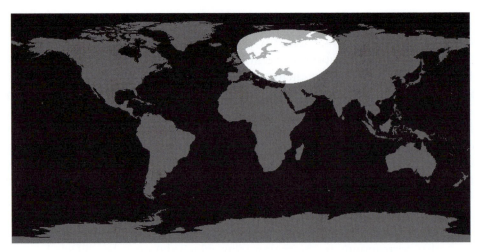

Figure 11.2. Typical station mask for a satellite altitude of 700 km.

are all situated on land and, as figure 11.2 suggests, this cannot be sufficient to provide complete coverage over the oceans. Partly to solve this problem, but also to reduce the number of receiving stations required to give complete global coverage, an alternative approach is to store the data on board the satellite, in a tape recorder or solid-state memory. The data can then be transmitted to a receiving station when the satellite is within view. If a whole orbit's-worth of data (ascending and descending) is stored, and downlinked at the end of the orbit, the data will be collected in a time of about 100 min and must be transmitted in a time of about 12 min, so the data transmission rate must be about eight times the rate at which they are collected. This fact, and the difficulty of providing very large amounts of reliable on-board data storage, can limit this mode of operation such that an instrument can collect data for only a fraction of the orbit.

Finally, we mention the use of relay satellites for increasing the useful coverage of a network of receiving stations. These are satellites in geostationary orbit, and the data collected by the remote sensing satellite are collected by whichever relay satellite is in view and then rebroadcast to Earth. The later Landsat satellites use relay satellites called TDRSS (Tracking and Data Relay Satellite System).

Whichever of these methods is adopted, we must give some consideration to the volumes of data involved, in order to calculate the transmission rate and data storage requirements. For an imaging sensor of swath width w and rezel area A, carried by a sensor whose ground-track speed is v, the mean rate at which rezels are viewed is vw/A. If n bits of data and k spectral bands (or equivalent, e.g. polarisation states) are recorded from each rezel, the *minimum* data rate must be

$$\frac{knvw}{A}$$

bits per second. Inserting some typical numbers into this equation, we find data rates for satellite-borne sensors that range from 10 kb s^{-1} (1 kb = 1024 bits) for a radar altimeter to 10 Mb s^{-1} (1 Mb = 1024 kb) for an imaging radar or high-resolution optical imaging system. The actual data rates will be higher than those calculated according to the formula, by a factor dependent on the instrument's design and the degree of oversampling. For example, a typical imaging radar will in fact generate data at of the order of 100 Mb s^{-1}. We can see, therefore, that spaceborne remote sensing systems have the potential to generate terabits of data per day (1 Tb $\approx 10^6$ Mb), and although at present most of these data are not permanently stored, the requirements on data storage are still large. Table 11.1 summarises the capacities of the main data storage media.[2]

11.3   Image processing

As was mentioned earlier, image processing is generally considered to consist of the three steps: preprocessing, image enhancement and classification. Roughly speaking, these steps involve, respectively, the removal of systematic errors in the data, increasing their intelligibility (perhaps by the removal of random errors, or by redisplaying the data) as a representation of the object being sensed, and extracting meaningful patterns from the data. As will be apparent from this brief description, and more so from the more detailed descriptions that follow, the distinctions between these steps are not clear-cut. This is one of the justifications for regarding image processing as a single coherent subject, whose aim is to extract meaningful, preferably quantitative,

Table 11.1. Storage media for digital data

Typical approximate values of the storage capacities (given in B = bytes = 8 bits) are given. For comparison, an uncompressed Landsat-7 ETM image requires about 400 MB. The approximate volumes needed to store 1 TB of data are also given. For comparison, the volumes of a typical desk drawer, a floor-to-ceiling storage rack and an entire office are of the order of 10^{-2}, 1 and 10 m^3, respectively

Medium/device	Storage capacity (GB)	Volume needed to store 1 TB (m^3)
Floppy disc	0.001	10
0.5-inch tape	0.2	10
Compact disc	0.7	10^{-2}
Removable magnetic cartridge	1	10^{-1}
Magneto-optical disc	5	10^{-1}
8-mm tape cartridge	10	10^{-1}
Hard disc	50	10^{-1}

[2] For comparison, the equivalent figures for an aerial photographic negative might be of the order of 1 GB, with of the order of 10^{-2} m^3 required to store 1 TB.

patterns from the detected data. Another justification is of course the very wide applicability of such techniques outside the field of remote sensing.

Most image processing is carried out on digital data, since this is the format in which most data are supplied, and since it is much easier to perform all but the simplest operations on digital data that are held in computer memory. We will therefore consider a one-band image to be a two-dimensional array of numbers (figure 11.3), each of which represents the intensity of the radiation reaching the sensor from one element (rezel) of the Earth's surface. The figure shows an image consisting of $m \times n$ pixels, such that the location of any pixel can be specified by (i, j), i being the column number and j the row number. The value $D(i, j)$ is an integer, variously called the pixel value, digital number (DN) or *grey level* of the pixel (i, j). For a multiband image, the array is three-dimensional, as shown in figure 11.4. In this case, the pixel (i, j) is represented by the pixel values $D(i, j, k)$, where $k = 1$ corresponds to band 1, $k = 2$ to band 2, and so on. There is no intrinsic difficulty in modifying these definitions to include continuous data such as photographic images, but it will be more convenient to continue to assume that we are dealing with digital data.

11.3.1 Preprocessing

As described at the beginning of this section, the preprocessing stage removes systematic errors from the data. The most important operations are the correction of radiometric and geometric errors, namely calibration of the detected signal and registration of the image data to true surface positions. We should also include under this heading the initial stages of processing synthetic aperture radar data (see chapter 9), which require substantial 'unscrambling' to convert them from a set of amplitudes and phases into a radar image.

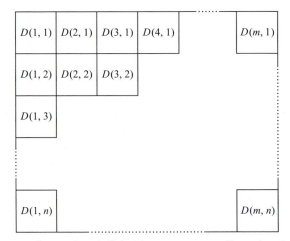

Figure 11.3. Structure of a one-band digital image as a two-dimensional array.

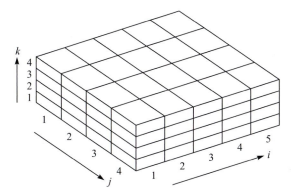

Figure 11.4. Structure of a multiband image as a three-dimensional array. Each cell contains a single value and is indexed by the values of (i, j, k). where i is the column number, j the row number and k the band number.

However, the details of SAR processing are beyond our scope here, and we will not add to the remarks made on the subject in chapter 9.

11.3.1.1   Radiometric correction

The data will in some cases need to be calibrated, although this is unusual and difficult for photographic images. For optical and infrared systems, there are effectively two steps in this process. The first is to establish the relationship between the pixel value recorded by the sensor and the relevant physical property (normally the spectral radiance) of the radiation incident on the sensor. This aspect of calibration is usually achieved, at least in part, through the design of the sensor itself, which will often include internal calibrators or be designed to make periodic observations of, for example, the Sun. In the case of satellite imagery, nominal calibration data are often provided to users by the operating agency, usually in the form of at-satellite radiances corresponding to the minimum and maximum pixel values (e.g. to pixel values of 0 and 255 for 8-bit data).

This first step in radiometric correction is to convert the pixel values into 'at-satellite radiances'. The second step is correction for atmospheric propagation effects, to obtain 'at-surface radiances'. The means by which this can be achieved have been discussed in sections 6.2.6 and 6.3.6. In some cases, for example where it is necessary to compare optical images that have been acquired under very different conditions of illumination, it can be desirable to correct the data for the effects of illumination geometry and even for the distance from the Sun to the Earth. Figure 11.5 illustrates a typical illumination geometry for a nadir-viewing sensor. If the surface scattering can be assumed to be Lambertian (see section 3.3.2), the reflected at-surface radiance L (i.e. the radiance at the surface heading towards the sensor) can be written as

$$L = \frac{E_{\text{exo}}}{\pi d^2} \rho \cos \theta \qquad (11.2)$$

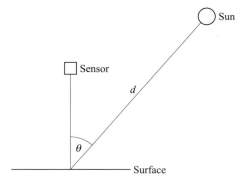

Figure 11.5. A sensor views a horizontal surface at nadir, the surface being illuminated by the Sun at a zenith angle of θ. The Sun's distance from the Earth is d astronomical units.

where E_{exo} is the mean exoatmospheric irradiance of the Sun at a distance of one astronomical unit,[3] d is the actual distance from the Sun to the area being sensed,[4] and ρ is the effective planetary reflectance of the surface. By using equation (11.2) to calculate ρ from L, d and θ, we obtain a measure of reflectance that is largely independent of the viewing conditions.

Active microwave systems are often calibrated against a target of known backscattering cross-section, such as a corner-cube reflector (three adjacent sides of a cube, forming an interior corner, constructed from a radar-reflective material.)

11.3.1.2 Geometric correction

Geometric correction involves relating the spatial coordinates in the image (i.e. the column and row coordinates (i, j) of a pixel) to the corresponding spatial coordinates on the Earth's surface. If data are available on the position and direction of view of the sensor at the time the image was acquired, this may be enough to establish the necessary relationship. In most cases, however, it is necessary to use ground control points (GCPs) to determine the relationship. The use of GCPs has already been mentioned very briefly in section 10.2. They are particularly important if several images are to be joined together in a mosaic, or if images of the same area, acquired perhaps at different times or with different sensors, need to be overlaid and compared.

The relationship between the coordinates of a pixel in the image and of the corresponding point on the Earth's surface can be expressed functionally. For example, if we denote the coordinates of a point in the image by (x_i, y_i)[5] and

---

[3] The astronomical unit, or AU, is defined as the semi-major axis of the Earth's orbit around the Sun, and is equal to 1.496×10^{11} m.

[4] This varies according to the time of year, reaching a maximum value of about 1.017 AU in early July and a minimum value of 0.983 AU in early January. Its value can be determined from standard astronomical tables.

[5] Note that we have temporarily abandoned the (i, j) column–row notation for coordinates in the image. The reason is that, for simplicity, we wish to treat the image coordinates as continuous variables in this section. We discuss below how to relate these continuous variables to the integer values.

the coordinates of the corresponding point on the surface by (x_s, y_s),[6] we can express the needed relationship quite generally as

$$x_s = f(x_i, y_i) \tag{11.3.1}$$

$$y_s = g(x_i, y_i) \tag{11.3.2}$$

The functions f and g will depend on the kind of relationship that we assume exists between image coordinates and surface coordinates. In simple cases, we can assume that the same functions f and g apply everywhere within the image. In more complicated cases, it may be necessary to find the appropriate forms of f and g locally.

The simplest useful model that has the form of equations (11.3) is the general linear model represented by equations (11.4):

$$x_s = a_1 + a_2 x_i + a_3 y_i \tag{11.4.1}$$

$$y_s = a_4 + a_5 x_i + a_6 y_i \tag{11.4.2}$$

Although much more complicated models, with many more parameters, are obviously possible, this model can allow for a shift in origin, rotation, different scales in the x and y directions, and some types of skew. Since it has six parameters (a_1 to a_6) and each GCP provides two pieces of information (an x-coordinate and a y-coordinate), in principle three GCPs are sufficient to determine this model. In practice, one would use many more than three (a good rule of thumb is to use ten times as many as the minimum), as a precaution against random errors. Determination of the appropriate values of the model parameters a_1 to a_6 is then carried out using a minimum least-squares method. Figure 11.6 shows a typical result of using GCPs to define the parameters of a transformation having the form of equation (11.4). The grid superimposed on the images has been derived from the GCPs using a least-squares fitting procedure. It represents the fact that the surface coordinates (x_s, y_s) are known for every point (x_i, y_i) in the image.

The example of figure 11.6 shows an image for which the transforming functions f and g, defined in equations (11.3), have been calculated. However, we will often wish to go one step further and actually change the geometry of the image so that it conforms to the chosen coordinate system (x_s, y_s). For example, in the case of figure 11.6 this would correspond to distorting the image so that the superimposed grid became square, with the grid lines running horizontally and vertically. In general, we can define such a transformation as follows:

$$i = F(i', j') \tag{11.5.1}$$

$$j = G(i', j') \tag{11.5.2}$$

where (i, j) are the pixel coordinates in the untransformed image, (i', j') are the pixel coordinates in the transformed image, and the functions F and G

---

[6] These could be, for example, longitude and latitude, or grid coordinates.

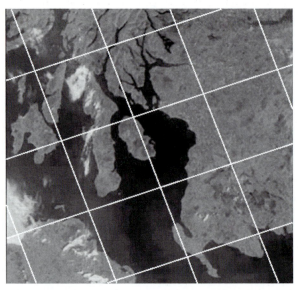

Figure 11.6. A typical satellite image after geometric correction (schematic). The lines are at 50-km intervals on the British National Grid system.

can be derived from the functions f and g in equations (11.3). Equations (11.5) provide a 'recipe' for finding the pixel coordinate (i, j) in the original untransformed image that should be copied into a location (i', j') in the new image. The pixel coordinates are only defined for integer values of the column and row numbers i and j, but equations (11.5) can yield non-integer values. If this happens, how do we choose the pixel value from the original image to copy into the new image? This is the problem of *resampling* an image.

The simplest resampling technique is nearest-neighbour resampling. Here, we simply identify the pixel in the original image that is spatially nearest to the calculated position (i, j), and copy its pixel value to the location (i', j') in the new image. This approach has the merit that it does not alter any of the pixel values; on the other hand, it can produce resampled images that are visually somewhat displeasing since they tend to have jagged edges. Smoother results are produced by interpolation, in which the pixel value copied into the new location (i', j') is an appropriately weighted average of the pixel values in the neighbourhood of the calculated location (i, j). The commonest forms are bilinear interpolation, which uses a 2×2 pixel neighbourhood, and bicubic interpolation, which uses a 4×4 pixel neighbourhood. Detailed discussions of these and other interpolation techniques can be found in almost any work on digital image processing, for example Richards (1993). However, interpolation is generally undesirable if subsequent processing of the data will make quantitative use of the pixel values, since these are to some extent corrupted by the averaging process.

Before leaving the subject of geometric correction, we should make a final remark about aerial photography. We saw in section 5.5 how the relief displacement in a pair of stereophotographs can be used to deduce the surface relief of the area common to both photographs. The relief displacement is itself a form of geometric distortion in the photographs. It can be removed if the relief and the viewing geometry are known, using equations (5.8) and (5.9) to define the relationship between the coordinates of a point on the surface and the coordinates of the corresponding point in the image. A photograph to which this procedure has been applied is termed an *orthophotograph* (see e.g. Avery and Berlin, 1992).

11.3.2 *Image enhancement*

Improvements to the image can be divided into those that operate on individual pixels without reference to their spatial context, and those that also make use of spatial information. The first type can generally be referred to as contrast modification, and the second as spatial filtering.

11.3.2.1 *Contrast modification*

In order to discuss image contrast and its modification, we first need to introduce the concept of the image histogram. This is simply a graph or table showing the number $h(d)$ of pixels in an image (or a subregion of an image) having pixel value d. Contrast modifications involve changing the shape of the histogram by reassigning one pixel value to another. This can be represented by a graph of the 'transfer function', showing how the output pixel value is related to the input pixel value. These ideas are illustrated in figure 11.7. The figure shows a contrast modification in which the transfer function has a constant gradient that is greater than unity. This is referred to as a linear contrast stretch, and it has the effect of expanding the range of pixel values occupied by the data. In the example of figure 11.7, the input data occupy a rather compressed range of pixel values, with no values at all occurring outside the range 27 to 175. The contrast stretch has therefore been chosen to expand this range to fill the whole range of pixel values (in this example, 0 to 255) that is available.[7] This will also obviously have the advantage of enhancing subtle tonal variations in the image (see figure 11.8). These can be enhanced even further by increasing the gradient of the transfer function, but this would lead to some 'clipping' of the data. For example, if the slope of the transfer function were 5, a range of input pixel values of only $255/5 = 51$ would lead to a range of 255 in the output values.

The linear contrast stretch is merely the simplest of a range of possible contrast modifications. In general, these can be thought of as attempts to

---

[7] Figure 11.7c also shows a characteristic result of increasing the contrast of an image. Because the range of pixel values has been expanded, some pixel values in the new image are not represented – the histogram values are zero.

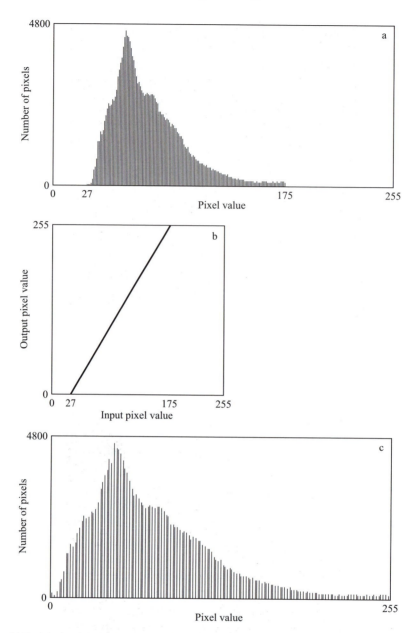

Figure 11.7. (a) An image histogram before contrast modification; (b) a transfer function relating input and output pixel values; (c) the image histogram after applying the transfer function (b) to the original image.

give the image histogram some desired form. We might, for example, wish to modify the contrast of one image so that its histogram matches that of another image, before attempting to make a mosaic from the two images.

Suppose we have an image with N_1 pixels and histogram $h_1(d)$, where d can take any value between 0 and n_1, and we want to find a transfer function $f(d)$ that will transform the histogram into $h_2(d)$, where d can now take any value

(a)

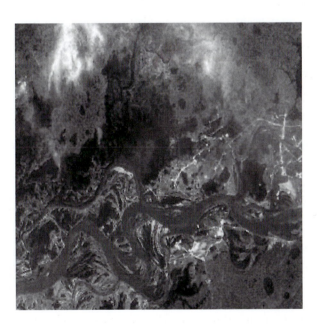

(b)

Figure 11.8. Linear contrast stretch. (a) Original image (with image histogram as shown in figure 11.7a); (b) after application of the transfer function shown in figure 11.7b. (MSU-E image of oil wells in the Ob' basin, © R&D Center ScanEx, Moscow. Reproduced by permission from http://scanex.ss.msu.ru.)

between 0 and n_2. The new image has N_2 pixels. Normally, n_2 will equal n_1 (e.g., if we are dealing with 8-bit data, both values will be 255), but we might, for example, wish to scale 10-bit input data into 8-bit output data. Similarly, N_2 will normally equal N_1, although we might wish to match the histograms of two different images. Whatever the values of n_1 and n_2, N_1 and N_2, the required transfer function can be calculated as follows. First, scaled cumulative histograms are calculated:

$$g_1(d) = \frac{n_1}{N_1} \sum_{j=0}^{d} h_1(i) \qquad (11.6.1)$$

$$g_2(d) = \frac{n_2}{N_2} \sum_{j=0}^{d} h_2(i) \qquad (11.6.2)$$

Thus $g_1(d)$ has a maximum value of n_1 when $d = n_1$, and similarly for $g_2(d)$. Next, we calculate the inverse function of g_2, $g_2^{-1}(d)$. This is defined by

$$g_2^{-1}(g_2(d)) = g_2(g_2^{-1}(d)) = d \qquad (11.7)$$

Finally, the required transfer function is calculated from

$$f(d) = g_2^{-1}(g_1(d)) \qquad (11.8)$$

As an example of this method, let us consider the very common contrast modification known as *histogram equalisation*. Here, the aim is to produce a histogram that is 'flat'; that is, all pixel values occur with equal frequency. For simplicity, we will assume that $n_1 = n_2 = n$, $N_1 = N_2 = N$. The required output histogram thus has $h_2(d) = N/(n+1)$ for all values of d, and hence, from equation (11.6.2),

$$g_2(d) = \frac{n}{n+1}(d+1)$$

The inverse function, defined by equation (11.7), is therefore

$$g_2^{-1}(d) = \frac{n+1}{n}d - 1$$

and hence the required transfer function is

$$f(d) = \frac{n+1}{N} \sum_{i=0}^{d} h_1(d) - 1$$

This is a suitably scaled version of the cumulative histogram of the original image. Figure 11.9 shows the image histogram before and after equalisation, and the transfer function. The histogram in figure 11.9c does not look particularly flat. This is a consequence of the fact that the transfer function defined by equation (11.8) cannot be realised precisely using integers. However, if the histogram were recalculated in 'bins' of, for example, 16 pixel values, this digitisation effect would be much less apparent and the histogram would

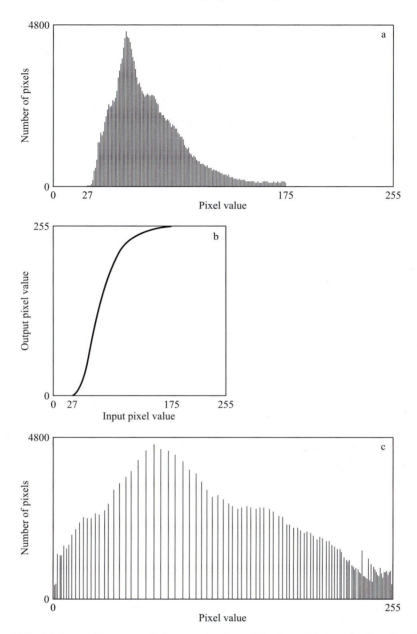

Figure 11.9. (a) Image histogram before contrast modification; (b) transfer function for histogram equalisation; (c) image histogram after equalisation.

appear much flatter. Figure 11.10 shows the appearance of the image of figure 11.8 after histogram equalisation.

Finally, we note that contrast modifications of the type discussed in this section are sometimes applied only to the look-up table (LUT) used to generate the display of an image, and not to the image data themselves. In this way, the operator of the image processing system is assisted in identifying features of interest in the image, but the original image data are unchanged.

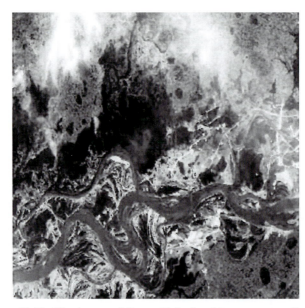

Figure 11.10. The image of figure 11.8 after histogram equalisation.

11.3.2.2   Spatial filtering

The contrast modifications discussed in the previous section do not necessarily alter the image data, merely the way they are displayed. Spatial filtering, on the other hand, does change the image data. The pixel value at a given location is modified according to the pixel values in its neighbourhood.

The simplest type of spatial filtering is spatial averaging, normally applied to reduce random noise or speckle in the data. The most obvious way of accomplishing this is to replace a given pixel value with the average pixel value for that pixel and its neighbours. This can be represented diagrammatically by a grid of boxes, each box representing a pixel, the central box representing the pixel to be processed, and the number in each box representing the weight of the contribution made by that pixel to the total sum (figure 11.11). If the total of all these weights is unity, the average pixel value (brightness) of the image will be unchanged.

Variants of this type of spatial averaging operator can have weights that decrease with distance from the centre, although the sum of the weights must still be unity if the operation is required not to change the average brightness of the image. However, all spatial averaging operations, as well as their desirable property of reducing noise, have the generally undesirable effect of blurring the image. For example, sharp edges will be smoothed out, and the contrast between a single bright or dark pixel against its neighbours will be reduced. For this reason, spatial averaging can also be described as a smoothing operation.

Many other types of spatial filter can be considered. For example, a sharpening filter has the opposite effect to a smoothing filter, narrowing the widths of boundaries between regions of different brightness, increasing the contrast

$\frac{1}{9}$	$\frac{1}{9}$	$\frac{1}{9}$
$\frac{1}{9}$	$\frac{1}{9}$	$\frac{1}{9}$
$\frac{1}{9}$	$\frac{1}{9}$	$\frac{1}{9}$

Figure 11.11. Uniform 3×3 smoothing filter.

between single pixels and their backgrounds, and (undesirably) increasing the prominence of noise in the image. One approach to the design of sharpening filters can be represented symbolically as

$$k\mathbf{I} + (1-k)\mathbf{A}$$

where $\mathbf{I}$ is the 'identity operator' (i.e. the operator that, when performed on the image, leaves it unchanged; this operator has a '1' in the central box and '0's everywhere else), $\mathbf{A}$ is an averaging (smoothing) operator, and k is some number greater than unity that defines the degree of sharpening. For example, if we take $k = 2$ and $\mathbf{A}$ as in figure 11.11, we obtain the sharpening filter shown in figure 11.12.

Another common type of sharpening filter can be represented as

$$\mathbf{I} - k\mathbf{L}$$

where $\mathbf{L}$ is the Laplacian operator, whose 3×3 representation is shown in figure 11.13a, and k is greater than zero. Figure 11.13b shows the corresponding sharpening filter for $k = 1/4$.

Instead of sharpening an image to enhance edges, we may sometimes wish to perform an edge-detection operation, in which edges are emphasised but from which uniform areas are removed. This can clearly be carried out using an operator

$$\mathbf{I} - \mathbf{S}$$

where $\mathbf{S}$ is a sharpening operator. Thus, the Laplacian operator of figure 11.13a is an edge-detection filter (it is in fact a digital implementation of the isotropic second derivative), as is any operator $\mathbf{I} - \mathbf{A}$. These are both isotropic edge-detection filters, meaning that they can detect an edge in any orientation. Non-isotropic filters can also be defined to detect edges in specific orientations.

$-\frac{1}{9}$	$-\frac{1}{9}$	$-\frac{1}{9}$
$-\frac{1}{9}$	$\frac{17}{9}$	$-\frac{1}{9}$
$-\frac{1}{9}$	$-\frac{1}{9}$	$-\frac{1}{9}$

Figure 11.12. A sharpening filter.

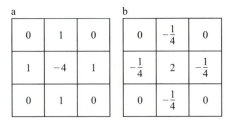

Figure 11.13 (a) Laplacian operator **L**; (b) sharpening filter **I** − **L**/4

Common examples of these filters are the Roberts and Sobel filters, illustrated in figures 11.14 and 11.15, respectively.

The filters that we have just been discussing are all examples of spatial *convolution operations*. Such an operation can be expressed mathematically as

$$d'(i,j) = \sum_{k=a}^{b} \sum_{l=c}^{d} w(k,l)d(i+k,j+l) \tag{11.9}$$

where $d(i,j)$ is the original pixel value at (i,j) and $d'(i,j)$ is its transformed value, and $w(k,l)$ defines the array of weights and is called the *kernel*. This kernel is defined over the range of integers $k = a$ to b and $l = c$ to d (a and c are equal to -1, b and d to $+1$, in the 3×3 example of figure 11.11). An alternative procedure for carrying out such an operation is through the use of Fourier transforms. In section 2.3 we introduced and discussed the concept of the Fourier transform, which relates the dependence $f(t)$ of some variable f on the time t to an equivalent description $a(\omega)$ in terms of the angular frequency ω. An analogous pair of transforms can be defined in the spatial domain. Since images are two-dimensional, it is most convenient to consider the two-dimensional Fourier transforms. These can be written as

$$a(\mathbf{q}) = \frac{1}{(2\pi)^2} \int_{-\infty}^{\infty} \int_{-\infty}^{\infty} f(\mathbf{x}) \exp(-i\mathbf{q} \cdot \mathbf{x}) \, d\mathbf{x} \tag{11.10.1}$$

$$f(\mathbf{x}) = \int_{-\infty}^{\infty} \int_{-\infty}^{\infty} a(\mathbf{q}) \exp(i\mathbf{q} \cdot \mathbf{x}) \, d\mathbf{q} \tag{11.10.2}$$

where **x** is the vector (x, y) describing a position in the image, and $f(\mathbf{x})$ is the image brightness (or whatever variable is to have the Fourier transform applied

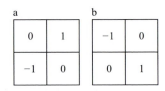

Figure 11.14. Roberts filters for edge detection in two orthogonal directions.

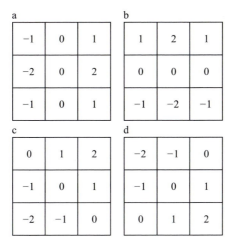

Figure 11.15. Sobel filters for directional edge-detection.

to it) at that position; **q** is a spatial frequency, expressed as a vector (q_x, q_y), and $a(\mathbf{q})$ is the corresponding description of the image in terms of its spatial frequency components (i is the square root of -1).

In fact, since we are considering digital images, we use the discrete Fourier transform (DFT) rather than the spatial Fourier transforms defined by equations (11.10). If the image is defined in a rectangular array of width m pixels, so that the column number j ranges from 0 to $(m-1)$, and height n pixels, with the row number k ranging from 0 to $n-1$,[8] the DFT is also defined in an array of $m \times n$ pixels, with

$$a(j', k') = \frac{1}{mn} \sum_{j=0}^{m-1} \sum_{k=0}^{n-1} f(j, k) \exp\left[-2\pi i\left(\frac{jj'}{m} + \frac{kk'}{n}\right)\right] \qquad (11.11.1)$$

The inverse transform is then

$$f(j', k') = \sum_{j=0}^{m-1} \sum_{k=0}^{n-1} a(j, k) \exp\left[2\pi i\left(\frac{jj'}{m} + \frac{kk'}{n}\right)\right] \qquad (11.11.2)$$

The term at spatial frequency zero appears at coordinates $(0, 0)$ in the DFT. The frequency sampling interval in the x-direction is m^{-1} cycles per pixel (and correspondingly n^{-1} in the y-direction). However, the maximum x-component of the spatial frequency in the DFT is not 1 cycle per pixel, as might be expected, but $1/2$ cycle per pixel. This occurs at $j' = m/2$, and larger values of j' correspond to negative spatial frequencies, with $j = m - 1$ corresponding to a spatial frequency of $-m^{-1}$ cycles per pixel. This is yet another example of the phenomenon of aliasing, discussed in section 10.3.4.5.

---

[8] In fact, practically all DFTs are calculated using the fast Fourier transform (FFT) algorithm, which is most conveniently performed when the image is square, with m and n equal to one another and to a power of 2. The book by Bracewell (1978) provides much more detail.

The potential advantage of considering spatial filtering operations in terms of Fourier transforms is that a convolution in the spatial domain is equivalent to a multiplication in the frequency domain. If the spatial extent of the convolution kernel is very large (i.e. the 'box' of the kind shown in figures 11.11 to 11.15 is large), the extent of the corresponding Fourier transform will be small (this is an extension of the idea discussed in equation (2.18)), and it may be computationally more efficient to calculate the DFT of the image, multiply by the appropriate filter function, and then retransform back into the spatial domain.

Figure 11.16 shows the Fourier transforms of some of the spatial filters described earlier in this section. Figure 11.16a shows that the smoothing filter of figure 11.11 retains the lower spatial frequencies but suppresses the higher ones. For this reason, an alternative name for such a filter is a low-pass filter. Similarly, the sharpening filter can be regarded as a high-boost filter, since it preserves the lower spatial frequencies and increases the amplitudes of the higher frequencies. The Laplace filter (figures 11.13a and 11.16c) suppresses the lowest spatial frequencies and is hence a high-pass filter. Finally, we note that the Sobel filter of figure 11.15c is antisymmetric, causing its Fourier transform (figure 11.16d) to be imaginary.

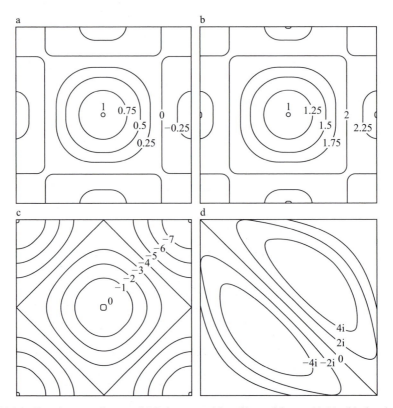

Figure 11.16. Fourier transforms of (a) the smoothing filter of figure 11.11; (b) the sharpening filter of figure 11.12; (c) the Laplace filter of figure 11.13a; (d) the Sobel filter of figure 11.15c. The region plotted is from $-1/2$ to $+1/2$ cycles per pixel in the x and y directions.

Figures 11.17 to 11.20 illustrate the effect of these filters on the image of figure 11.10.

Finally, we should mention two more important classes of spatial filter. The first of these are the *non-linear* filters. These cannot be represented as convolution operations with the form of equation (11.9). Perhaps the simplest example of a non-linear filter is the median filter, in which the central pixel of an $N \times N$ box is replaced by the median value of all the pixels in the box. This has some advantages with respect to the simple averaging filter of figure 11.11, notably the fact that it will preserve sharp edges in an image. Many other non-linear filters have been devised to take into account specific characteristics of the image, for example filters to reduce speckle in SAR images (although in fact the median filter is also rather effective in this role).

The last type of spatial filter that we shall discuss is the destriping filter. This is intended to correct the effects of poor row-to-row calibration in some scanning systems, notably the Landsat MSS system. We noted in section 6.2.2 that the Landsat TM system scans 16 lines of data simultaneously. The MSS system used a similar approach, scanning six lines simultaneously using six different detectors per spectral band. Since these detectors did not have identical calibrations, the result was to produce a characteristic 'striping' or 'banding' in the images that have a period of 6 pixels. Most destriping algorithms work by adjusting the pixel values in a single line of pixels so that their mean and standard deviation match the mean and standard deviation of some reference pixels. They can thus be thought of as a form of local contrast modification. For example, a typical MSS destriping algorithm processes the image in blocks

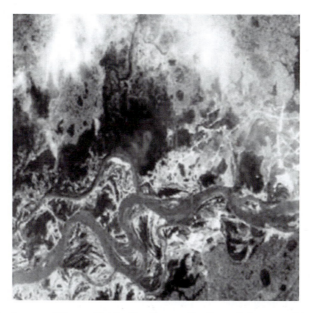

Figure 11.17. The image of figure 11.10 after the application of the smoothing filter shown in figure 11.11.

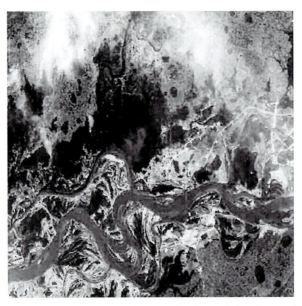

Figure 11.18. The image of figure 11.10 after application of the sharpening filter shown in figure 11.12. Note that although edges have been enhanced, noise in the image has also been emphasised.

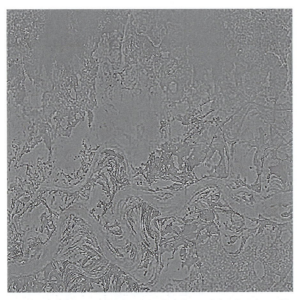

Figure 11.19. The image of figure 11.10 after application of the Laplace filter of figure 11.13a. A constant value of 128 has also been added to the pixel values so that both negative and positive values can be shown. The filter is an isotropic edge-detector.

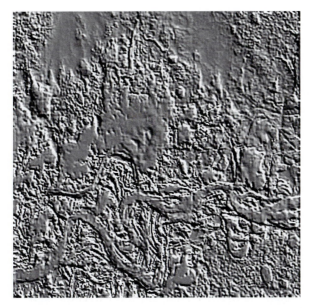

Figure 11.20. The image of figure 11.10 after the application of the Sobel filter of figure 11.15c. A constant value of 128 has also been added to the pixel values so that both negative and positive values can be shown. The filter is a directional edge-detector.

100 pixels wide and 6 pixels in the along-track direction. Linear contrast stretches are applied to the second to fifth rows of pixels so that their means and standard deviations match those of the first row. However, alternative approaches are also possible through the use of Fourier transforms, in which periodic noise can be detected in the frequency domain and hence reduced by suitable filtering.

11.3.3   Band transformations

Band transformations are transformations of multiband (e.g. multispectral) images. In a normal band transformation, the same operation is performed on each pixel in the image. This operation consists of generating a new pixel value from some mathematical combination of the pixel values of the various bands of the image.

One of the simplest types of band transformation is calculation of a *vegetation index*. This is an operation performed on a red and a near-infrared band of an optical-infrared image. As we discussed in section 3.5.1, the reflectance of green-leaved vegetation is low in the red part of the spectrum as a consequence of absorption by chlorophyll, and high in the near-infrared region, mostly as a result of multiple scattering in the mesophyll. Reflectance measurements in these regions are thus strongly correlated with the amount of *photosynthetically active radiation* (PAR) absorbed by the plant material, and vegetation indices are designed to exploit this fact.

A vegetation index is calculated from the reflectance r_r in the red band (typically 0.6 to 0.7 μm) and the reflectance r_i in the near-infrared band (typically 0.8 to 1.0 μm). In principle, these reflectances should be calibrated and corrected for atmospheric propagation effects, but in practice it is sometimes acceptable to use the original (uncorrected) pixel values. The simplest vegetation index is the *ratio vegetation index* (RVI), which is defined as

$$\text{RVI} = \frac{r_i}{r_r}$$

This is, however, somewhat unsatisfactory, since it diverges to infinity if the red reflectance is zero. Much more widely used is the *normalised difference vegetation index* (NDVI), defined as

$$\text{NDVI} = \frac{r_i - r_r}{r_i + r_r} \tag{11.12}$$

This is mathematically better behaved, since it can only take values between -1 and $+1$. Many modifications of this concept have been proposed in an attempt to generate an index that is proportional to the (suitably defined) amount of plant material. For example, one common modification is to take account of the reflectance properties of bare soil. Figure 11.21 and plate 9 show examples of the NDVI.

A more sophisticated approach to estimating the amount of plant material uses more than just the two spectral bands of the vegetation indices. The prototype of such transformations is the *Kauth–Thomas transformation*, which was originally applied to Landsat MSS images of agricultural areas in the USA. Denoting the pixel values in bands 4 to 7 [9] of an MSS image by d_4 to d_7, respectively, the Kauth–Thomas transformation can be written as

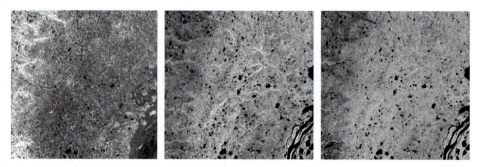

Figure 11.21. Left: band 3 (red) of part of a Landsat TM image; centre: band 4 (near-infrared); right: NDVI. The image covers an area 21 km × 21 km in the Nenets Autonomous Okrug, Russia.

---

[9] For historical reasons, the MSS bands of Landsats 1 to 3 were numbered 4 to 7, although they were renumbered as 1 to 4 for Landsats 4 and 5. Their wavelengths are 0.5–0.6, 0.6–0.7, 0.7–0.8 and 0.8–1.1 μm, respectively.

$$b = 0.433d_4 + 0.632d_5 + 0.586d_6 + 0.264d_7 + 32 \qquad (11.13.1)$$
$$g = -0.290d_4 - 0.562d_5 + 0.600d_6 + 0.491d_7 + 32 \qquad (11.13.2)$$
$$y = -0.829d_4 + 0.522d_5 - 0.039d_6 + 0.194d_7 + 32 \qquad (11.13.3)$$
$$n = 0.223d_4 + 0.012d_5 - 0.543d_6 + 0.810d_7 + 32 \qquad (11.13.4)$$

The new variable b is called the 'brightness', associated mainly with variations in the soil background. As can be seen from equation (11.13.1), it is really just a weighted average of the four pixel values. The variable g is the 'greenness', associated with variations in green vegetation. Since band 5 of the MSS is a red band, and bands 6 and 7 are near-infrared bands, we can see from equation (11.13.2) that the greenness variable is similar to a vegetation index. The third variable, y, is the 'yellowness', and is associated with the yellowing of senescent variation, and the fourth variable, n, is possibly associated with variations in atmospheric conditions. Since it appears not to be directly associated with vegetation, it has been given the name 'nonesuch'. The Kauth–Thomas transformation is also known as the *tasselled-cap transformation*, from the shape of the surface traced out in the three-dimensional b–g–y space as vegetation ripens and then senesces.

11.3.3.1   The principal and canonical components transformations

In general, if we write d_i to represent the pixel value in band i, where i ranges from 1 to N for N-band data, we can define N linear combinations of the bands as follows:

$$d_1' = a_{11}d_1 + a_{12}d_2 + \cdots + a_{1N}d_N$$
$$d_2' = a_{21}d_1 + a_{22}d_2 + \cdots + a_{2N}d_N$$

and so on. These N equations can be written more compactly in vector-matrix notation as

$$\mathbf{d}' = \mathbf{A}\mathbf{d} \qquad (11.14)$$

where $\mathbf{d}$ is a column vector containing the N original pixel values d_1 to d_N, $\mathbf{d}'$ is the corresponding vector after transformation, and $\mathbf{A}$ is the matrix of coefficients a_{ij}. (In fact, the Kauth–Thomas transform described in the previous section is an example of such a set of linear combinations, apart from the fact that they all have the extra value of 32 added to them for practical convenience.)

The principal components of a multiband image are the set of linear combinations of the bands that are both independent of, and also uncorrelated with, one another. The concept is easiest to understand in the case where there are only two bands of data. Figure 11.22 illustrates this case. In figure 11.22a, the pixel values are plotted in the two-dimensional d_1–d_2 space, and it is clear that there is significant correlation between the two bands. However, by making suitable linear combinations d_1' and d_2' from d_1 and d_2, this correlation can be removed. This is shown in figure 11.22b, and figure 11.22a also shows the d_1' and d_2' axes plotted on the original data. It is clear that, in order to meet the

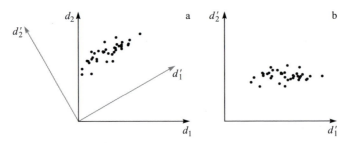

Figure 11.22. Principal components of two-band data.

requirement that the transformed bands d_1' and d_2' should be independent of one another, the corresponding axes should be at right angles to one another. Thus, we can see that the transformation is just a rotation in the two-dimensional pixel value space. The angle of the rotation is determined by the correlation between the values of d_1 and d_2.

The transformation matrix $\mathbf{A}$ in equation (11.14) is calculated from the *covariance matrix* (or variance–covariance matrix) of the original data. For N-band data, this is an $N \times N$ matrix $\mathbf{C}$ in which the element $\mathbf{C}_{ij}$ is defined as

$$\mathbf{C}_{ij} = \langle (d_i - \langle d_i \rangle)(d_j - \langle d_j \rangle) \rangle \qquad eqno(11.15)$$

where d_i is the pixel value in band i, as usual, and the angle brackets $\langle \rangle$ denote an average over all pixels. An element $\mathbf{C}_{ii}$ on the leading diagonal of the matrix is the variance σ_i^2 of the pixel values in band i, and an off-diagonal term $\mathbf{C}_{ij}$ (with $j \neq i$) is related to the correlation coefficient ρ_{ij} between bands i and j through

$$\mathbf{C}_{ij} = \rho_{ij}\sigma_i\sigma_j \qquad (11.16)$$

The matrix $\mathbf{A}$ is calculated from the eigenvectors of the matrix $\mathbf{C}$. There are N eigenvectors, each satisfying a vector–matrix equation of the form

$$\mathbf{C}\mathbf{x} = \alpha\mathbf{x} \qquad (11.17)$$

where $\mathbf{x}$ is the eigenvector and α is its corresponding eigenvalue. By convention, the eigenvectors are normalised so that $\mathbf{x} \cdot \mathbf{x} = 1$, and arranged in order such that the first eigenvector has the largest eigenvalue, the second eigenvector has the next largest eigenvalue, and so on. With this definition, the matrix element a_{ij} is just the jth element of the ith eigenvector.

A principal components transformation (PCT), also known as a *Karhunen–Loève transformation* or a *Hotelling transformation*, involves replacing the N bands d_1 to d_N of a multiband image by the corresponding principal components d_1' to d_N'. In the case of imagery having more than three bands, it will often occur that the first three principal components will contain a large percentage (e.g. 95% or more) of the total image variance. Thus, the effective dimensionality of the image data has been reduced by the transformation. By assigning these three components to, say, the red, green and blue channels of a colour display, an image can be displayed in only three bands that contains most of the information present in the original multiband image. It may also happen that one or more of the principal components will correlate with some physically meaningful variable, as is the case with visible–near-infrared images of vegetated areas.

Plate 10 (left) shows a four-band image of Stavanger, Norway (in fact, of course, the plate can show only three of the four bands). The four principal components of this image are shown in figure 11.23. In this case, the variances contained in the principal components, expressed as a fraction of the total variance, are 97.0%, 2.6%, 0.2% and 0.2%, respectively. It is clear that the first principal component responds mostly to land/water differences, but also to vegetation differences. The fourth principal component is mostly noise. Plate 10 (right) shows the result of combining the first three principal components into a red–green–blue composite image.

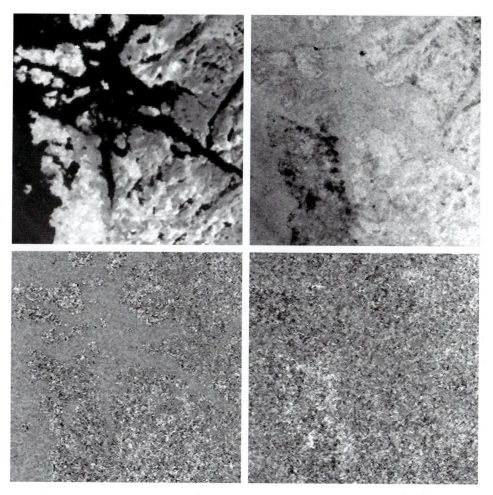

Figure 11.23. The four principal components of the four-band image shown in plate 10. Linear contrast stretches have been applied to each of these components.

The *canonical components transformation* (CCT) is a similar idea to the principal components transformation. In the latter, the principal components are arranged so that the image variance is greatest in the first principal component. In the former, the transformation takes account of clustering in the data, and the components are chosen such that the separability (see section 11.3.4.2) of the clusters is greatest in the first canonical component. We will not discuss this further here. The interested reader is referred to, for example, the book by Richards (1993) for more details.

11.3.4   Image classification

Image classification is the process of making quantitative decisions from image data, grouping pixels or regions of the image into classes intended to represent different physical objects or types. The output of the classification process may be regarded as a thematic map rather than as an image. The majority of classification techniques use mainly the radiometric data (pixel values) present in the image, with little or no reference to spatial variation, and these techniques will be described first. These techniques can be thought of as follows. Suppose we have an n-band image, and the pixel values in each band can take k different values (e.g. $k = 256$ for 8-bit data). The number of possible coordinates in the n-dimensional pixel-value space is k^n, a number that can very easily exceed a million (it is over 16 million for the case where $n = 3$ and $k = 256$). However, it is extremely unlikely that the image represents a million or more different classes of data, or that we could make use of the information if it did. What we require is some simplification of the data in the n-dimensional pixel-value space, identifying a volume within this space as representing a single class of data.

11.3.4.1   Density slicing and pseudocolour display

Density slicing is a particularly simple classification technique, applied to a single-band image or to one band of a multiband image. A range of input pixel values (a 'slice' of the image) is mapped to a single pixel value in the output image. Several such ranges can be defined, of course. If each of these is mapped to a different colour in the output image, the result is called a *pseudocolour image* (see plate 9). Density slicing is commonly used where the pixel values have a direct relationship to a physical variable. For example, the pixel values in a calibrated thermal infrared image of an ocean area may correspond directly to the sea-surface temperature, or one might wish to slice up a vegetation index image into different ranges of the vegetation index. Another common use of density slicing is to generate image masks, for example land/water masks, which can be used to define areas of the image to which subsequent processing is to be applied.

11.3.4.2   Multispectral classification

Multispectral classification can be approached from two fundamentally different directions. The first of these, *supervised classification*, uses information about the known distribution of classes to initiate the process. From field investigations or ancillary data (e.g. maps), the image class is already known in some areas of the image. This knowledge is used to define training data – that is, a statistical description of the range of pixel values – for each of the classes of interest. The entire image is then examined, pixel by pixel, to determine to which of the classes, if any, a pixel belongs. This process obviously requires some kind of rule to decide to which class a pixel belongs. The simplest is the *box classifier* or *parallelepiped classifier*. In this case, *n*-dimensional (*n* is the number of bands in the image) boxes are defined so that they enclose all, or a fixed, high proportion, of the training data for each class. The rule is then very simple: if a pixel is within a particular box, it is assigned to the corresponding class. This is illustrated schematically for two-band data in figure 11.24. Although the box classifier is exceptionally easy to apply, figure 11.24 shows two of its disadvantages. While the pixel at *a* should obviously be assigned to class *A*, the pixel at *b* could be assigned to either *A* or *B*, and the pixel at *c* cannot be assigned at all.

More sophisticated classification algorithms use a *discriminant function*, which quantifies how well a pixel 'fits' a given class. For example, the *Euclidean distance algorithm* calculates the distance, in the *n*-dimensional pixel-value space, from a given pixel to the mean position (centroid) of the training data for each class. The smallest distance determines the class to which the pixel will be assigned. Thus, referring again to figure 11.24, pixels *a* and *b* will both be assigned to class *A*, and pixel *c* will be assigned to class *C*. The *maximum likelihood algorithm* in effect models the probability distributions for

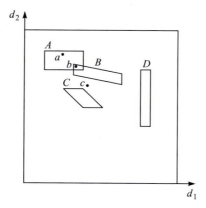

Figure 11.24. Schematic illustration of the box classifier for two-band data. d_1 and d_2 are the pixel values in bands 1 and 2, respectively, and the large square shows the total 'volume' of pixel-value space that is available. The boxes *A*, *B*, *C* and *D* have been constructed around the training data for the corresponding classes; the points *a*, *b* and *c* are three pixels to be classified.

each class, using the training data, from which it is possible to estimate the likelihood that a given pixel belongs to a particular class. The most probable assignment can then be made. This approach also has the advantage that a probability threshold can be imposed, so that no classification at all will be made if the maximum likelihood of belonging to any class is below some critical value.

The opposite approach to supervised classification is *unsupervised classification*. In this case, the entire image is first analysed without reference to any training data. The aim of this analysis is to identify distinguishable clusters of data in the *n*-dimensional pixel-value space. Once these clusters have been identified, they can be associated with physical classes using training data.

The first step in an unsupervised classification is to cluster the data. Various clustering algorithms exist. One of the most widely used is the *isodata algorithm*, also known as the *K-means* or *migrating means* algorithm. This is an iterative procedure, in which the user first specifies the number of clusters to be found. The algorithm assigns nominal centre coordinates to each of these clusters in the *n*-dimensional pixel-value space. Next, every pixel in the image is then assigned to the appropriate cluster, using a discriminant function such as the Euclidean distance. The centre coordinates of the clusters are then recalculated from the mean values of the pixels assigned to them, and the process is repeated. It is normally terminated when fewer than some specified proportion (for example, 2%) of the pixels have their assignments changed.

Figure 11.25 illustrates the operation of the isodata algorithm. In figure 11.25a, the four cluster means *A* to *D* have been assigned arbitrarily, and the consequent assignment of the pixels to the clusters is shown by the rectangular boxes. In figure 11.25b, the cluster means have 'migrated' to their new positions, giving the new assignments shown by the rectangular boxes. In fact, in this very simple example, the process has converged completely after only one iteration, since no further changes in the assignment of pixels to the clusters occur with subsequent iterations.

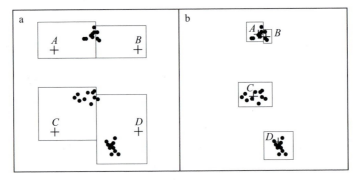

Figure 11.25. Schematic illustration of the operation of the isodata clustering algorithm. The square boxes represent the *n*-dimensional pixel-value space, the dots are the coordinates of individual pixels, and the crosses are the cluster means. (a) The initial cluster means, and (b) the cluster means after one iteration.

In essence, supervised classification forces the image classes to correspond to physical classes that are defined by the user, but without a guarantee that these classes will be statistically distinct. Unsupervised classification, on the other hand, forces the image classes to be statistically distinct, but does not guarantee that they will correspond to the physical classes required by the user. *Hybrid classification* combines both approaches. In this technique, an unsupervised classification is first carried out to determine the number of distinguishable clusters present in the image. These are then compared with the training data, and clusters are split, merged or deleted as necessary. It may also be necessary to modify the definitions of the physical classes, for example by merging two physical classes that cannot be discriminated in the image. The process is normally an iterative one.

An important aspect of this process is the ability to quantify the extent to which two clusters are statistically separable. Various measures of separability can be used, all of which are defined from the probability distributions $p_1(\mathbf{d})$ and $p_2(\mathbf{d})$ of the two clusters in the n-dimensional pixel-value space. (The vector $\mathbf{d}$ defines the coordinates of a pixel in this space.) For example, the *divergence d* is defined as

$$d = \int (p_1(\mathbf{d}) - p_2(\mathbf{d})) \ln \frac{p_1(\mathbf{d})}{p_2(\mathbf{d})} d\mathbf{d} \qquad (11.18)$$

where the integral is carried out over the entire pixel-value space. This is zero for identical distributions, and infinite for distributions that do not overlap at all and hence are completely separable. The *transformed divergence*

$$d^* = 2\left(1 - \exp\left(-\frac{d}{8}\right)\right) \qquad (11.19)$$

is also zero for identical distributions, but has a maximum value of 2 for non-overlapping distributions. A third commonly used measure of separability is the *Jeffries–Matusita distance*

$$J = \int \left(\sqrt{p_1(\mathbf{d})} - \sqrt{p_2(\mathbf{d})}\right)^2 d\mathbf{d} \qquad (11.20)$$

which, like the transformed divergence, ranges from 0 for identical distributions to 2 for non-overlapping distributions. As a rough guide for real data, the divergence d should exceed about 10 for reasonable separability.

As a final remark before leaving the subject of multispectral classification, we can observe that although we have been assuming throughout this section that a pixel is characterised by its coordinates in the n-dimensional pixel-value space, this is not a fundamental necessity for image classification. Other properties can be used to characterise a pixel. These could be, for example, radar backscatter coefficients in different polarisation states, texture parameters (see section 11.3.4.3), or single-band pixel values from different dates. The essential aspect is that the pixel is characterised by more than one variable, and so can be described in a multidimensional *feature space*.

11.3.4.3 *Texture classification*

Up to this point, we have discussed image classification on a pixel-by-pixel basis, and have given no consideration to the spatial context of a pixel. Spatial considerations can be applied in a very simple manner *after* a multispectral (in general, multifeature) classification has been performed, for example by applying a *majority filter* to the classified image. Here, a small neighbourhood, typically 3×3 pixels, is investigated around each pixel in the classified image. The classification of the central pixel in this neighbourhood is reassigned to whichever image class is most strongly represented in the neighbourhood. This procedure usually reduces the level of noise in the classification. However, the spatial context of a pixel can also be used to define the features on which the classification itself is based. This is the topic of *image texture*.

Although a precise definition of image texture is rather difficult to formulate, it can be loosely defined as structure in the spatial variation of the pixel values. Everyday experience, and especially the consideration of black-and-white photographs, tells us that image texture is an important aspect of the way in which the human brain interprets a scene. Our task is to define quantitative measures of image texture that correspond to its perception by the brain, or at least have some utility in differentiating between different regions of an image. Many texture measures have been proposed. The simplest is the variance of the pixel values (or, similarly, the range between the maximum and minimum values) within a small neighbourhood of the pixel in question. This does not, however, provide any information on the spatial scale of the variations. Scale-dependent information can be obtained in various ways, for example by calculating the Fourier transform of a small neighbourhood in the image and then extracting the coefficients for different spatial frequencies. However, one of the commonest approaches to the quantification of image texture is through the use of the *grey-level co-occurrence matrix* (GLCM), also known as the *spatial dependency matrix*.

If the pixel values in the image are drawn from a set of N integers (e.g. $N = 256$ for 8-bit data), the GLCM is an $N \times N$ square matrix $\mathbf{P}$. It is calculated as follows. First, we choose some spatial separation to define the position of a pixel relative to some reference pixel. For example, the vector $(1, 0)$ would define the pixel immediately to the right of the reference pixel. Next, we define a neighbourhood of pixels within which the GLCM is to be calculated. The matrix element $\mathbf{P}_{ij}$ is then the number of times that pixel value i occurs in a pixel within the neighbourhood *and* pixel value j occurs at the chosen separation from this first pixel. The elements are normally expressed as proportions of the total; that is, the sum of all the matrix elements is 1. Once the GLCM has been calculated, various texture parameters can be derived from it. Some of the commonest are the *energy*, defined as

$$\sum \mathbf{P}_{ij}^2$$

the *entropy*, defined as

$$\sum \mathbf{P}_{ij} \ln \mathbf{P}_{ij}$$

and the *contrast*, defined as

$$\sum |i - j|^a \mathbf{P}_{ij}^b$$

where a and b are often taken as 2 and 1, respectively. In all of these definitions, the sum is taken over all the matrix elements.

Since image texture is a spatial property, any texture parameter must necessarily be defined by examining variations of the pixel values within some finite neighbourhood of the image. The texture parameter associated with a particular pixel is calculated from the pixel values in a 'window' of the image centred on that pixel, and this has the effect of degrading the spatial resolution. The choice of window size is therefore important: if it is too small, too few pixels will be available to give a statistically meaningful measure of texture, whereas if it is too large, the resolution of the 'texture image' will be unnecessarily degraded.

11.3.4.4   *Error matrices and classification accuracy*

An *error matrix*, also known as a *confusion matrix*, is a method of quantifying the performance of a classification. It is a square matrix $\mathbf{E}$ of $N \times N$ elements, where N is the number of classes in the classified image. The element $\mathbf{E}_{ij}$ is the number of pixels known to belong to class i and classified as belonging to class j. Thus, the elements on the leading diagonal, $\mathbf{E}_{ii}$, correspond to correctly classified pixels, and all the off-diagonal elements correspond to erroneous classifications. As an example, figure 11.26 shows a hypothetical error matrix for a four-class classification. From the figure, we see that $\mathbf{E}_{21} = 40$, for example, meaning that 40 pixels known to belong to class 2 (bare ground) were erroneously classified as belonging to class 1 (water).

The error matrix can be used to calculate various performance figures for the classification. The *user's accuracy* (also called the *consumer's accuracy*) for class i is defined as

	Known class			
	Water	Bare ground	Crops	Forest
Water	187	40	7	0
Bare ground	11	246	12	9
Crops	0	21	239	39
Forest	0	0	140	49

Classification

Figure 11.26. An error matrix.

$$\frac{\mathbf{E}_{ii}}{\displaystyle\sum_{j=1}^{N}\mathbf{E}_{ji}}$$

so the example of figure 11.26 gives user's accuracies of 79.9%, 88.5%, 79.9% and 74.1% for classes 1 to 4 (water to forest), respectively. The *producer's accuracy* for class i is

$$\frac{\mathbf{E}_{ii}}{\displaystyle\sum_{j=1}^{N}\mathbf{E}_{ij}}$$

which has values of 94.4%, 80.1%, 60.1% and 50.5% for classes 1 to 4, respectively. Single-parameter classification accuracies can also be specified from the error matrix. The simplest and most widely used of these is just the proportion of pixels that are correctly classified, that is,

$$\frac{\displaystyle\sum_{i=1}^{N}\mathbf{E}_{ii}}{\displaystyle\sum_{i=1}^{N}\sum_{j=1}^{N}\mathbf{E}_{ij}}$$

For our example data, this has a value of 72.1%.

It is highly unlikely that the correct classification will be known for every pixel in an image (otherwise there would have been no need to perform the classification in the first place), and some care must be exercised in choosing the pixels from which the error matrix is calculated. The proportions of pixels known to belong to the various image classes should correspond as closely as possible to the proportions for the image as a whole. A random sampling strategy is usually preferred as a way to try to achieve this.

11.3.5   *Detection of geometric features*

The recognition and classification of objects in an image on the basis of their shape is a high-level operation in image processing. The edge-detection filters discussed in section 11.3.2.2 are clearly simple examples of shape detection, and other convolution operators can be used to detect particular shapes, in an approach usually referred to as *template matching*. The convolution kernel is designed in such a way that, if it is applied to an object having the desired shape, it will produce a maximum output. As a simple example, linear features that are a single pixel wide can be detected using the four directional templates shown in figure 11.27. One possible method of using these templates would be to apply each of them to the original image data, then to add the squares of the outputs, and finally to threshold the result.

An alternative approach to the detection of objects having known shape, size and orientation, but unknown position, is through the use of Fourier

-1	-1	-1
2	2	2
-1	-1	-1

-1	2	-1
-1	2	-1
-1	2	-1

-1	-1	2
-1	2	-1
2	-1	-1

2	-1	-1
-1	2	-1
-1	-1	2

Figure 11.27. Templates (convolution kernels) for detection of linear features that are a single pixel in width.

transforms. This can be illustrated by considering the problem represented in figure 11.28. The two images 11.28a and 11.28b contain the same features with the same relative positions, but there is an unknown spatial displacement between them that we wish to determine. For example, image 11.28a could be a subregion of an aerial photograph of a group of icebergs, and image 11.28b could be the same subregion of another aerial photograph acquired from the same vantage point, but some time later, the problem being to determine the motion of the icebergs during that time. Clearly, one could in principle define a template corresponding to figure 11.28a and move it around over figure 11.28b to see where it best fits, but the template would be large and the procedure therefore time-consuming. Alternatively, one simply calculates the Fourier transforms of the two images and multiplies them together. The resulting product, when retransformed back from the spatial frequency domain into the spatial domain, is just the correlation function of the two images, and it will have a maximum at the image coordinates corresponding to the displacement between the two images.

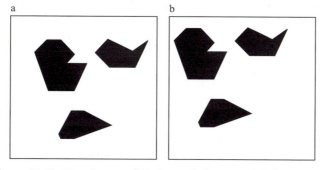

Figure 11.28. Two images differing only by a spatial displacement.

A third approach to the detection of objects on the basis of their shape is through the use of *Hough transforms*. These can usefully be applied even when the size and orientation of the object is unknown, as long as they can be suitably parametrised. The simplest example of the use of the Hough transform is in the detection of straight lines, illustrated in figure 11.29.

Figure 11.29a shows an image of width and height 100 pixels. The image contains a straight line, and P_1 to P_3 are three representative pixels on this line. The position and orientation of the line can be specified by the parameters r and θ, where r is its perpendicular distance from the origin and θ is the angle between the perpendicular and the x-axis, as shown in the figure. The coordinates (x, y) of any point on the line will satisfy the equation

$$x \cos\theta + y \sin\theta = r \tag{11.21}$$

The purpose of the Hough transform is to determine the values of r and θ most appropriate to a group of pixels such as those represented by P_1 to P_3. Figure 11.29b shows a plot in r–θ space. The curve labelled P_1 is the set of all points in this space that satisfy equation (11.21) for the (x, y) coordinates of the point P_1, and similarly for the other two points. As expected, the three curves in figure 11.29b intersect at a single point, corresponding to the values of r and θ shown in figure 11.29a.

In practice, the Hough transform is calculated using an 'accumulator array'. This is a two-dimensional array representing r–θ space, and initially all its members are set to zero. The procedure scans the image to identify the pixels to be transformed (these will often be derived by performing a line- or edge-detection on the original image). For the (x, y) coordinates of every such pixel, the contents of the accumulator array at (r, θ) are incremented by 1 for every combination of r and θ that satisfies equation (11.21). Once this procedure has been completed, the accumulator array is scanned to find the maxima that correspond to straight lines in the original image.

Other Hough transforms can be defined for shapes that can be parametrised analogously to equation (11.21). For example, a circle can be defined through the three parameters a, b and r:

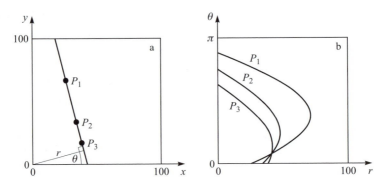

Figure 11.29. Illustration of the Hough transform for straight lines.

$$(x - a)^2 + (y - b)^2 = r^2$$

where a and b are the coordinates of the centre and r is the radius. The accumulator array in this case is three-dimensional. A fuller discussion of Hough transforms can be found in the book by Sonka et al. (1993).

As a final approach to the detection of geometric features, we mention the use of image segmentation. The aim of this process is to identify homogeneous regions within the image, typically after classification so that all the pixels within a region belong to the same class. Contiguity is then defined by identifying the boundary of the region, for example using an edge-detection algorithm, and then performing morphological operations on the boundary to ensure that it forms a closed loop with no gaps or dangling ends. Once this procedure has been carried out, shape factors (e.g. the parameters of the best-fitting ellipse, the length of the boundary, the circularity, etc.) can be calculated for the region and used to characterise its shape.

11.3.6 Data compression

In section 11.2 we discussed the data volumes necessary to specify images, and noted that these can easily be many megabytes in size. This can obviously raise problems of storage and transmission, and methods by which the data can be compressed are of particular interest. The subject of data compression might more logically have been presented in section 11.2, but it is more convenient to discuss it now, following on from the discussion of classification. The reason for this is that the classification process itself is a form of data compression, since fewer bits are required to specify a classified image than the original image from which it was derived. In general, methods of data compression can be divided into *reversible* (or lossless, or loss-free) processes from which all the original information can be retrieved, and *irreversible* (lossy) processes that entail some loss of information. Clearly, a multispectral classification is an irreversible compression.

Reversible data compression methods can be divided into two types: those that operate on a pixel-by-pixel basis, and those that make use of the spatial uniformity present in an image. *Huffman coding* is a common example of the former, used for example in the widespread GIF [10] (graphics interchange format) format. In this case, the binary numbers used to specify pixel values are recoded into sequences of 0s and 1s (not strictly binary numbers, since leading zeros are now significant), such that frequently occurring pixel values are assigned short sequences and infrequent values are assigned long ones. Table 11.2 shows a simple example of a Huffman code, for data originally specified using three bits. The first column shows the original pixel value, which is always a 3-bit binary number (corresponding to integers between 0 and 7). The second column shows the number of pixels that have the relevant pixel

---

[10] GIF is a trademark of the Compuserve Corporation.

Table 11.2. An example of a Huffman code

Pixel value	Number of pixels	Number of bits	Coded value	Number of bits
000	100	300	10000	500
001	200	600	101	600
010	300	900	001	900
011	800	2400	11	1600
100	700	2100	01	1400
101	200	600	000	600
110	100	300	10001	500
111	100	300	1001	400
Totals	2500	7500		6500

value, from which we see that there are 2500 pixels in total, with pixel value 3 (011) being the most commonly occurring. The third column shows the number of bits required to specify all the pixels that have a given pixel value, and of course the total number of bits required is just $2500 \times 3 = 7500$. The fourth column shows how the pixel values are recoded into sequences of 0s and 1s using a Huffman code. For example, pixel value 3 (011) is recoded to the sequence 11, requiring only two bits, whereas pixel value 6 (110), which occurs much less frequently, is given the much longer code 10001. The last column shows the number of bits needed to specify all the pixels that have a given recoded value, from which we can see that the total number of bits needed to specify the whole image has been reduced from 7500 to 6500, a saving of 13.3%.

There are two essential considerations in designing and using a Huffman code. The first is the rather obvious point that the code must be stored along with the encoded data. The second consideration is that the decoding process should be unambiguous. For example, if four consecutive pixels in the original image have pixel values of 1, 2, 7 and 6, their binary representation is 001010111110 before coding, and 101001100110001 after coding. Since the number of bits needed to specify a pixel value is no longer constant, it must be possible to determine from this sequence of bits how to reconstitute the original data. By referring to table 11.2, we can confirm that this is indeed true in this case. To guarantee an unambiguous code requires that it be generated by a suitable algorithm. We should also note that the encoded data have practically no redundancy, meaning that if any data are lost (e.g. through corruption of the medium on which they are stored), it may be difficult or impossible to retrieve subsequent data. For this reason, it is safer to encode the image on a line-by-line basis because, despite the fact that this slightly increases the data volume, it does at least mean that loss of data in one line of the image will not affect other lines.

The Huffman code is nearly optimal, in the sense that it reduces the data volume to almost the minimum that is theoretically possible. This theoretical

minimum is the information content of the data. If each pixel value can take
one of M possible values, such that the value i occurs n_i times, the information
content I can be defined as

$$I = N \log_2 N - \sum_{i=1}^{M} n_i \log_2 n_i \qquad (11.22)$$

bits, where

$$N = \sum_{i=1}^{M} n_i$$

is the total number of pixels. In the example of table 11.2, $M = 8$ and
$N = 2500$, and applying equation (11.22) to the data we find that $I \approx 6369$
bits. Thus, the data volume after compression using the Huffman code, to 6500
bits, is only 2% larger than this theoretical minimum.

The other main type of reversible data compression employs the spatial
information in the image. It is evident that if every pixel in a 1000×1000
pixel image has a pixel value that, while specified as an 8-bit number, actually
has the same value (say 42), the whole image can be precisely specified in very
few bits – as indeed we have just done in the preceding statement. *Run-length
encoding* consists of specifying a pixel value, and then the number of places
within the sequence of data (the length of the run) for which this value is
repeated. Its scope for compressing the data volume obviously depends on
the prevalence of homogeneous areas within the image, and it therefore
tends to be more suitable for compressing classified images than original
data unless the latter have been smoothed to reduce the level of noise. The
PCX (PC Paintbrush) and Macintosh PICT formats are common examples of
run-length encoded formats. A more sophisticated version of run-length encoding uses *tesseral addressing* to specify the area and location of square regions
within the image. However, whatever spatially based scheme is used, it is
apparent that little or no saving in data volume can be made if the image is
noisy, since in this case pixel values will change frequently, resulting in short
'runs' of data.

Irreversible (lossy) data compression can take a large number of forms,
many of which are obvious. Images can be cropped to remove uninteresting
areas, or sub-sampled to reduce their spatial resolution and volume at the same
time. An alternative to sub-sampling is to form the Fourier transform of the
image and then crop it so that it contains only the lower spatial frequencies. As
an example, if the spatial resolution is halved using this technique, the dimensions of the Fourier transform array will also be halved, so that the number of
pixels in the Fourier transform is one quarter of the number in the original
image. On the other hand, the Fourier transform actually needs two numbers
for each pixel, to hold the real and imaginary parts, respectively, so the data
volume has in fact only been halved. More efficient schemes make use of
transforms that are similar in function to the Fourier transform, but recognise
that the input (spatial domain) data will always be real. Examples of such

transforms include the *Hadamard transform*, the *Hartley transform* and the *discrete cosine transform* (DCT). The DCT is used in the widespread **JPEG** image format. In the usual implementation of this format, the image data are processed in blocks of 8 × 8 pixels. The DCT is calculated for each block, and sufficient terms are retained so that the regenerated image will appear reasonably accurate to the human eye. Very large compression factors can be achieved using this method, and it is often used for the storage and transmission of 'quick-look' images.

PROBLEMS

1. How many uniformly spaced geostationary satellites would be needed to ensure that all points on the Earth's surface with latitudes less than 66.5° can be seen at an elevation angle of at least 10°?

2. (a) Estimate appropriate values, corresponding to those in table 11.1, for printed paper as a data storage medium.
(b) Estimate the potential data storage requirements from a five-year space-borne remote sensing mission.

3. Identify the nature of the spatial filters whose kernels are given below:

(a)			(b)			(c)		
1	3	1	1	3	1	−1	−3	−1
3	−8	3	3	6	3	−3	16	−3
1	3	1	1	3	1	−1	−3	−1

4. A (small) image consists of 64 pixels arranged in an 8 × 8 grid. The pixel values I_1 and I_2 in two spectral bands are shown:

BAND 1

```
50  39 34 42 55 85 77 75
51  31 26 20 46 66 77 61
67  18 22 15 20 40 71 75
28  26 24 24 34 62 80 70
67  74 47 24 29 28 72 83
64  81 75 34 35 32 56 39
78  66 70 62 55 25 60 61
77  73 53 32 60 50 47 45
```

BAND 2

```
52  50 37 42 55 30 37 25
55  40 69 80 45 70 33 65
61  72 72 78 82 45 32 69
37  80 75 75 80 68 34 65
32  41 50 78 71 78 30 23
37  27 30 45 72 65 60 50
40  42 30 65 51 64 63 58
45  35 64 45 55 58 50 55
```

(a) Construct histograms for I_1 and I_2. Do the pixels appear, on this basis, to be divisible into a number of classes?

(b) Perform a cluster analysis on the data. How many classes are represented in the image, and how many distinct regions does it contain? Use both manual clustering and (if you have a computer and some spare time) the isodata algorithm to identify the clusters.

5. A single band of an image has a histogram with a Gaussian distribution of standard deviation σ. Show that the theoretical limit to which the data can be compressed using Huffman coding is

$$\log_2 \sigma + 2.05$$

bits per pixel. You may assume that σ is large enough for the digitisation noise to be ignored. You may find the following integrals useful:

$$\int_{-\infty}^{\infty} \exp(-x^2/2\sigma^2)\, dx = \sigma\sqrt{(2\pi)}; \qquad \int_{-\infty}^{\infty} x^2 \exp(-x^2/2\sigma^2)\, dx = \sigma^3 \sqrt{(2\pi)}$$

The Global Positioning System

A1.1 Introduction

The Global Positioning System (GPS) is a satellite-based positioning system operated by the United States Department of Defense. It has been operational since 1993, and although it is primarily intended for military use, it is, subject to one or two provisos, also available for world-wide civilian use. It is now widely used in remote sensing – in field work, to determine the position of training areas for image classification (see section 11.3.4.2), and also by some remote sensing satellites, for precise determination of the satellite's own position. This appendix therefore presents a brief introduction to GPS and its capabilities. A much fuller discussion is given by, for example, Leick (1995).

The fundamental idea behind GPS is not a new one. It is a radio-positioning system in which timing signals are transmitted at known times from a number of radio beacons at known locations. By measuring the times at which these signals are received, the distances to the various beacons can be calculated, and hence the position of the receiver can be deduced. What sets GPS apart from its predecessors is that the beacons are carried on satellites, providing genuinely global coverage.

A1.2 Space segment

The 'space segment' of the GPS system consists of 24 satellites in circular orbits around the Earth. The semi-major axis (radius) of these orbits is about 26 600 km and the inclination is 55°, giving them a nodal period (see equation (10.13)) of 43 082 s, or exactly half a sidereal day. This means that the sub-satellite track of a given satellite retraces itself exactly, although approximately 4 min earlier, every day. The satellites are arranged in six regularly spaced orbital planes, as shown in figure A1.1. As figure A1.1 demonstrates, the altitude and number of the satellites ensures that a comparatively large number, typically eight to twelve, are visible from any location on the Earth's surface at any time.

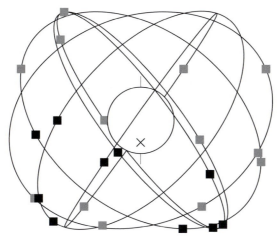

Figure A1.1. Typical configuration of GPS satellites in six orbital planes around the Earth. The squares show the positions of the satellites; those shown in black are visible from the location on the Earth's surface marked with a cross.

The satellites transmit at frequencies of 1575.42 MHz (this frequency is designated L1) and 1227.60 MHz (L2). The L1 carrier signal is modulated with a unique timing signal at 1.023 MHz, with a repeat period of 1 ms. This is the signal used for basic, non-military, positioning called SPS (standard positioning service). Both the L1 and L2 carrier signals are (also) modulated at 10.23 MHz for more precise positioning (PPS, or precise positioning service), and at 50 Hz to carry essential 'housekeeping' data identifying the satellite and its orbital parameters, the time, and so on.

A1.3   Determination of position

To determine an unknown position in three dimensions, it is in principle necessary to measure the distances from only three known positions. Although the satellites themselves carry extremely accurate atomic clocks, most GPS receivers are equipped only with clocks controlled by quartz crystals, and this means that the absolute time at the receiver is not known with sufficient accuracy. It is therefore necessary to observe at least *four* satellites, to solve for all four variables (the three spatial coordinates of the receiver, and the time). In practice, receivers track as many satellites as are available, and use a least-squares method to determine the most probable values of the variables.

The simplest and least accurate mode of operation employs a single GPS receiver using the SPS transmissions. The accuracy with which the propagation time from a satellite to the receiver can be measured is typically 5 ns, corresponding to an uncertainty of about 1.5 m in the distance. However, ionospheric and tropospheric delays (see section 8.3.5) introduce an additional uncertainty of the order of 4 m. When the geometry of the 'constellation' of satellites, and the least-squares procedure for modelling the observed data, are taken into account, the accuracy in a position determination[1] is of the order of 25 m. This can be further degraded by the use of 'selective availability', a deliberate dithering of the satellites' clocks, if it is felt necessary to deny the full accuracy of the system to an enemy. When selective availability is implemented, the stated accuracy of the SPS service is such that there is a 95% probability that a measured position will be within 100 m of its true value.

The accuracy of a single-receiver SPS measurement can of course be improved by averaging. The correlation time for the position errors is about 4 min,[2] so that repeated measurements at intervals much less than 4 min are not statistically independent. If n measurements are made at intervals of about 4 min or greater, and averaged, the uncertainty is reduced by a factor of $\sqrt{n}$, so that, for example, the average of 100 measurements (which would require at least 6.7 h) would reduce the radius of the 95% probability circle from 100 m to 10 m.

Considerably higher accuracies are available using *differential GPS*. This technique requires (at least) two receivers, and relies on the fact that many of the error terms show a very high degree of spatial correlation. Expressed simply, we can say that if two GPS receivers A and B are located not too far apart (say 10–100 km at most), the position error measured by A at a given instant (i.e. the difference between the measured and true positions) will be almost identical to that measured by B at the same instant, provided that both receivers are tracking the same set of satellites. This means that, although the absolute positions of A and B may only be known to an accuracy of 100 m, the position of B relative to A can be determined to a much greater accuracy. The simplest differential technique uses the *pseudoranges* measured by the receivers. The pseudorange is the propagation time from the satellite to the receiver, converted to a distance by multiplying by the speed of light. It differs from the true range because of receiver errors and atmospheric and ionospheric propagation delays. Differential correction of GPS positions using pseudoranges can achieve an accuracy of the order of 1 m. However, very much greater accuracies can be achieved by tracking the phase of the carrier signal itself. The wavelength of the L1 carrier is about 190 mm, and the phase can be determined to within about one hundredth of a wavelength, so phase-tracking differential GPS can achieve accuracies of a few millimetres.

---

[1] From this point onward, positional accuracies refer only to the horizontal component, for example the latitude and longitude. The accuracy in the vertical direction is poorer by a factor of about 2.

[2] This figure applies to the horizontal component. For the vertical component, the correlation time is somewhat longer, at about 10 min.

Data tables

This appendix collects together some of the more commonly needed general data in remote sensing. Some of them also appear in the main text; however, it is felt that it will be a convenience also to present them here.

A2.1 Physical constants

c	Speed of light *in vacuo*	$2.9979 \times 10^8 \, \text{m s}^{-1}$ (defined as $299\,792\,458$ m s^{-1})
h	Planck constant	$6.6261 \times 10^{-34} \, \text{J s}$
e	Charge on the proton	$1.6022 \times 10^{-19} \, \text{C}$
m_e	Mass of the electron	$9.1094 \times 10^{-31} \, \text{kg}$
u	Atomic mass unit	$1.6605 \times 10^{-27} \, \text{kg}$
m_0	Permeability of free space	$1.2566 \times 10^{-6} \, \text{H m}^{-1}$ (defined as $4\pi \times 10^{-7}$ H m^{-1})
ε_0	Permittivity of free space	$8.8542 \times 10^{-12} \, \text{F m}^{-1}$
Z_0	Impedance of free space	$3.7673 \times 10^2 \, \Omega$
G	Gravitational constant	$6.6726 \times 10^{-11} \, \text{N m}^2 \, \text{kg}^{-2}$
R	Gas constant	$8.3145 \, \text{J K}^{-1}\text{mol}^{-1}$
N_A	Avogadro number	$6.0221 \times 10^{23} \, \text{mol}^{-1}$
k	Boltzmann constant	$1.3807 \times 10^{-23} \, \text{J K}^{-1}$
σ	Stefan–Boltzmann constant	$5.6705 \times 10^{-8} \, \text{W m}^{-2} \text{K}^{-4}$

A2.2 Units

The table lists some units in common use in remote sensing, together with their SI equivalents

1 ångström (Å)	$10^{-10} \, \text{m}$
1 inch	$25.4 \, \text{mm}$
1 foot	$304.8 \, \text{mm}$
1 statute mile	$1.609 \, \text{km}$
1 nautical mile	$1.852 \, \text{km}$
1 astronomical unit (AU)	$1.496 \times 10^{11} \, \text{m}$

1 hectare	$10^4\,\text{m}^2$
1 acre	$4.05 \times 10^3\,\text{m}^2$
1 pound	$0.454\,\text{kg}$
1 knot	$0.514\,\text{m}\,\text{s}^{-1}$
1 bar	$10^5\,\text{Pa}$
1 atmosphere	$1.013 \times 10^5\,\text{Pa}$
1 torr $= 1\,\text{mm Hg}$	$133\,\text{Pa}$
1 electron-volt (eV)	$1.602 \times 10^{-19}\,\text{J}$
1 calorie	$4.187\,\text{J}$
1 gauss	$10^{-4}\,\text{T}$
1 gamma	$10^{-9}\,\text{T}$

A2.3   Illuminance at the Earth's surface

Illuminance is the photometric quantity that is analogous to the radiometric quantity called irradiance. Its definition is discussed in section 5.3. The following table gives typical values of the illuminance at the Earth's surface.

Condition	Illuminance (lux)
Noon, clear day	10^5
Noon, overcast	10^4
Sunrise and sunset	500
Sun 5° below horizon	5
Full moon, clear sky	0.5
Full moon, overcast	0.05
Clear starlit sky	0.005

A2.4   Properties of the Sun and Earth

Sun's radius	$6.96 \times 10^8\,\text{m}$
Sun's mass	$1.99 \times 10^{30}\,\text{kg}$
Total radiated solar power	$3.85 \times 10^{26}\,\text{W}$
Sun's black-body temperature	$5770\,\text{K}$
Earth's equatorial radius	$6\,378\,135\,\text{m}$
Earth's polar radius	$6\,356\,775\,\text{m}$
Semi-major axis of Earth's orbit around the Sun	$1.496 \times 10^{11}\,\text{m}$
Earth's mass	$5.976 \times 10^{24}\,\text{kg}$
Earth's angular velocity about its polar axis	$7.292 \times 10^{-5}\,\text{s}^{-1}$
Sidereal day	$86\,164.09\,\text{s}$
Tropical year (equinox to equinox)	$31\,556\,926\,\text{s}$
Sidereal year (fixed star to fixed star)	$31\,558\,150\,\text{s}$
GM	$3.986 \times 10^{14}\,\text{m}^3\,\text{s}^{-1}$
Standard gravitational acceleration at Earth's surface (*g*)	$9.807\,\text{m}\,\text{s}^{-2}$
Earth's dynamical form factor (J_2)	1.0826×10^{-3}

Mean global albedo	0.35
Land area	$1.49 \times 10^{14} \text{ m}^2$
Ocean area	$3.61 \times 10^{14} \text{ m}^2$
Mean land elevation	860 m
Mean ocean depth	3.9 km

Mean annual atmospheric temperature at sea level:

at the equator	27°C
at 30° N or S latitude	20°C
at 60° N or S latitude	–2°C
at the poles	–25°C

A2.5   Position of the Sun

Since much environmental remote sensing is dependent on solar illumination, it is important to be able to calculate the Sun's position in the sky at a given date, time and position on the Earth's surface. This will permit, for example, the hours of darkness and solar elevation angles to be calculated.

The relevant variables describing the Sun's position relative to the Earth are the *solar declination δ* and the *equation of time E*. It will sometimes also be necessary to know the distance D between the centres of the Earth and the Sun. The declination is the Sun's angle above or below the Earth's equatorial plane. The equation of time is the difference in position between the true Sun and a fictitious Sun (the 'mean Sun') that appears to travel uniformly across the sky. It is normally expressed in minutes of time. For a position on the Earth's surface having longitude λ (with longitudes east of the Greeenwich meridian being positive), the Sun's *hour angle H* is given, in degrees, by

$$H = 15T - 180 + \lambda + \frac{E}{4} \tag{A2.1}$$

where T is the Universal Time (Greenwich Mean Time) in hours, λ is measured in degrees and E in minutes of time. It may be necessary to add or subtract 360° from equation (A2.1) to generate a value that is in the range −180° to +180°.

For a position on the Earth's surface having latitude φ (defined such that north latitudes are positive) and from which the solar hour angle is H, the Sun's elevation angle a is given by

$$\sin a = \sin \delta \sin \phi + \cos \delta \cos \phi \cos H \tag{A2.2}$$

and its azimuth angle A (north = 0°, east = 90°, etc.) is given by

$$\cos A = \frac{\sin \delta - \sin \phi \sin a}{\cos \phi \cos a} \tag{A2.3}$$

The ambiguity in equation (A2.3) is resolved by noting that if the hour angle H is negative, the azimuth A must be less than 180°, while if H is positive, A must be greater than 180°.

The values of δ, E and D are tabulated in a number of official publications, or they can be calculated from standard astronomical formulae. The following formulae are simple approximations that use only the day number d within the year (i.e. January $1 = 1$, February $1 = 32$, etc.) In all cases, the formulae assume that trigonometrical quantities are expressed in degrees. Equation (A2.4) gives the value of E in minutes, to an accuracy of about 1 min. Equation (A2.5) is accurate to about 1°. Equation (A2.6) gives the value of D in astronomical units, accurate to about 0.001 AU.

$$E \approx 9.9 \sin(1.97(d - 80)) - 7.7 \sin(0.986(d - 3)) \qquad \text{(A2.4)}$$

$$\sin \delta \approx 0.3987 \sin(0.986(d - 80)) \qquad \text{(A2.5)}$$

$$D \approx 1 - 0.0167 \cos(0.986(d - 3)) \qquad \text{(A2.6)}$$

Example

Suppose we wish to estimate the Sun's position from 18°S, 32°E at 12:00 UT on 1 November 1999. The *Astronomical Almanac* gives $\delta = -14.37°$ and $E = +16.40$ min for this date and time. Equation (A2.1) gives the Sun's hour angle $H = 36.10°$, and substitution of this value into equation (A2.2) gives the solar elevation as 55.19°. From equation (A2.3), $\cos A = +0.0102$. Taking the inverse cosine gives $A = 89.42°$ or 270.58°, and since H is positive, the correct value is $A = 270.58°$.

Using the approximations of equations (A2.4) and (A2.5), we obtain $\delta = -15.43°$ and $E = +16.64$ min, and hence $H = 36.16°$, $a = 55.33°$ and $A = 268.74°$. The solar distance is given by equation (A2.6) as 0.992 AU.

References

Avery, T.E. and Berlin, G.L. (1992). *Fundamentals of Remote Sensing and Airphoto Interpretation* (5th edition). New York: Macmillan Publishing Company.

Barrett, E.C. and Curtis, L. F. (1992). *Introduction to environmental Remote Sensing* (3rd edition). London: Chapman and Hall.

Beckmann, P. and Spizzichino, A. (1963). *The scattering of electromagnetic waves from rough surfaces.* Oxford: Pergamon Press.

Berk, A., Bernstein, L.S. and Robinson, D.C. (1989). *MODTRAN: A moderate resolution model for LOWTRAN 7.* Hanscom Air Force Base, Massachusetts: Air Force Geophysics Laboratory.

Bracewell, R.N. (1978). *The Fourier Transform and its applications* (2nd edition). Tokyo: McGraw-Hill Kogakusha Ltd

Campbell, J.B. (1996). *Introduction to Remote Sensing* (2nd edition). London: Taylor and Francis.

Chen, H.S. (1985). *Space Remote Sensing Systems: an introduction.* Orlando: Academic Press.

Chetty, P.R.K. (1988). *Satellite technology and its applications.* Blue Ridge Summit, Pennsylvania: Tab Books.

Colwell, R.N. (ed.) (1983). *Manual of Remote Sensing* (2nd edition), 2 volumes. Falls Church, Virginia: American Society of Photogrammetry.

Cracknell, A.P. (1981). *Remote sensing in meteorology, oceanography and hydrology.* Chichester: Ellis Horwood.

Cracknell, A.P. and Hayes, L.W.B. (1991). *Introduction to Remote Sensing.* London: Taylor and Francis.

Curran, P.J. (1985). *Principles of Remote Sensing.* London and New York: Longman.

Deirmendjian, D. (1969). *Electromagnetic scattering on spherical polydispersions.* New York: American Elsevier Co.

Deschamps, P.Y. and Phulpin, T. (1980). Atmospheric correction of infrared measurements of sea surface temperature using channels at 3.7, 11 and 12 μm. *Boundary-Layer Meteorology,* **18**, 131–143

Drewry, D.J. (1983). *Antarctica: Glaciological and geophysical folio.* Cambridge: Scott Polar Research Institute.

Drury, S.A. (1998). *Images of the Earth: a guide to Remote Sensing.* Oxford: Oxford University Press.

Elachi, C. (1987). *Introduction to the physics and techniques of Remote Sensing*. New York: John Wiley.

Eppler, D.T., Farmer, L.D., Lohanick, A.W., Anderson, M.R., Cavalieri, D.J., Comiso, J., Gloersen, P., Garrity, C., Grenfell, T.C., Hallikainen, M., Maslanik, J.A., Mätzler, C., Melloh, R.A., Rubinstein, I. and Swift, C.T. (1992). Passive microwave signatures of sea ice. In Carsey, F.D. (ed), *Microwave Remote Sensing of Sea Ice*, Washington, DC: American Geophysical Union.

Feynman, R.P., Leyton, R.B. and Sands, M. (1964). *Lectures on Physics*. Reading, Massachusetts: Addison-Wesley.

Forshaw, M.R.B. et al. (1983). Spatial resolution of remotely sensed images. *International Journal of Remote Sensing*, **4**, 497.

Goetz, A.F.H. (1979). *Preliminary Stereosat mission description*. Jet Propulsion Laboratory report no. 720-33. Pasadena, California: Jet Propulsion Laboratory.

Graham, L.C. (1974). Synthetic Interferometer Radar for Topographic Mapping. *Proceedings of the IEEE*, **62**, 6.

Hapke, B. (1993). *Theory of reflectance and emittance spectroscopy*. Cambridge: Cambridge University Press.

Harris, R. (1987). *Satellite Remote Sensing: an introduction*. London: Routledge and Kegan Paul.

Hecht, E. (1987). *Optics* (2nd edition). Reading, Massachussetts: Addison-Wesley.

Henderson-Sellers, A. (1984). *Satellite sensing of a cloudy atmosphere: observing the third planet*. London: Taylor and Francis.

Houghton, J.T. (1986). *The physics of atmospheres*. Cambridge: Cambridge University Press.

Jepsky, J. (1985). Airborne laser profiling and mapping. *Lasers and Applications*, **4**, 95.

Joss, J. and Gori, E.G. (1978). Shapes of raindrop size distributions. *Journal of Applied Meteorology*, **17**, 1054–1061.

Justice, C.O. et al. (1985). Analysis of the phenology of global vegetation using meteorological satellite data. *International Journal of Remote Sensing*, **6**, 1271.

Kneizys, F.X., Shettle, E.P., Abreu, L.W., Chetwynd, J.H., Andreson, G.P., Gallery, W.O., Selby, J.E.A. and Clough, S.A. (1988). *User's guide to LOWTRAN 7*. Hanscom Air Force Base, Massachusetts: Air Force Geophysics Laboratory.

Kramer, H.J. (1996). *Observation of the Earth and its environment* (3rd edition). Berlin: Springer-Verlag.

Laws, J.O. and Parsons, D.A. (1943). The relation of raindrop size to intensity. *Transactions of the American Geophysical Union*, **24**, 452–460.

Leick, A. (1995). *GPS Satellite Surveying* (2nd edition). New York: Wiley-Interscience.

Lillesand, T.M. and Kiefer, R.W. (1994). *Remote Sensing and Image Interpretation*. New York: John Wiley.

Lipson, S.G., Lipson, H. and Tannhauser, D.S. (1995). *Optical Physics* (3rd edition). Cambridge: Cambridge University Press.

Long, M.W. (1983). *Radar reflectivity of land and sea*. Dedham, Massachusetts: Artech House Inc.

Longair, M.S. (1984). *Theoretical concepts in physics*. Cambridge: Cambridge University Press.

MacDonald, R.A. (1995). CORONA: Success for space reconnaissance, a look into the Cold War, and a revolution for intelligence. *Photogrammetric Engineering and Remote Sensing*, **61**, 689–720.

Marshall, G.J., Dowdeswell, J.A. and Rees, W.G. (1994). The spatial and temporal effect of cloud cover on the acquisition of high quality Landsat imagery in the European Arctic sector. *Remote Sensing of Environment*, **50**, 149–160.

Marshall, J.S. and Palmer, W.M.K. (1948). The distribution of raindrops with size. *Journal of Meteorology*, **5**, 165–166.

Massonnet, D. (1993). Geoscientific applications at CNES. In Schreier, G. (ed.), *SAR Geocoding: Data and Systems*. Karlsruhe: Herbert Wichmann Verlag.

Mather, P.M. (1987). *Computer processing of remotely-sensed images: an introduction*. Chichester: John Wiley.

Newton, D.C. (1989). Letter. *New Scientist*, 14 January, p. 72.

Omar, M.A. (1975). *Elementary solid state physics*. Reading, Massachusetts: Addison-Wesley.

Rees, W.G. (1988). Foreword to *Radio echo-sounding as a glaciological technique* (comp. A.D. Macqueen). Cambridge: World Data Centre 'C' for glaciology.

Rees, W.G. (1992). Orbital subcycles for Earth remote sensing satellites. *International Journal of Remote Sensing*, **13**, 825–833.

Rees, W.G. (editor) (1999). *The Remote Sensing data book*. Cambridge: Cambridge University Press.

Rees, W.G. and Satchell, M.J.F. (1997). The effect of median filtering on synthetic aperture radar images. *International Journal of Remote Sensing*, **18**, 2887–2893.

Richards, J.A. (1993). *Remote Sensing Digital Image Analysis* (2nd edition). Berlin: Springer-Verlag.

Robinson, I.S. (1994). *Satellite Oceanography*. Chichester: Wiley-Praxis.

Ryerson, R.A. (editor) (1998). *Manual of Remote Sensing* (3rd edition), 2 volumes. Bethesda, Maryland: American Society for Photogrammetry and Remote Sensing.

Sabins, F.F. (1996). *Remote Sensing: principles and interpretations*. New York: W.H. Freeman.

Schanda, E. (1986). *Physical fundamentals of Remote Sensing*. Berlin: Springer-Verlag.

Schowengerdt, R.A. (1997). *Remote Sensing: Models and Methods for Image Processing* (2nd edition). New York: Academic Press Inc.

Scorer, R.S. (1986). *Cloud investigation by satellite*. Chichester: Ellis Horwood.

Smith, D.E. et al. (1985). A global geodetic reference frame from LAGEOS ranging (SL5.1AP). *Journal of Geophysical Research*, **90**, 9221.

Sonka, M., Hlavac, V. and Boyle, R. (1993). *Image processing, analysis and machine vision*. London: Chapman and Hall.

Stewart, R.H. (1985). *Methods of satellite oceanography*. Berkeley: University of California Press.

Stiles, W.H. and Ulaby, F.T. (1980). The active and passive microwave response to snow parameters. 1: wetness. *Journal of Geophysical Research*, **85**, 1037–1044.

Swain, P.N. and Davis, S.M. (1978). *Remote Sensing: the quantitative approach*. New York: McGraw-Hill.

Tsang, L., Kong, J.A. and Shin, R.T. (1985). *Theory of microwave remote sensing*. New York: John Wiley.

Ulaby, F.T., Moore, R.K. and Fung, A.K. (1981, 1982, 1986) *Microwave Remote Sensing*, 3 volumes. Reading, Massachussetts: Addison-Wesley.

Van de Hulst, H.C. (1957, 1981) *Light scattering by small particles*. New York: John Wiley. Republished New York: Dover Publications.

Wahl, T., Eldhuset, K. and Asknes, K. (1986). SAR detection of ships and ship wakes. In *SAR Applications Workshop* (ESA Special Publication SP-264), p. 61. Paris: European Space Agency.

Warren, D. and Turner, J. (1988). Cloud track winds from polar orbiting satellites. In *Proceedings of the 1988 International Geoscience and Remote Sensing Symposium*, (ESA SP-284), p. 549. Paris: European Space Agency.

Závody, A.M., Mutlow, C.T. and Llewellyn-Jones, D.T. (1995). A radiative transfer model for sea surface temperature retrieval for the along-track scanning radiometer. *Journal of Geophysical Research*, **100** (C1), 937–952.

Zebker, H.A. and Goldstein, R.M. (1986). Topographic mapping from interferometric SAR observations. *Journal of Geophysical Research*, **91**, B5.

Hints and solutions to numerical problems

Chapter 2. Electromagnetic waves in free space

1. The y-component of the wave described in the problem can be obtained from equation (2.2) by transforming x to y, y to z and z to x, and setting $E_0 = E$. Applying the same transformations to equation (2.3) gives

$$B_z = \frac{E}{c}\cos(\omega t - kx)$$

The z component of the new wave can similarly be obtained from equation (2.9) by transforming x to y, y to z and z to x, and setting $E_0 = 2E$. Applying these transformations to equation (2.10) gives

$$B_y = -\frac{2E}{c}\cos(\omega t - kx)$$

The flux density is found by substituting $E_0 = \sqrt{5}E$ into equation (2.7). It is 6.63 kW m^{-2}.

2. Using the examples of Stokes vectors given in section 2.2, the detected power is 1 unit for random polarisation, 2 units for linearly x-polarised radiation, 0 for linearly y-polarised radiation, and 1 unit for all other fully polarised states except elliptical. For the example of elliptically polarised radiation listed in section 2.2, the detected power is 1.6 units.

3. Use the fact that

$$\exp(x) \approx 1 + x$$

for $|x| \ll 1$.

4. From equation (2.17), the Fourier transform is given by

$$a(\omega) = \frac{1}{2\pi} \int\limits_{-\infty}^{\infty} \exp\left(-\frac{(t - t_0)^2}{2\sigma^2} - i\omega t\right) dt$$

Change the variable to

$$z = \frac{t - t_0}{\sigma\sqrt{2}} + \frac{i\omega\sigma}{\sqrt{2}}$$

so that the integral becomes

$$a(\omega) = \frac{\sigma\sqrt{2}}{2\pi} \int\limits_{-\infty}^{\infty} \exp(-z^2)\, dz\, \exp\left(-\frac{i\omega t_0}{2} - \frac{\omega^2\sigma^2}{2}\right)$$

This proves the result. We note that a Gaussian function of width (standard deviation) σ has been transformed into another Gaussian function of width $1/\sigma$. This is an example of the 'uncertainty principle', discussed in chapter 2, relating the width of a function in the time domain to the width of the corresponding representation in the frequency domain. The $\exp(-i\omega t_0/2)$ part of the Fourier transform is a phase factor arising from the fact that $f(t)$ was centred not on $t = 0$ but on $t = t_0$.

5. The formula for the ratio is

$$\frac{e^{hf/kT_2} - 1}{e^{hf/kT_1} - 1}$$

where $T_1 = 300\,\mathrm{K}$ and $T_2 = 6000\,\mathrm{K}$. (a) When $f = 1\,\mathrm{GHz}$, the Rayleigh–Jeans approximation gives the ratio as $\approx T_1/T_2 = 0.050$. (b) When $f = 1000\,\mathrm{GHz}$, the Rayleigh–Jeans approximation is no longer quite good enough. Evaluating the ratio directly gives $(\exp(0.00800) - 1)/(\exp(0.16005) - 1) = 0.046$. (c) At a wavelength of $1\,\mu\mathrm{m}$ it is still possible to evaluate the ratio directly: $(\exp(2.4007) - 1)/(\exp(48.01) - 1) = 1.41 \times 10^{-20}$. (d) At $0.1\,\mu\mathrm{m}$, the ratio $\approx \exp(24.01)/\exp(480.1) = \exp(-456) \approx 10^{-198}$. Calculation of (c) and (d) shows that thermal emission can almost always be ignored with respect to reflected solar radiation, where present, at visible and near-infrared wavelengths.

Chapter 3. Interaction of electromagnetic radiation with matter

1. Substitute equation (3.12.2) into (3.12.1) and rearrange to give a quadratic equation in m^2:

$$m^4 - \varepsilon' m^2 - \frac{\varepsilon''^2}{4} = 0$$

This can be solved to give the required result. Similarly, substitute equation (3.12.1) into (3.12.2) to give a quadratic equation in κ^2.

2. Substitute the result for κ from problem 3.1 into equation (3.15), introducing the variables $\lambda_0 = 2\pi c/\omega$ and $x = \varepsilon''/\varepsilon'$, to give

$$l_a = \frac{\sqrt{2}\lambda_0}{4\pi\sqrt{\varepsilon'}} \frac{1}{\sqrt{\sqrt{1 + x^2} - 1}}$$

The ratio of the approximate expression for l_a to the accurate expression can thus be written as

$$\frac{\sqrt{2}}{x}\sqrt{\sqrt{1 + x^2} - 1}$$

It is now straightforward to show that this ratio does not fall below 0.99 provided that x does not exceed about 0.28.

3. From the data given, the dielectric constant at $100\,\mathrm{MHz}$ is $88.2 - 719i$. This gives $\kappa = 17.8$ (see problem 3.1), and hence (from equation (3.15)) $l_a = 13\,\mathrm{mm}$. Repeating the

calculation for a frequency of 100 kHz gives $\kappa = 600$ and $l_a = 0.4$ m. Note that when $\varepsilon' \ll \varepsilon''$ and $\varepsilon'' = \sigma/2\pi\varepsilon_0 f$ (as here), we have $l_a \approx (c^2\varepsilon_0/4\pi\sigma f)^{1/2}$. Significant absorption lengths in sea water are only possible at very low frequencies (which is why VLF has to be used for radio communication with submarines.)

4. From equation (3.32.1) we find that the amplitude reflection coefficient for perpendicularly polarised radiation is $-0.972 + 0.007i$, while for parallel polarised radiation equation (3.32.3) gives $-0.016 + 0.117i$. These coefficients can be written in complex exponential form as $r\,e^{i\phi}$, where the values of r are 0.972 and 0.118, respectively, and the values of ϕ are 3.134 and 1.707 radians, respectively. Defining the x-axis of the reflected radiation to correspond to the perpendicularly polarised component, we find from the definition of the Stokes vector $\mathbf{S}$ in equation (2.13) that $\mathbf{S} = [0.959, 0.931, 0.033, -0.227]$ relative to the incident radiation. This is practically 100% polarised (the incidence angle is close to the Brewster angle), predominantly linearly polarised but with a significant circularly polarised component.

5. Use equation (3.33.1) and the equation preceding (3.43). It is convenient to change the variable to $\mu = \cos\theta$, in which case the expression for the diffuse albedo becomes

$$2\int_0^1 \left(\frac{\mu - \sqrt{n^2 + \mu^2 - 1}}{\mu + \sqrt{n^2 + \mu^2 - 1}}\right)^2 \mu\,d\mu$$

This can be evaluated to the expression given in the problem. As n tends to 1, the albedo tends to zero (no dielectric contrast), and as n tends to infinity the albedo tends to 1 (perfect reflection).

6. Put

$$R = A\cos\theta_0\cos\theta_1$$

where A is a constant to be determined. Substituting into equation (3.39) gives

$$r(0,0) = 2\pi A \int_0^{\pi/2} \cos^2\theta_1 \sin\theta_1\,d\theta_1 = \frac{2\pi A}{3}$$

so that $A = 3/2\pi$. Hence, from equation (3.41) the diffuse albedo is given by

$$6\int_0^{\pi/2} \cos^2\theta_0 \sin\theta_0\,d\theta_0 \int_0^{\pi/2} \cos^2\theta_1 \sin\theta_1\,d\theta_1 = \frac{2}{3}$$

as required.

7. For the stationary-phase model, condition (3.56.1) gives $\cos(\theta) > 1.58/\pi$ and so $\theta < 59.8°$. Condition (3.56.2) gives $L/\lambda > 6/2\pi = 0.95$ and condition (3.56.3) gives $L/\lambda > \sqrt{(17.3/4\pi)} = 1.17$. Thus condition (3.56.2) is ineffective, and from equation (3.55), $m < \sqrt{2}/(2 \times 1.17) = 0.60$. For the scalar model, condition (3.58.1) gives $L/\lambda > 1/(2 \times 0.18) = 2.78$, condition (3.58.2) gives $L/\lambda > 6/2\pi = 0.95$ and condition (3.58.3) gives $L/\lambda > \sqrt{(17.3/4\pi)} = 1.17$. The 'active' conidition is thus that $L/\lambda > 2.78$, hence $m < \sqrt{2}/(2 \times 2.78) = 0.25$.

8. Solve the coupled differential equations (3.61.1) and (3.61.2) by trying a solution

$$F_- = \exp(\mu z)$$
$$F_+ = R \exp(\mu z)$$

9. Following the argument presented for cloud on page 77, we estimate the scattering cross-section of a single ice crystal as $8 \times 10^{-7} \, \text{m}^2$. The number density of the ice crystals is $2 \times 10^8 \, \text{m}^{-3}$, so the scattering coefficient γ_s is of the order of $200 \, \text{m}^{-1}$ and the ratio γ_s/γ_a is of the order of 2×10^5. Since this is very much greater than 1, we expect virtually all of the radiation that enters the snowpack to be reflected out of it. The attentuation length is roughly $1/\gamma_s \approx 5 \, \text{mm}$, so provided the snowpack is a few centimetres deep it will be optically thick.

Chapter 4. Interaction of electromagnetic radiation with the Earth's atmosphere

1. Combining equations (4.10) and (4.11) gives

$$\frac{\Delta f_{\text{Doppler}}}{\Delta f_{\text{pressure}}} = \frac{fRT}{\sigma N_A pc}$$

Substitute $R = kN_A$ and $c = f\lambda$.

2. Model the attenuation coefficient as

$$\gamma_a = \gamma_{a,0} \, \exp\left(-\frac{z}{z_0}\right)$$

where $\gamma_{a,0}$ is the attenuation coefficient at sea level and z_0 is the scale height. Thus the optical thickness is given by

$$\tau = \int_0^\infty \gamma_a \, dz = \gamma_{a,0} z_0$$

and hence the scale height is 2.0 km.

3. Assume the water droplets are spherical, with radius a and scattering cross-section πa^2. Equation (4.15) gives the visibility v as

$$v = \frac{16 a \rho_w}{3\rho}$$

where ρ_w is the density of water and ρ is the density of the fog. Substituting the appropriate values gives $a \approx 19 \, \mu\text{m}$.

4. Combining equations (4.16) and (3.84) gives

$$\gamma_s = \frac{36\pi^2 \tau \varepsilon_p}{(\varepsilon_\infty + \varepsilon_p + 2)^2 c \rho_w} \rho f^2$$

Substituting the data, and recalling that a scattering coefficient in m^{-1} is converted to dB km^{-1} by multiplying by 4.34×10^3, gives the required result. A possible criterion for considering the transparency or otherwise of a cloud is a total attenuation of, say, 3 dB.

5. For raindrops of radius a, equations (4.18) and (4.19) become

$$\gamma_s = \pi a^2 N$$

and

$$\rho = \frac{4}{3}\pi a^3 \rho_w N$$

respectively, and the rain rate R is given by

$$R = \frac{4}{3}\pi a^3 N v$$

where v is the sedimentation speed. Solving for the two rain rates gives

$$R = 1\,\text{mm h}^{-1}: a = 0.5\,\text{mm}, N = 60\,\text{m}^{-3}, v = 8\,\text{m s}^{-1}$$

$$R = 100\,\text{mm h}^{-1}: a = 1.2\,\text{mm}, N = 320\,\text{m}^{-3}, v = 13\,\text{m s}^{-1}$$

Chapter 5. Photographic systems

1. Use the results of section 2.6 to estimate the fraction of the Sun's irradiance between wavelengths $0.51\,\mu\text{m}$ and $0.61\,\mu\text{m}$: The mean exoatmospheric irradiance is $1.37\,\text{kW m}^{-2}$ and the fraction between the wavelength limits is approximately 0.127. Multiplying these two figures, and multiplying the result by the factor K in equation (5.1), gives an exoatmospheric illuminance of 118 klx. The optical thickness of the atmosphere can be taken as 0.3 for vertical propagation (see figures 4.5 and the discussion of aerosols in section 4.3), which should be increased by a factor of $1/\sin(45°)$ to 0.42 to allow for the oblique propagation. Thus we expect the illuminance at the surface to be roughly 77 klx, although for a horizontal surface this should be further reduced by a factor of $\sin(45°)$ to allow for the oblique illumination. This gives a final figure of about 55 klx. Figure 5.21 shows an illuminance of about 60 klx, so our estimate is a remarkably good one.

2. Scale of the negative $= 1/100\,000$ (equation (5.4)); coverage $= 2.5 \times 3.5\,\text{km}$ (equation (5.5)); ground resolution $= 1.25\,\text{m}$ (equation (5.6)); height accuracy for relief displacement $= 29\,\text{m}$ (equation (5.11)); height accuracy for stereophotography $= 36\,\text{m}$ assuming a base-height ratio of 0.28 (which would give a vertical exaggeration factor of about 2).

3. (a) Denote the four specified points by A, B, C and D, respectively. The principal point P is where the projections of the lines DC and BA meet, since these are the images of vertical lines. Simple coordinate geometry gives the coordinates of P as (81.1, 79.6) mm. (A diagram is helpful to see what is going on, but is unlikely to give the position of P with sufficient accuracy.)

   (b) Use equation (5.10). For the vertical that is imaged as AB, we have $h' = AB = 12.18\,\text{mm}$ and $x' = AP = 53.57\,\text{mm}$, so taking $H = 212\,\text{m}$ gives $h = 39.3\,\text{m}$. Repeating the calculation for CD as a check gives $h' = 11.60\,\text{mm}$ and $x' = 51.03\,\text{mm}$, and hence $h = 39.3\,\text{m}$ in this case too.

   (c) The scale of the negative at ground level is $f/H \approx 1/2409$. The length $AC = 11.0\,\text{mm}$ corresponds to the width of the building at this scale, so the width is 26.5 m.

Chapter 6. Electro-optical systems

1. Consider the diagram below.

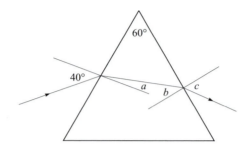

We have

$$\sin a = \frac{\sin 40°}{n}$$

$$b = 60° - a$$

$$\sin c = n \sin b$$

and the total deviation of the ray is $c - 20°$. For $n = 1.601$, we obtain $a = 23.671°$, $b = 36.329°$ and $c = 71.524°$, giving a deviation of $51.524°$. For $n = 1.569$, the corresponding deviation is $46.655°$, so the range of deviation is $4.89°$.

2. From equation (2.23), the spectral irradiance at the surface can be written as

$$E_\lambda = L_{\lambda,S} \, \Delta\Omega \, \cos \theta$$

where $L_{\lambda,S}$ is the spectral radiance from the Sun, $\Delta\Omega$ is the solid angle subtended by the Sun, and θ is the incidence angle of the radiation. Equation (3.43) shows that the BRDF of the surface is $1/\pi$ in all directions, so from equation (3.35) the reflected spectral radiance is

$$L_{\lambda,\text{out}} = \frac{L_{\lambda,S} \, \Delta\Omega \, \cos \theta}{\pi}$$

Setting

$$L_{\lambda,S} = \frac{2hc^2 \varepsilon_S}{\lambda^5 (\exp[hc/\lambda k T_S] - 1)}$$

where ε_S and T_S are, respectively, the Sun's emissivity and temperature, and

$$L_{\lambda,\text{out}} = \frac{2hc^2}{\lambda^5 (\exp[hc/\lambda k T_b] - 1)}$$

where T_b is the brightness temperature of the reflected radiation, gives

$$\exp[hc/\lambda k T_b] - 1 = \frac{\pi}{\varepsilon_S \, \Delta\Omega \, \cos \theta} (\exp[hc/\lambda k T_S] - 1)$$

Substitute $\Delta\Omega = 6.76 \times 10^{-5}$ sr, $\varepsilon_S = 0.99$ and $T_S = 5800$ K to obtain $T_b \approx 330$ K and $4 \, \mu$m and 150 K at $10 \, \mu$m.

3. Differentiate equation (6.5) with respect to z and equate to (6.6), to give the thermal diffusion equation

$$\frac{\partial^2 T}{\partial z^2} = \frac{C\rho}{K} \frac{\partial T}{\partial t} \tag{i}$$

This is most easily solved using complex exponential notation. Substitute the expression

$$T = A \exp(i[\omega t - kz]) \qquad \text{(ii)}$$

into the diffusion equation to find

$$k^2 = \frac{i\omega C\rho}{K}$$

and hence

$$k = (1 - i)\sqrt{\frac{\omega C\rho}{2K}} \qquad \text{(iii)}$$

Substituting equation (ii) into (6.5) gives

$$F = ikKA \exp(i[\omega t - kz])$$

so if we set $F_0 = ikKA$ we obtain the solutions

$$F = F_0 \exp(i[\omega t - kz])$$

and

$$T = \frac{F_0}{ikK} \exp(i[\omega t - kz])$$

Substituting for k from (ii), and taking the real parts of these expressions, gives equations (6.7) and (6.8).

4. For convenience, set $\exp(-\tau) = f$ and $\sec(\theta) = n$, so that equations (6.14) and (6.15) become

$$T_{b1} = fT_{b0} + (1 - f)T_a \qquad \text{(i)}$$

and

$$T_{b2} = f^n T_{b0} + (1 - f^n)T_a \qquad \text{(ii)}$$

respectively. These can be solved for f to give

$$f = \left(\frac{T_{b2} - T_a}{T_{b1} - T_a}\right)^{\frac{1}{n-1}} \qquad \text{(iii)}$$

Substituting the data gives $f = 0.861$ and hence the optical thickness $\tau = 0.15$. Inserting this value of f into equation (i) gives $T_{b0} = 297.0\,\text{K}$.

5. Substitute the expression for $\gamma(h)$ into equation (6.23), change the height variable to $z = (h - h_0)^{1/2}$ and recall that

$$\int_0^\infty \exp(-\beta z^2)\, dx = \frac{1}{2}\sqrt{\frac{\pi}{\beta}}$$

Putting $\tau = \ln 10$, $R = 6378\,\text{km}$ and substituting the given values of β and γ_0 gives $h_0 \approx 7.1\,\text{km}$.

Chapter 7. Passive microwave systems

1. Use table 7.1 (recalling that D is specified in decibels) and equations (7.5) and (7.9) to show that the effective area of a rectangular antenna is approximately ab and that of a circular paraboloid is approximately πd^2.

2. The discussion in section 7.3.1 shows that the phase shift between adjacent antenna elements is $kd \sin \theta_0$ when the beam has been steered through an angle θ_0. Since addition of integer multiples of 2π to the phase makes no difference, it follows that the antenna will also respond to the direction θ, where

$$\sin \theta = \sin \theta_0 + \frac{n\lambda}{d}$$

and n is any integer. (a) In this case, $\lambda/d = 1.5$, so provided that $\sin \theta_0 > 0.5$, the antenna will have a second response corresponding to $n = -1$. (b) In order to avoid these multiple responses, λ/d must be at least 2. In this case, when $\sin \theta_0$ has its maximum possible value of $+1$, the next largest value of $\sin \theta$ is less than -1 and thus impossible.

3. The observed brightness temperature will be an area-weighted average of the brightness temperatures of the different species present in the IFOV (this is a consequence of the Rayleigh–Jeans law). Assuming fractions f_w, f_f and f_m of the three species, and that $f_m = 1 - f_f - f_w$ (no other species present), we have

$$80 f_w + 252 f_f + 200 (1 - f_f - f_w) = 119 f_w + 253 f_f + 168 (1 - f_f - f_w) = 180$$

Solution: $f_w = 30\%, f_f = 32\%, f_m = 38\%$.

4. If the fraction of the IFOV occupied by ice is f, the antenna temperature is

$$253f + 119(1 - f) = 119 - 134f$$

Setting this equal to $119 - 0.9$ gives $f = 0.00672$ as the minimum detectable value of f. The IFOV (solid angle) is $\lambda^2/A_e = 0.00022$ sr, so the area of the IFOV on the surface $= 140 \, \text{km}^2$. Thus, the smallest detectable floe area is $140 \times 0.00672 \, \text{km}^2 = 0.9 \, \text{km}^2$.

5. Substitute

$$\tau' = \int_0^z \gamma \, dz = \tau(1 - \exp(-\beta z))$$

into equation (3.74) to obtain the expression for $a(z)$. We then have

$$\ln(a(z)) = \ln \tau + \ln \beta - \beta z - \tau \exp(-\beta z)$$

Differentiate this with respect to z and equate to zero to find the value of z at which $a(z)$ is a maximum. To show that the effect of changing τ (at fixed β) is to shift $a(z)$ along the z-axis without change of scale, we have to show that $a(z - z_0, \tau^*) = a(z, \tau)$ for all values of z. This can be simplified to

$$\ln \tau - \tau \exp(-\beta z) = \ln \tau^* + \beta z_0 + \tau^* \exp(-\beta(z - z_0)) \qquad \text{(i)}$$

so equating the exponentials gives

$$\tau^* \exp(\beta z_0) = \tau \qquad \text{(ii)}$$

Substituting equation (ii) into (i) confirms the equality, and the result is established. It implies that, provided the simple model of the absorption coefficient is valid, the vertical resolution of the sounding technique is independent of the height being sounded (e.g. see figure 6.22).

Chapter 8. Ranging systems

1. Figure 8.5 shows a dry-atmosphere zenith correction of about 2.35 m for a wavelength of 1 μm, so for a path that passes through $1 - 0.26 = 0.74$ atmospheres at an angle of $45°$, the estimated correction is $2.35 \times 0.74/\cos(45°) \approx 2.46$ m. Figure 8.6 shows a water vapour correction of about 0.35 m per metre of precipitable water, or about 0.018 m for a vertical path through the entire atmosphere. If we assume that all of the water vapour will be found below 10 000 m, the only correction needed is to multiply by $1/\cos(45°)$ to allow for the oblique path, so the correction is approximately 0.025 m.

2. Let the radar be located at height H above the mean surface, and consider a scatterer at radial distance r from the nadir point and height h above the mean surface. The distance from the radar is approximately

$$H + \frac{r^2}{2H} - h$$

so a signal received from this scatterer at time t must have been emitted at time

$$t - \frac{2}{c}\left(H + \frac{r^2}{2H} - h\right)$$

If we can write the number of scatterers between r and $r + dr$ and between h and $h + dh$ as

$$kr\, f(h)\, dr\, dh$$

(the $r\, dr$ part accounts for the area of surface and the $f(h)\, dh$ part for the distribution of scatterers with height), the received power at time t is given by

$$P_r = k \int_{r=0}^{\infty} \int_{h=-\infty}^{\infty} P_t\left(t - \frac{2H}{c} - \frac{r^2}{Hc} + \frac{2h}{c}\right) r\, f(r)\, dr\, dh$$

Change the r variable to

$$s = t - \frac{2H}{c} - \frac{r^2}{Hc}$$

so that the expression for P can be rewritten as

$$P_r = \frac{kHc}{2} \int_{s=-\infty}^{t-2H/c} \int_{h=-\infty}^{\infty} P_t\left(s + \frac{2h}{c}\right) f(h)\, ds\, dh$$

This can now be differentiated with respect to s to give the required result.

3. From equation (8.14), the time-differential of the return pulse is the convolution of transmitted pulse with surface height distribution converted to a time distribution through the factor $c/2$. Thus we have two Gaussians, with widths to $1/e$ points of 3.00 ns and 6.67 ns. The convolution of these is another Gaussian, with width to $1/e$ points of $(3.00^2 + 6.67^2)^{1/2} = 7.31$ ns. This is the answer we seek.

4. First, find the Fourier transform of the original chirp signal:

$$a(\omega) = \frac{1}{2\pi} \int_{-T/2}^{T/2} \exp\left(i\left(\omega_0 t' + \frac{\Delta\omega}{2T} t'^2\right)\right) \exp(-i\omega t')\, dt'$$

The signal that emerges from the delay-line is therefore, in frequency representation,

$$a'(\omega) = \frac{1}{2\pi} \int_{-T/2}^{T/2} \exp\left(i\left(\omega_0 t' + \frac{\Delta\omega}{2T} t'^2 - \frac{\omega_0 T}{\Delta\omega}\omega - \frac{T}{2\Delta\omega}\omega^2\right)\right) \exp(-i\omega t') \, dt'$$

To retransform into the time domain will require the following operation:

$$f(t) = \int_{-\infty}^{\infty} a'(\omega) \exp(i\omega t) \, d\omega$$

To evaluate this integral, we first rearrange it:

$$f(t) = \frac{1}{2\pi} \int_{-T/2}^{T/2} \exp\left(i\omega_0 t' + \frac{\Delta\omega}{2T} t'^2\right)\right) \int_{-\infty}^{\infty} \exp\left(i\left(\left[t - t' - \frac{\omega_0 T}{\Delta\omega}\right]\omega - \frac{T}{2\Delta\omega}\omega^2\right)\right) d\omega$$

The second integral, over ω, can be evaluated by completing the square. It is

$$\sqrt{\frac{2\pi i\,\Delta\omega}{T}} \exp\left(-\frac{i\,\Delta\omega}{2T}\left(t - t' - \frac{\omega_0 T}{\Delta\omega}\right)^2\right)$$

so inserting this result into the first integral and simplifying it gives

$$f(t) = \frac{1}{2\pi}\sqrt{\frac{2\pi i\,\Delta\omega}{T}} \exp\left(i\left(-\frac{\omega_0^2 T}{2\Delta\omega} + \omega_0 t - \frac{\Delta\omega}{2T} t^2\right)\right) \int_{-T/2}^{T/2} \exp\left(i\frac{\Delta\omega t}{T} t'\right) dt'$$

The integral over t' is just $T\mathrm{sinc}(t\Delta\omega/2)$, so this is the modulating function. It is centred at $t = 0$ and falls to zero when $t = \pm 2\pi/\Delta\omega$. The only one of the phase terms (to the left of the integral) that is not negligible is the term $\exp(i\omega_0 t)$, which is the carrier wave term.

Chapter 9. Scattering systems

1. Use equation (9.4) with $\lambda = 5\,\mathrm{cm}$, $P_t = 4\,\mathrm{kW}$, $R = 800\,\mathrm{km}$, $\eta = 1$, $\sigma^0 A = 10^{-2.5} \times 10^3 = 3.16\,\mathrm{m}^2$ to obtain $G = 5.07 \times 10^4$. From equations (7.5), (7.6) and (7.9), the antenna's effective area $A_e = \lambda^2 G/4\pi = 10.1\,\mathrm{m}^2$.

2. Let the wind speed be v, direction ψ (measured eastwards from north). The data give

$$0.8v - 30 + (3.5 - 0.1v)\cos 2\psi = -22.9$$

for the north-looking observation, and

$$0.8v - 30 + (3.5 - 0.1v)\cos(\pi - 2\psi) = -21.1$$

for the east-looking observation. The second equation can be simplified to

$$0.8v - 30 - (3.5 - 0.1v)\cos 2\psi = -21.1$$

The two equations can be solved to give $v = 10.0$, $\cos 2\psi = -0.36$, so $\psi = 55.5°$, 235.5°, 124.5° or 304.5°, namely two possible directions with a 180° ambiguity in each. A third observation could remove one of these two directions, but not the 180° ambiguity (which is present in the model, though not in the slightly more sophisticated one given in equation (9.6)).

3. First consider a single component, with amplitude a and phase ϕ. Writing this in complex exponential notation, and then taking the real part x, gives $\langle x \rangle = 0$ (by symmetry) and $\langle x^2 \rangle = a^2/2$. If we now add a very large number N of such components, the central limit theorem shows that the probability distribution for the sum X of all the xs will be Gaussian, with zero mean and variance $Na^2/2$. In other words,

$$p(X) = \frac{1}{a\sqrt{N\pi}} \exp\left(-\frac{X^2}{Na^2}\right)$$

Clearly, an almost identical argument applies to the imaginary parts y, giving

$$p(Y) = \frac{1}{a\sqrt{N\pi}} \exp\left(-\frac{Y^2}{Na^2}\right)$$

The amplitude $R = (X^2 + Y^2)^{1/2}$ of the signal thus has the probability distribution

$$p(R) = \frac{2R}{a^2 N} \exp\left(-\frac{R^2}{Na^2}\right)$$

This is a Rayleigh distribution, as required.

4. Assume that

$$a_M = a \exp(i[\phi + 2kr_M])$$

and

$$a_S = a \exp(i[\phi + 2kr_S])$$

where a and ϕ are real constants that may vary from pixel to pixel. By comparing the phase angles of a_M and a_S, we find that $r_M - r_S = -0.011, +0.011, +0.007, +0.004$ m for pixels 1 to 4, respectively, although integer multiples of 0.025 m (half a wavelength) can be added to or subtracted from these values. By direct calculation, $r_M - r_S = -40.087$ m for pixel 1. To find the coordinates (x, z) of the scatterer that generates pixel 2, we note that $r_M - r_S$ is greater than for pixel 1 by $(0.022 + 0.025\,n)$ m, where n is an integer, and r_M is greater than for pixel 1 by 15 m. The easiest way to solve this is to make linear approximations of the equations:

$$\Delta r_M = 0.40082x - 0.91616z$$
$$\Delta(r_M - r_S) = -9.6115 \times 10^{-5}x - 4.2056 \times 10^{-5}z$$

where Δ indicates the difference from pixel 1. Inserting $\Delta r_M = 15$ m and $\Delta(r_M - r_S) = 0.022 + 0.025\,n$ m gives $x \approx -186 - 218n$ m, $z \approx -98 - 96n$ m. Only the solution with $n = -1$ satisfies the condition on the surface slope, so pixel 2 is located at $(x, z) = (32, -2)$ m. Repeating the analysis for pixels 3 and 4, we obtain positions of $(73, 0)$ m and $(105, -3)$ m, respectively.

Chapter 10. Platforms for remote sensing

1. Since we require only to calculate the velocity increments, without consideration of how these are achieved, we may consider a body (rocket, satellite, etc.) of fixed mass m. To prove equation (10.2), consider the total energy (kinetic plus potential) of the body at A, just after it has been given velocity Δv_1. The kinetic energy is

$$\frac{1}{2}m\,\Delta v_1^2$$

and the gravitational potential energy is

$$-\frac{GMm}{R_E}$$

The body comes to rest at B, so its kinetic energy is zero and its potential energy is

$$-\frac{GMm}{R}$$

Equating the total energies at A and B gives the required result for Δv_1.

To prove equation (10.3), we need only to calculate the velocity v of a body in a circular orbit of radius R. This is easily done by equating the gravitational force

$$\frac{GMm}{R^2}$$

to the centripetal force

$$\frac{mv^2}{R}$$

To prove equations (10.4) and (10.5), we need to find the elliptical transfer orbit between the Earth's surface and the desired circular orbit of radius R. If the velocity of the body when it is grazing the Earth's surface is Δv_1, the total energy is

$$\frac{1}{2}m\,\Delta v_1^2 - \frac{GMm}{R_E}$$

as before. When the body reaches B, conservation of angular momentum dictates that its velocity must be given by

$$\frac{R_E}{R}\Delta v_1$$

so its total energy is given by

$$\frac{1}{2}m\frac{R_E^2}{R^2}\Delta v_1^2 - \frac{GMm}{R}$$

Equating the two energies gives equation (10.4), and equation (10.5) is just the difference between the required circular velocity $v\,(=(GM/R)^{1/2})$ and the velocity $R_E\,\Delta v_1/R$ of the elliptical orbit at B.

2. Equation (10.10) shows that the largest angular error will be approximately $2e$. This proves the result.

3. We are considering the descending (southbound) part of satellite's orbit. For definiteness, let us assume that the orbit crosses the Greenwich meridian in the southbound direction at 09:30 GMT. From equations (10.11), with $b = +52°$ and $i = 98.2°$, we obtain $\phi = 127.236°$ and $l = l_0 - 169.371°$. Thus, the satellite crossed latitude $+52°$ at a time $(52.764/360)$ of an orbit earlier, hence at $09:30 - (52.764/360) \times (16/233)$ days $= 09:30 - 14.5$ min $= 09:15.5$ GMT. However, the satellite was at a different longitude, so the local time was different. We know that during these 14.5 min the satellite moved $10.629°$ west. We also know that the Earth rotated $360 \times 14.5/1440 = 3.625°$ east, so relative to the Earth's surface the satellite moved $14.254°$ west. Thus, at 09:15.5 GMT the sub-satellite point was at longitude $14.504°$ east, where the local time was $09:15.5 + 14.254/15\,\text{h} = 10:12.5$.

4. The longitudinal spacing of *spatially* adjacent orbital tracks is $360/502 = 0.717°$. In exactly three days, the satellite makes $3 \times 502/35 = 43.03$ orbits. The nearest integer to this is 43, so we calculate the time taken to make 43 orbits. This is $43 \times 35/502$ days $= 2.998008$ days, during which time the satellite travels $360 \times 2.998008 = 1079.283°$ west. This is equivalent to travelling $0.717°$ east. Thus, every (nearly) three days the satellite moves to the spatially adjacent suborbital track, which is the definition of a drifting subcycle. The drift is eastwards.

Chapter 11. Data processing

1. Putting $\theta = 10°$ and $h = 35{,}800$ km into equation (11.1) gives $\phi = 71.43°$. A small circle of angular radius ϕ with its centre at latitude 0 and longitude 0 will pass through a point at latitude b, longitude l, where $\cos \phi = \cos b \cos l$, so setting $b = 66.5°$ gives $l = 37.0°$. This satellite will therefore provide coverage of longitudes between $-37°$ and $+37°$, or a range of $74°$. Thus five satellites, covering $360/5 = 72°$ each, will be adequate to cover the whole of the specified area.

2. (a) Reasonable estimates for printed paper are that a (fairly large) book might have a storage capacity of 10 MB, with a volume of 500 m^3 being required to store 1 TB.

(b) At a typical data rate of 1 MB s^{-1}, a five-year mission could generate 150 TB of data. Even with the highest density of data storage, this would currently require several cubic metres.

3. All of the filters are isotropic; that is, they do not have any directionality. The change of sign in (a) shows that it will reduce low spatial frequencies relative to higher frequencies. The sum of the weights is not zero (it is 8), so it is an edge-enhancing or sharpening filter rather than one that will remove uniform areas of the image. It is not normalised, so unless the output is divided by 8 it will increase the image brightness. (b) This is an averaging (smoothing) filter with tapered weights, again unnormalised. Filter (c), like (a), increases high spatial frequencies relative to low ones. In this case, however, the sum of the weights is zero, so we see that this filter will entirely suppress uniform areas. It is therefore an edge-detection filter. (Note that such filters cannot be normalised since the total weight is zero.) In fact, filter (b) is an approximation to a Gaussian smoothing filter, and (a) and (c) have been derived from it.

4. (a) Single-band histograms show marginal evidence for two classes in I_1, but no evidence in I_2.

(b) On the other hand, a cluster plot shows three distinct clusters. Class 1 (say) has I_1 less than 40 and I_2 greater than 60; class 3 has I_1 greater than 60 and I_2 less than 50; and anything else belongs to class 2. Classifying the image on this basis gives:

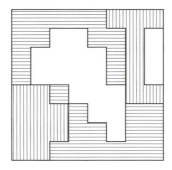

where no shading = class 1, horizontal shading = class 2, vertical shading = class 3. There are six or seven distinct regions, depending on whether the single class-2 pixel joined by one corner to a larger class-2 region is classed as a separate region or not.

5. Set

$$n_x = A \exp\left(-\frac{(x - x_0)^2}{2\sigma^2}\right)$$

and substitute into equation (11.22). The total number of pixels is

$$N = \int_{-\infty}^{\infty} n_x \, dx = A\sigma\sqrt{2\pi} \tag{i}$$

The second term in equation (11.22) is

$$\frac{1}{\ln 2} \int_{-\infty}^{\infty} n_x \ln n_x \, dx = \frac{A\sigma\sqrt{2\pi}}{\ln 2} \ln A - \frac{A\sigma^3\sqrt{2\pi}}{2\sigma^2 \ln 2} \tag{ii}$$

so the minimum total number of bits needed to specify the image is

$$\frac{A\sigma\sqrt{2\pi}}{\ln 2} \ln\left(\sigma\sqrt{2\pi}\right) + \frac{A\sigma\sqrt{2\pi}}{2 \ln 2}$$

Dividing this expression by equation (i) to find the average number of bits per pixel gives the required result.

Index

absorption
   coefficient, 45, 64, 74, 102, 108, 165, 173, 191
   cross-section, 70, 101–2
   length, 37, 63, 86, 164, 184
   lines, 79, 91–5, 189, 214
   resonant, 96
accuracy, classification, 301–2
across-track direction, 224
active systems, 4
ADEOS satellites, 172
Advanced Millimetre-wave Atmospheric Sounder, 190
Advanced Very High Resolution Radiometer, 153, 160
aerosols, atmospheric, 98–100, 108, 130
   measurement of, 146, 159, 163, 173, 174
AFS index, 114, 117
agriculture, applications of remote sensing in, 2, 4, 132, 144, 151, 221, 242
air
   mean molar mass, 89
   refractive index, 41, 87
aircraft, 246–8
AIRS, 172
albedo, 48–9, 86, 155
   measurement of, 3, 77
aliasing, 18, 195, 206, 266
Alissa, 214
Almaz satellites, 227
along-track direction, 217, 223, 224
Along-Track Scanning Radiometer, 157, 161
altimetric orbits, 264–6
AMAS, 190
ambiguity
   distance, 240
   height, 240
   range, 195, 237
Ångström relation, 100
angular frequency, 10
anode, 135
antenna, 175–81, 198, 202, 210, 213, 214–16, 223–5
   dipole, 178
   dish, 178
   monopole, 178
   phased, 183

rectangular, 178
   temperature, 176
   Yagi, 178
anthocyanins, 79
apogee, 251, 258
archaeology, applications of remote sensing in, 4, 132, 156
area, effective, 179, 191, 215
argon, atmospheric, 87, 88
Argon satellite programme, 131
Aristotle, 1
array, phased 182–3, 191
ASA number, 113
ASCAT, 222
ascending node, 253, 255, 261
astronomical unit, 276, 313
astronomy, radio, 175
atmosphere
   composition, 87–9, 134, 157, 166–9, 172–3
   density, 88, 190
   dry, 41–2, 193, 197–8, 209, 212
   microwave brightness temperature, 186–8
   optical thickness, 96–100, 129
   pressure, 88–90
   scale height, 89, 99, 165
   temperature distribution, 3, 90
   temperature sounding, 3, 163–6, 189
   turbulence, 99, 106–8
   upwelling illuminance, 129–30
   windows, 4–5, 96–7
atmospheric correction, 6, 68, 87, 145–7, 152, 157–8, 163, 182, 184, 186–8, 196–8, 209–10, 212, 292
Atmospheric Infrared Sounder, 172
atom, hydrogen, 91–2
ATSR, 157, 161
attenuation, 65, 72–3, 96–8, 100–2, 108, 145, 156, 163, 186, 211
AU, 276, 313
autocorrelation function, 56
averaging filter, 284–5
AVHRR, 153, 160
azimuth
   defocussing, 237
   direction, 224, 227
   shift, 234–6

335